Nikolay S. Sidorenkov

The Interaction Between Earth's Rotation and Geophysical Processes

Related Titles

Bull, W. B.

Tectonic Geomorphology of Mountains

A New Approach to Paleoseismology

328 pages
2007
Hardcover
ISBN: 978-1-4051-5479-6

Lynch, A. H., Cassano, J. J.

Applied Atmospheric Dynamics

290 pages
2006
E-Book
ISBN: 978-0-470-86175-2

Bohren, C. F., Clothiaux, E. E.

Fundamentals of Atmospheric Radiation

An Introduction with 400 Problems

490 pages with 184 figures
2006
Softcover
ISBN: 978-3-527-40503-9

Leeder, M., Perez-Arlucea, M.

Physical Processes in Earth and Environmental Sciences

336 pages
2006
Softcover
ISBN: 978-1-4051-0173-8

Potter, T. D., Colman, B. R. (eds.)

Handbook of Weather, Climate and Water

1000 pages
2003
Hardcover
ISBN: 978-0-471-21490-8

Nikolay S. Sidorenkov

The Interaction Between Earth's Rotation and Geophysical Processes

WILEY-VCH

WILEY-VCH Verlag GmbH & Co. KGaA

The Author

Dr. Nikolay S. Sidorenkov
Hydrometcenter of Russia
Bolshoy Predtechensky 11 -13
123242 Moscow
Russ. Federation

Cover
Grafik Design Schulz, Fußgönheim

Library of Congress Card No.: applied for

British Library Cataloguing-in-Publication Data
A catalogue record for this book is available from the British Library.

Bibliographic information published by the Deutsche Nationalbibliothek
The Deutsche Nationalbibliothek lists this publication in the Deutsche Nationalbibliografie; detailed bibliographic data are available on the Internet at http://dnb.d-nb.de.

Printed in the Federal Republic of Germany
Printed on acid-free paper

Printing and Binding: betz-Druck GmbH, Darmstadt

ISBN: 978-3-527-40875-7

Contents

The Interaction Between Earth's Rotation and Geophysical Processes. Nikolay S. Sidorenkov

ISBN: 978-3-527-40875-7

Preface

The book addresses the nature of the Earth's rotation instabilities and associated geophysical processes. The spectrum of the Earth's rotation instabilities that comprise variations in the length of the day, polar motion, and precession and nutation of the rotation axis in inertial space, includes periods of several hours to thousands of years. Instabilities of the Earth's rotation are related to various geophysical processes such as terrestrial, oceanic and atmospheric gravitational and thermal tides; redistribution of air and water masses; variations in the angular momentum of the atmosphere and ocean; air mass exchange between the summer and winter hemispheres; mechanical interaction between the atmosphere, the ocean and the solid Earth; the quasibiennial wind oscillation in the equatorial stratosphere; the El Niño–Southern Oscillation, multiyear atmospheric and oceanic waves, atmospheric circulation epoch change, climate variations, evolution of ice sheets, and so forth.

All these processes are discussed in the book, their nature and the mechanism of their influence on the rotation of the Earth described as far as possible.

There are several books on the Earth's rotation published. The publications mainly address either the celestial-mechanical or astrometry and geodetic problems of determination of the Earth's rotation instabilities or the observation data processing methods. In contrast to these, our monograph covers the physical aspects of the nonuniformity of the Earth's rotation and polar motion and nutation. In terms of the research area, our book is closest to the book by Munk and MacDonald (1960), but this was issued about 50 years ago. Since then the study of all aspects of the Earth's rotation and adjacent areas has phenomenally progressed, so a new book seems long overdue.

The author has studied the Earth's rotation instabilities and related geophysical problems for about 45 years and has received a number of fundamental scientific results. Among them are a concept of translational–rotational motion of continua, theories of zonal atmospheric circulation and seasonal variations in the Earth's rotation rate, excitation mechanisms of the Chandler wobble and annual polar motion, methods of calculation of global water exchange and hydrometeorological forecasting based on the Earth's rotation parameters, a concept of multiyear and

The Interaction Between Earth's Rotation and Geophysical Processes. Nikolay S. Sidorenkov

ISBN: 978-3-527-40875-7

decadal fluctuations in the Earth's rotation rate, and others. The author has discovered: diurnal nutation of the atmospheric angular momentum vector with a wide spectrum of oscillations; an interhemispheric thermal engine in the atmosphere; interannual oscillations of the Earth–ocean–atmosphere system; multiyear waves in the ocean and atmosphere; superharmonics of the Chandler period in phenomena of the El Niño–Southern Oscillation and quasibiennial atmospheric oscillations; correlations between the decadal fluctuations in the Earth's rotation, on the one hand, and the changes in the ice mass in Antarctica, variations of atmospheric circulation epochs and global air temperature variations, and so forth, on the other.

The results made it possible to understand the nature of many peculiarities of the Earth's rotation. When interpreting the Earth's rotation instabilities, the author frequently faced situations when observation data were radically contrary to generally accepted concepts. In those cases, a criterion of true was the concordance of a model with observation data rather than with abstract mathematical theorems, theories and conclusions. That is why some of the author's models and estimations conflict with the fixed notions and have not been recognized yet (a concept of translational–rotational motion of continua, a theory of zonal atmospheric circulation, models of macroturbulent transport of the angular momentum, the El Niño–Southern Oscillation, tidal impacts on atmospheric processes, and so forth.).

The problems addressed in the book lie at the interface between astronomy, physics of the Earth, physics of atmosphere and ocean, climatology, glaciology, and so forth. The subject of the research is dealt with all the areas of geosciences. All materials in the book are presented in detail, so that the book could be accessible even to nonspecialists and some specialists may probably find this approach elementary. We had great difficulties in mathematical notations because of a variety of geophysical parameters under study. So, different parameters are sometimes denoted by identical symbols. The author apologizes in advance for such inconveniences.

Study of any natural phenomenon is confined, as a rule, to its observation, analysis, interpretation, and use in solving scientific and practical problems. In accordance with this approach, the book logically expounds the following: the results of calculation of parameters of the Earth's rotation instabilities (Chapter 3), the lunisolar tides and their effects on the Earth's rotation (Chapter 5), the influence of atmospheric and hydrospheric processes on the Earth's rotation, more focus being given to the nature of these phenomena (Chapters 6–11). The studies described in Chapters 5–11 could be difficult to understand without a general knowledge about the Earth's motion and the theory of estimation of the Earth's rotation instabilities. Hence, a brief account of these subjects is given in the first three chapters. The closing chapter (12) addresses the use of the geodynamic laws revealed by the author in hydrometeorological forecasting. Tables of data on the rotation and some global processes are given in the Appendix. A list of the abbreviations used is given in the Appendix as well.

The Earth's rotation instabilities are correlated with many characteristics of natural processes in all frequency ranges. It can be argued from the author's multiyear experience that the Earth's rotation variations can be used as a good validation test for various geophysical models because these variations are a unique index to many processes in all spheres of the Earth, including the biosphere. Various problems can be solved using the Earth's rotation parameters. The reader will find some methods in this book and can derive others from studying how some particular characteristics available to him/her are related to the parameters of the Earth's rotation given in the Appendix.

The author is deeply obliged to Michael Efroimsky for his assistance in the publication of the book, L.P. Kuznetsova, I.V. Ruzanova and B.M. Shubik for their help in translation of the text into English, and to G.L. Averina for her help in the manuscript preparation. Many results were obtained thanks to the support of the Russian Foundation for Basic Research (Projects 02-02-16178a, 06-02-16665a).

We would be grateful if readers would send us their remarks or point out any mistakes or slips. Our address is: Hydrometcentre of Russian Federation, B. Predtechensky pereulok, 11–13, Moscow, 123242 Russia. E-mail: sidorenkov@mecom.ru

1
Introduction

The Earth's rotation accounts for the alternation of day and night, the daily cycle of solar radiation influx, formation of diurnal and semidiurnal tidal waves and finally causes diurnal variations in all characteristics of the atmosphere, hydrosphere and biosphere. The revolution of the Earth around the barycenter of the Earth–Moon system and the revolution of the Earth–Moon system around the Sun modulate the amplitudes of the diurnal oscillations of the solar radiation influx and atmospheric tides, and in the end define the variability of terrestrial processes over periods of up to several years.

The Sun revolves around the barycenter of the Solar System along compound curves of the fourth order (conchoids of a circle), so-called "Pascal's limacons". The curvature of the Sun's trajectory constantly changes and the Sun moves with varying acceleration. Being a satellite of the Sun, the Earth revolves around it and also moves with the Sun around the Solar System's barycenter. Like the Sun, the Earth undergoes all varying accelerations. Similar to the lunisolar tides, the accelerations disturb processes in the Earth's shells, producing decadal fluctuations in the latter.

Movements in the Earth's shells are observed mainly from the earth surface. Reference systems for description of the movements are tied to the Earth as well. Different points of the earth surface move with different velocities and varying accelerations. For this reason any movement looks rather complicated in a reference system tied to the Earth. Newton's laws are valid in such a reference system provided that so-called inertial forces, the Coriolis force and centrifugal force, are taken into account. The Coriolis force and centrifugal force are caused by the movement of the terrestrial reference system in an inertial system rather than by the interaction of bodies. Terrestrial processes are formed under the action of many forces. Among them the inertial forces connected with the Earth's rotation play a key role. Their contribution to atmosphere dynamics is especially significant. As a result of the Earth's rotation, the direction of movement of air masses deflects to the right in the Northern hemisphere and to the left in the Southern hemisphere; the cyclonic and anticyclonic vortices arise; systems of western winds and east winds (trade winds) are formed in the middle latitudes and in the equatorial latitudes, respectively; zones of higher pressure are formed in the subtropical latitudes and zones of lower pressure, near to the polar circles. The centrifugal force makes level surfaces

The Interaction Between Earth's Rotation and Geophysical Processes. Nikolay S. Sidorenkov

ISBN: 978-3-527-40875-7

(equigeopotential surfaces) stretch out along the equatorial axis and compress along the polar axis, as a result, these surfaces tend to form ellipsoids of rotation. Owing to the fact that the reference system is noninertial, atmospheric transfer processes seem so complicated that for the sake of their interpretation geophysical hydrodynamics has accepted the concept of negative viscosity, which contradicts to physical laws.

Bodies and particles in continua move along elliptic gravity potential surfaces and everywhere gravity is vertically directed to the center of the Earth. Gravity force tends to adjust moving bodies and particles in continua to a direction of the local gravity vertical. As a result, all bodies and particles of geophysical continua move in a translational–rotational manner. An exact description of their motion requires not only momentum conservation equations but also angular momentum conservation equations.

The Earth's rotation around its axis gives a basis for celestial and terrestrial reference systems in astronomy, serves as a natural standard of time and allows the universal time scale to be defined. The Earth's rotation is characterized by the vector of instantaneous angular velocity, which can be decomposed into three components: one component along the mean axis of rotation and two others, in the perpendicular plane. The first component defines the instantaneous velocity of the Earth's rotation around its mean axis, or the length of day, and the other two the coordinates of the instantaneous pole. The vector of the angular velocity of the Earth's rotation does not remain constant. Change in the vector's first component is manifested in nonuniformity of the Earth's rotation, and the two other in the motion of the poles.

Polar motion is the movement of the rotation axis in the body of the Earth measured relative to the Earth's crust. But the Earth's rotation axis also moves relative to the inertial celestial reference system and undergoes precession and numerous nutations.

Instabilities of the Earth's rotation (nonuniformity of rotation, polar motion, precession and nutation) distort the coordinates of celestial objects and complicate the universal time scale. The distortions can be taken into account only if peculiarities of the Earth's rotation are known and there is a theory of the Earth's rotation nonuniformity, polar motion, and precession and nutations. Nowadays, astronomical measurement accuracy requirements are becoming increasingly stringent in connection with the necessity of solving a number of scientific and applied problems in astronomy, geodesy, space research and so forth. Therefore, the study of the Earth's rotation is of great importance to modern astrometry, geodesy and geophysics.

Traditionally, the Earth's rotation instabilities are studied by astrometry. Astronomical methods register rotation instabilities. By their nature, the Earth's rotation instabilities are purely geophysical phenomena. They are related to processes in geospheres and depend on the structure and physical properties of the Earth's shells. The Earth's rotation instabilities reflect geophysical processes and give irreplaceable information on the latter, serving as natural integral characteristics of them and associated phenomena. Studying instabilities of the Earth's rotation broadens our knowledge in various areas of Earth sciences. Data on the Earth's rotation instabilities serve as criteria that can be used to verify some theories and models in geophysics, geology, space science, and so forth.

Doubts concerning constancy of the Earth's rotation rate arose after E. Halley discovered the secular acceleration of the Moon in 1695. The idea of secular slowing down of the Earth's rotation under the effect of tidal friction was first proposed by I. Kant in 1755. Nowadays, it is universally recognized that the secular slowing down of the Earth's rotation really exists and is caused by the tidal friction. The value of the secular slowing down is only discussed (Yatskiv *et al.*, 1976).

Simon Newcomb first suggested irregular fluctuations in the Earth's rotation rate in 1875. Their existence was ultimately proved at the beginning of the twentieth century. During the last hundred years, deviations in the length of day from the average value reached $\pm 45 \times 10^{-4}$ s.

Evidence of polar motion was also obtained then. Seth C. Chandler discovered a 14-month period of the latitude variations in 1891. The International Latitude Service (ILS) was established in 1899 for the purpose of monitoring the North Pole's motion. The main components of the polar motion are the Chandler motion whose amplitude is about 160 ms of arc, the annual motion, whose amplitude is about 90 ms of arc, and the secular motion toward North America with a velocity of about 10 cm/year.

In the 1930s, quartz clocks allowed seasonal variations of the Earth's rotation rate to be discovered. A more uniform scale of the Atomic Time was created in 1955 and parameters of seasonal variations began to be determined quite confidently. The length of day was established to have annual and semiannual variations with amplitudes of 37×10^{-5} s and 34×10^{-5} s, respectively.

Until the 1980s, estimations of polar motion and nonuniformity of the Earth's rotation were based on optical astrometric observations of latitude variations and the universal time variations. The observations were nonuniform and had various systematic errors. Reanalysis of the optical astrometric data in the Hipparch system, performed under the direction of J. Vondrak (Vondrak, 1999), partly eliminated these shortcomings and the data could be used in studying long-period instabilities of the Earth's rotation.

In the late 1970s, new engineering complexes were introduced: very long baseline interferometer (VLBI), global positioning system (GPS), satellite laser ranging (SLR), lunar laser ranging (LLR), Doppler orbitography and radio navigation (DORIS service) and new methods of monitoring the Earth's rotation instabilities with unprecedented accuracy. Instead of traditional astrooptical time and latitude estimations, scientists began to observe extragalactic radio sources and satellites of the Earth and process the results of the measurements (time and geometrical delays) to produce corrections to the universal time, the coordinates of the Earth's pole, and corrections to precession and nutation. Thanks to these methods, the resolution and accuracy of the estimation of rotation instabilities has increased 100-fold and are now $0''.0001$ of arc for the pole coordinates and nutation, and 0.000 005 s for corrections to Universal time UT1; which corresponds to several millimeters on the Earth surface. The time resolution of measurements reached several hours.

Regular rawinsounding of atmosphere by means of aerological station network started in the postwar years. The estimations based on these first, very limited data on the winds in atmosphere showed that seasonal variations in the Earth's rotation were mostly caused by redistribution of the angular momentum between the Earth and atmosphere (Pariiski, 1954; Munk and MacDonald, 1960).

The decadal fluctuations in the Earth's rotation rate, which are changes in the rotation rate with characteristic times of 2 to 100 years, are many times the seasonal variations. The fluctuations can be explained by extremely large increments of either the angular momentum of the atmosphere or the moment of inertia of the Earth. Therefore, it is believed that the decadal fluctuations in the Earth's rotation rate cannot be caused by geophysical processes on the Earth's surface (Pariiski, 1954; Munk and MacDonald, 1960). The fluctuations are usually considered to be related to the processes of interaction of the Earth's core and mantle (Hide, 1989).

Practically all variations in the Earth's rotation rate with periods of several days to two-three years (this range includes seasonal, quasibiennial and 55-day variations) are caused by changes in the atmospheric angular momentum (Munk and MacDonald, 1960; Lambeck, 1980; Sidorenkov, 2002a). Polar motion with a one-year period is mainly caused by seasonal redistribution of air masses between Eurasia and oceans. In the case of the Chandler wobble and nutation of the Earth's axis, the role of the atmosphere is still unclear and requires further study.

Although the mass and moment of inertia of the atmosphere is almost a million times less than those of the Earth and a hundred times less than those of the ocean, it appears that its contribution to the Earth's rotation instabilities with periods of several days to several years is prevailing. This paradoxical fact is explained by the high mobility of air. Whereas the characteristic velocity of movement within the Earth's mantle is 1 mm/year and the velocity of ocean currents is 10 cm/s, the velocity of wind in jet streams may exceed 100 m/s.

As a result of strong winds, changes in the atmospheric angular momentum considerably surpass variations in the angular momentum of the ocean and the liquid core. Energy estimations confirm the reliability of that conclusion as well. In fact, the Earth's rotation instabilities, on account of the law of angular momentum conservation, may be a consequence of movements with reversed sign in the shells surrounding the solid Earth: the atmosphere, hydrosphere, cryosphere, liquid core or the space. It is clear that the power of the energy sources exiting those movements should be not less than that of instabilities of the Earth's rotation. For the within-year and interannual nonuniformities of the Earth's rotation, the power is as follows:

$$\frac{dE}{dt} = C\omega\frac{d\omega}{dt} \approx 10^{14}\text{–}10^{15}\ \mathrm{W} \tag{1.1}$$

where E is the kinetic energy of the Earth's rotation, C is the polar moment of inertia, ω is the angular velocity and $d\omega/dt$ is the angular acceleration equal to 10^{-19}–$10^{-20}\ \mathrm{s}^{-2}$. The average powers of the energy sources are approximately as follows: atmospheric air movements – 2×10^{15} W, oceanic currents – about 10^{14} W, geomagnetic storms – 10^{12} W, auroras polaris – 10^{11} W, earthquakes – 3×10^{11} W, volcanoes – 10^{11} W, heat flows from the Earth's deep interior – 10^{13} W, interplanetary magnetic field and solar wind interacting with magnetosphere – less than 10^{12} W (Magnitskiy, 1965; Kulikov and Sidorenkov, 1977; Zharkov, 1983). The presented values indicate that only atmospheric air movements, and possibly currents in the ocean as well, are likely to cause the Earth's rotation instabilities. The power of other geophysical processes is small compared with the power of variations of the Earth's rotation. Note that such important, in terms of the Earth's rotation, effects as transport of water from the ocean to the continent (including the ice sheets of Antarctica and Greenland) and global redistribution of air masses would be impossible in the absence of atmospheric air movements. Bearing all the above in mind, as well as the fact that currents in the ocean are mostly generated by winds, we come to the conclusion of the paramount importance of atmospheric processes as far as the nature of the Earth's rotation instabilities is concerned.

Changes in the Earth's rotation rate are partly caused by changes in the moment of inertia of the Earth, which in turn results from tidal deformations. A theory of these oscillations is well developed (Woolard, 1959; Yoder, Williams and Parke, 1981; Wahr, Sasao and Smith, 1981). Therefore, the tidal oscillations are usually excluded from evaluation of the influence of various geophysical processes on the Earth's rotation.

The diurnal and semidiurnal atmospheric tides cause small changes in polar motion, nutation and the Earth's rotation rate. The most important effect is the direct annual nutation whose amplitude is about 0.1 ms of arc and excitation of free nutation of the core with amplitude ranging between 0.1 and 0.4 ms of arc. However, the excitation of the Earth's rotation instabilities by the diurnal and semidiurnal oceanic tides is approximately by two orders of magnitude greater than the corresponding influence of the atmospheric tides (Brzezinski *et al.*, 2002).

The book consists of thirteen chapters.

Chapter 1 describes the role of the Earth's rotation in dynamics of terrestrial processes, and gives a history of discovery and interpretation of the Earth's rotation instabilities. Also, the structure of the book is given here.

Chapter 2 acquaints the reader with motions of the Earth around the Sun and the barycenter of the Earth–Moon system. Compound motions of the Earth's rotation axis are described, and their geometrical interpretation given.

Chapter 3 addresses the motion of the geographical poles and variations of the angular rate of the diurnal Earth's rotation. A history of discovery of the motion of the Geographical North Pole and nonuniformity of the Earth's rotation rate is given in this chapter. Time series of instrumental observations of the North Pole's coordinates and the Earth's rotation rate are given. The results of mathematical analysis of the time series are presented, and the seasonal, multiyear and secular components are separated.

The theory of estimations of the Earth's rotation instabilities is described in Chapter 4. The differential equations are deduced for instabilities of rotation of an absolutely firm and perfectly elastic Earth under the action of exciting functions empirically calculated. The advantages and disadvantages of "balance method" and "method of the moment of forces" used to estimate various effects on the Earth's rotation instabilities are presented. Polar motion under the action of a harmonious exciting function is described. Basic equations of the theory of precession and nutation are deduced.

Chapter 5 addresses the lunisolar tides and their influence on the Earth's rotation instabilities. The derivation and decomposition of tidal potential is given in this chapter. The basic harmonics of zonal, diurnal and semidiurnal tides are given. A theory of tidal oscillations of the Earth's rotation angular velocity, polar motion, precession and nutation of the Earth axis is expounded. Introductory information on atmospheric tides is given.

The effects of seasonal redistribution of air masses on the Earth's rotation are addressed in Chapter 6. Detailed calculations of seasonal redistribution of air masses in the atmosphere are given. Components of the tensor of inertia of the atmosphere and the amplitude of their annual variations are calculated. Annual variations of the Earth's rotation rate and polar motion are estimated and compared with the calculations carried out by other authors.

The results of a study into the atmospheric angular momentum are discussed in Chapter 7. Data on zonal atmospheric circulation are analyzed. Series of components of the atmospheric angular momentum calculated by David Salstein (Atmospheric and Environment Research, Inc., USA) on the basis of the NCEP/NCAR reanalysis data from 1948 to the present time are described. The results of analysis of time series of the axial and equatorial components of the angular momentum of winds in the atmosphere are given. The existence of diurnal nutation of the vector of the angular momentum of atmospheric winds is shown. Components of lunisolar tides are separated. The contribution of variations of the angular momentum of winds to seasonal variations of the Earth's rotation rate and nutation is evaluated.

The zonal atmospheric circulation is described in Chapter 8. A concept of translational–rotational motion of geophysical continua is formulated. The origin of the zonal circulation and atmosphere superrotation is shown. A special theory of zonal circulation and subtropical maxima of pressure is developed. A new mechanism of seasonal variations of the angular momentum of the atmosphere and seasonal variations of the Earth's rotation is suggested.

Chapter 9 addresses the interrelation of the Chandler motion with oceanic variations known as the El Niño and La Niña, and with atmospheric oscillations manifested as the Southern Oscillation and Quasibiennial Oscillation of winds. Proofs of the existence of multiyear waves in the ocean and atmosphere are offered. The results of analysis of oceanic and atmospheric characteristics indicating that there is a connection between the variations in the ocean and atmosphere and Chandler variations of the Earth are described. A new model of excitation of free polar motion is presented.

Chapter 10 addresses the moments of forces of friction of wind and pressure on mountains. A theory of mechanical interaction of the atmosphere with the underlying surface is expounded. The results of calculations of the Earth's rotation rate by the method of moment of forces are described. A mechanism of the movement of lithosphere plates is proposed.

Chapter 11 discusses the geophysical processes that may be responsible for the decadal components (periods of 2 to 100 years) in the Earth's rotation rate fluctuations. The contribution of interplanetary magnetic field and solar wind is estimated; as well as the influence of glaciers and variations of the sea level on changes in the Earth's rotation. Variations in the mass of ice in Antarctica, Greenland and the ocean, which are necessary for explanation of the observed decadal instabilities of the Earth's rotation, are calculated. Data on the connection of the decadal fluctuations in the Earth's rotation rate with geomagnetic variations, decadal variations in atmospheric circulation and climate change are given. The nature of these connections is discussed.

Chapter 12 discusses how laws of tidal oscillations in the Earth's rotation can be used in hydrometeorological forecasting. Lunar cycles in variations of hydrometeorological characteristics are discovered. A technique of air-temperature forecasting is described. It is revealed that there is a connection between the extremality of natural processes and long-term variability of tidal forces. It is justified that tidal forces must be introduced into motion equations for global atmospheric and oceanic models in order to radically improve weather forecasting.

Unsolved problems of the Earth's rotation instabilities and prospects of further researches are discussed in Chapter 13.

The Appendix contains descriptions of surface spherical functions; the figure of the Earth, list of acronyms, tables of the annual values of the Earth's rotation velocities and secular polar motion, and the mass of ice in Antarctica, Greenland and the mass of water in the ocean, which are calculated from the above data, and indices of quasibiennial oscillations.

2
Motions of the Earth

2.1
Earth's Revolution

2.1.1
Introduction

The Sun is the main body of the Solar System. The Sun's diameter is 1 390 600 km, which corresponds to 109.1 diameters of the Earth. Its volume is equal to 1 301 200 volumes of the Earth, and the mass is 333 434 times greater than Earth's mass. Since the ratio between the Sun's and the Earth's volumes is much higher than their mass ratio, the Sun's average density is only 0.256 of the Earth's density. The Sun rotates with respect to one of its diameters with various velocities in different points rather than as a rigid body, all points of which have the same angular velocity. Thus, the points on the surface of the Sun's equatorial zone make one revolution with respect to the stars during 25.38 days, and the points laying further from the equator rotate more slowly. The further a point is from the equator the longer is the period of its rotation, and, consequently, the lower is its velocity. For example, the period of rotation of the near-pole points is 30 days or longer.

Eight planets, many minor planets, comets, meteoric streams, and meteoritic bodies move around the Sun. The Earth is the third planet distant from the Sun (after Mercury and the Venus). The Moon, its satellite, moves around the Earth at an average distance of 384 400 km from it. The diameter of the Moon is approximately 3.1 times smaller than the Earth's diameter. During one Earth revolution around the Sun the Moon makes about 13.5 revolutions, moving almost in the orbital plane of the Earth's motion around the Sun. Hence, the Moon is either ahead of the Earth (the last quarter) or behind it (the first quarter), between the Earth and Sun (the new moon) and further from the Earth (the full moon). Therefore, the lunisolar attraction forces acting on the Earth from the Moon and the Sun continuously change, thereby strongly complicating the Earth's motion.

The Earth makes two apparent motions with respect to the Sun: the daily motion (from east to west) as a result of the Earth's rotation around its axis and the annual motion (from west to east with a rate of about 1°/day) as a result of the Earth's orbital

The Interaction Between Earth's Rotation and Geophysical Processes. Nikolay S. Sidorenkov

ISBN: 978-3-527-40875-7

motion around the Sun. The trajectory of the apparent annual motion of the Sun among the stars, as seen by an observer on the Earth, passes near the ecliptic and among the constellations: Aries, Taurus, Gemini, Cancer, Leo, Virgo, Libra, Scorpio, Sagittarius, Capricornus, Aquarius, and Pisces. A more accurate determination of the ecliptic needed for our discussion is given below.

Let us consider now the plane passing through the Sun's center S, the mass center of the Earth–Moon system (the barycenter) T, and the velocity vector V of the barycenter moving around the Sun. It will be an instantaneous orbit plane of the center of mass of the Earth–Moon system. This plane does not keep its position unchanged in space (relative to the stars) but rotates around the center of the Sun S, due to the attraction of the Earth and the Moon (and, consequently, the barycenter) by planets.

The motion of the orbital plane of the mass center of the Earth–Moon system, which occurs due to the planets perturbation, can be divided into two motions: a slow displacement of the plane with a low velocity and a series of its very small periodical fluctuations.

The plane passing through the center of the Sun and possessing only the secular motion of the instantaneous orbital plane of the Earth–Moon system center of gravity is called the heliocentric ecliptic plane. The plane parallel to it and passing through the Earth's center is called the geocentric ecliptic plane. The intersection of the celestial sphere by the geocentric ecliptic plane is called *the ecliptic*.

Because the ecliptic is displaced as a result of the secular motion, it is accepted to refer it to a certain time or epoch to fix the position of the vernal equinox point. For example, one talks about the ecliptic of the adopted epoch J1950.0, J2000.0, and so forth. The movable ecliptic position for any instant can be given with respect to one of the motionless ecliptics.

Let us recall that the points of intersection of the celestial sphere with an imaginary axis of the Earth's rotation are called the celestial poles. The celestial pole found in the area of the Ursa Minor constellation is the North celestial pole, opposite to it is the South celestial pole. The great circle of the celestial sphere, which is perpendicular to the Earth's rotation axis and whose plane passes through the celestial sphere center, is called the celestial equator. The ecliptic is intersected with the celestial equator in two diametrically opposite points. One of these points, where the Sun moves along the ecliptic passes from the southern hemisphere of the celestial sphere to the northern one, is called the vernal equinox point. At present, it is in the Pisces constellation. The opposite point of intersection of the equator and the ecliptic is the autumnal equinox point. It is situated in the Virgo constellation. The ecliptic points that are equidistant from the vernal and autumnal equinox points are called the solstice points. In the Northern hemisphere of the celestial sphere (in the Taurus constellation) the point of the summer solstice is situated, and in the Southern hemisphere (in the Sagittarius constellation) is the point of the winter solstice. The Sun passes through the vernal equinox point approximately on March 21, the summer solstice – June 22, the autumnal equinox point – September, 23, and the winter solstice – December 22.

Two diametrically opposite points of the celestial sphere, equidistant from all ecliptic points, are called the poles of the ecliptic. The north ecliptic pole is situated in the Dragon constellation, the southern one – in the Dorado constellation.

According to the universal gravitation law, two mass points (or bodies) with masses M and m are mutually attracted with the force directly proportional to the product of their masses and inversely proportional to the square of distance R between them.

$$F = \gamma \frac{Mm}{R^2} \tag{2.1}$$

where $\gamma = 6673 \times 10^{-14}\,\mathrm{m^3\,kg^{-1}\,s^{-2}}$ is the coefficient of proportionality, or the gravitational constant. M in the product $\gamma M = 39\,860\,044 \times 10^7\,\mathrm{m^3\,s^{-2}}$ is the Earth's mass.

The point of mass, on which the external forces do not act, can be only at two states: at relative rest or a uniform and rectilinear motion. This is the concept of Newton's first law – the law of inertia.

The motion of a rigid body is more complicated. It can make the translation motion: translation and rotate simultaneously with respect to the center of mass. The center of inertia or the body center of mass (the material system) is the point in which the total mass of the body or the system is supposed to be concentrated and all external forces acting on the body points or system are applied to. The motion of the inertia center, or the center of mass follows the same laws as the motion of an individual point of mass.

The Earth makes a variety of motions of both the periodic and secular types. It rotates around its axis, doing a complete circuit in $23^{\mathrm{h}}\,56^{\mathrm{m}}\,04^{\mathrm{s}}$ (86 164.09 s) and revolves (that is, describes the translational motion) around the Sun with a period of 365.2422 days. Simultaneously with the motion around the Sun, the Earth's center of gravity rotates around the center of mass of the Earth–Moon system with a period of 27.3217 days. The Earth rotation axis makes the long-periodical motion (systematically divided into precession and nutation) with a period of 25 784 years, being the generating line of the almost circular cone. The Earth's body, in its turn, wobbles with respect to the axis of rotation, owing to which the Earth's geographic poles are displaced over its surface, describing complicated helical curves.

The Sun, in addition to the rotation around its axis (which does not influence the Earth's motion), rotates around the center of mass of the entire Solar System, the position of which depends on the planet's position for the given time. The orbital motion of the Sun basically consists of two almost circular motions with a radius of about 0.003 astronomical unit and the periods close to the Jupiter and the Saturn periods of revolution. Like many stars, the Sun also has its proper motion and displaces in the direction of the Hercules constellation with a velocity of 19.5 km/s, or 6×10^8 km/year. And, finally, the Solar System moves around the Galaxy center with a velocity of 250 km/s, doing the full circuit each 200 million years. The Earth, being the Sun's satellite, also participates in all these motions.

2.1.2
Orbit of the Earth's Center

All the bodies of the Solar System, the Earth including, move around the Sun along their own orbits. The mean distance between the Earth and the Sun is

149 600 000 km. It is called the astronomical unit and used as a standard for all astronomical and astronomy–geodetic calculations.

Let us consider at first a simplified problem of the Earth's motion. We will assume that the Solar System consists only of two bodies: the Sun and the Earth (without the Moon). The Sun center of gravity will be away from the mass center of the Sun–Earth system at a distance of $X = \frac{M_E}{M_S + M_E} \cdot A = 0.00003$ astronomical unit, $(1{:}332\,480 = 0.000\,030\,077)$, where M_S is the mass of the Sun, M_E is the mass of the Earth. The distance X is small in comparison with the astronomical unit; therefore we suppose also that the Earth moves around the motionless Sun. If we assume that the Sun and the Earth are spheres with a density decreasing to the periphery and the dimensions of the Earth and the Sun are small in comparison with the distance between them, then this problem can be solved as a problem of motion of two mass points mutually attracted by the universal gravitation law.

In this case, the Earth's translation motion around the Sun is described by two of Kepler's laws. The first law states that the orbit of each planet (the Earth) is an ellipse with the Sun at one focus of the ellipse. The second law defines that the line connecting the planet to the Sun (the Earth's radius-vector) sweeps equal areas in equal time intervals.

Like any ellipse, the orbit of the Earth's center of gravity in its motion around the Sun has two axes of symmetry – the major and the minor axes (Figure 2.1). The major axis passes through the ellipse focuses, that is, through the Sun S. Its intersection points with the orbit are called the apses.

The rate of change in the distance between the Earth and the Sun in apses is equal to zero; therefore, the Earth is closest to the Sun in one of the apses and as far as possible removed from the Sun in another apse. The point Π, where the Earth is the closest to the Sun, is called the perihelion; the most remote point A is the aphelion.

The Earth's position in the orbit is determined by the angle ΠSE reckoned from the perihelion towards in the direction of the Earth's motion up to its radius-vector SE. This angle is called the true anomaly and is designated as ϑ.

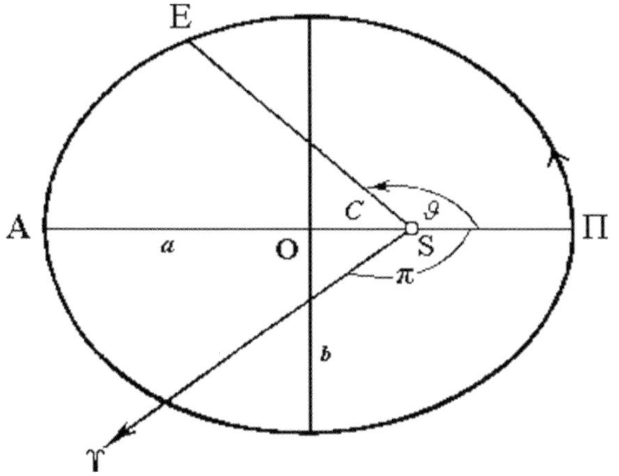

Figure 2.1 Earth's orbit.

The orbit of the Earth is determined by a and b semiaxes. But, as a rule, instead of the minor semiaxis b the orbit eccentricity is used, equal to

$$e = \frac{c}{a} = \frac{a - \Pi S}{a} = 0.01675 \tag{2.2}$$

where a is the orbit semimajor axis, c is the distance from the focus to the center of the ellipse, ΠS is the perihelion distance of the Earth from the Sun.

The distance between the Earth and the Sun continuously changes owing to the ellipticity of the Earth's orbit. The velocity of the Earth's motion along the orbit also changes: the further the Earth is from the Sun the lower is its velocity. In the most distant point of the orbit, Earth's velocity is $V_{min} = 29.27$ km/s; in the nearest point (that is, in the perihelion) $V_{max} = 30.27$ km/s. The product of the radius-vector $\boldsymbol{r}$ of the Earth's inertia center by its velocity is a constant value ($\boldsymbol{r}V_C = \text{const}$). It is impossible to observe directly the Earth's motion around the Sun. But the Earth's motion among the stars, which is observed from the Sun, will be the same as the Sun's motion among the stars, which is observed from the Earth. Therefore, in order to study the Earth's motion around the Sun it is necessary to observe the Sun's motion among the stars and other distant objects.

When the distance between the Earth and the Sun is the shortest (the Earth is in the perihelion) the velocity of the Sun's motion among the stars is the highest. It is possible to conditionally say that the Sun is in the perigee.

At the maximum distance between the Earth and the Sun, the speed of the Sun's motion among the stars is the lowest. An apparent deceleration caused by the motion of more remote objects (which is less visible than the motion of closer objects) perpendicular to the beam of sight is also added to the true slow displacement of the Sun. Therefore, the velocity of the Sun's motion among the stars changes during the year. To obtain the velocity of motion for some time during the year, the average velocity is taken (for example, the average arithmetic of the velocities in the perihelion and the aphelion). This average velocity is considered constant during the year; then one searches the correction for the given moment of the year. Since the true velocity within one half of the orbit is higher than the mean velocity and within another half of the orbit it is lower than the mean velocity (in the opposite points of the orbit, the corrections to the mean velocity are equal but of different sign), then it is easy to understand that the above correction changes during the year according to the sine law.

2.1.3 Motion of the Barycenter of the Earth–Moon System Around the Sun

Let us consider now the joint motion of the Earth and the Moon around the Sun. Unlike the case where the Earth's motion around the Sun alone was considered, here the Earth's orbit will not be an ellipse but a complicated curve whose points do not lie in one plane. The full solution of the problem of motion of the Earth–Moon system around the Sun is very difficult. Despite the efforts of the greatest mathematicians of the nineteenth and twentieth centuries, it was only possible to obtain solutions for some special cases rather than the general solution.

Let us assume that the Moon's motion is connected with the Earth's motion in such a way that their general center of mass moves around the Sun along an ellipse. This assumption does not quite correspond to the truth. However, since the distance to the Sun exceeds by many times the distance from the Moon to the Earth and the mass of the Sun is very large, the errors can be neglected in this case.

The position of the Earth–Moon system center of mass (further on – the barycenter) is given as usual

$$M_E x = M_M (a_M - x) \tag{2.3}$$

whence

$$x = \frac{M_M}{M_E + M_M} a_M \tag{2.4}$$

where M_M is the mass of the Moon, a_M is the distance between the Earth and the Moon.

The mean distance between the Moon and the Earth is 384 400 km, and the Earth's mass is larger by 81.30 times than that of the Moon. Proceeding from this, $x = 4670$ km, that is, the mass center of the Earth–Moon system is situated inside the Earth and is closer to its surface than to the center.

Over one month, the Earth's center of mass *E* describes an elliptic orbit around the barycenter O_l, this orbit being similar to the orbit of the Moon mass center *M* around the same barycenter; however, the first orbit is smaller in the ratio M_M/M_E and is turned in its plane by 180° (Figure 2.2). The terrestrial ellipse dimensions are smaller than the lunar ellipse dimensions by as many times as the Moon's mass M_M is

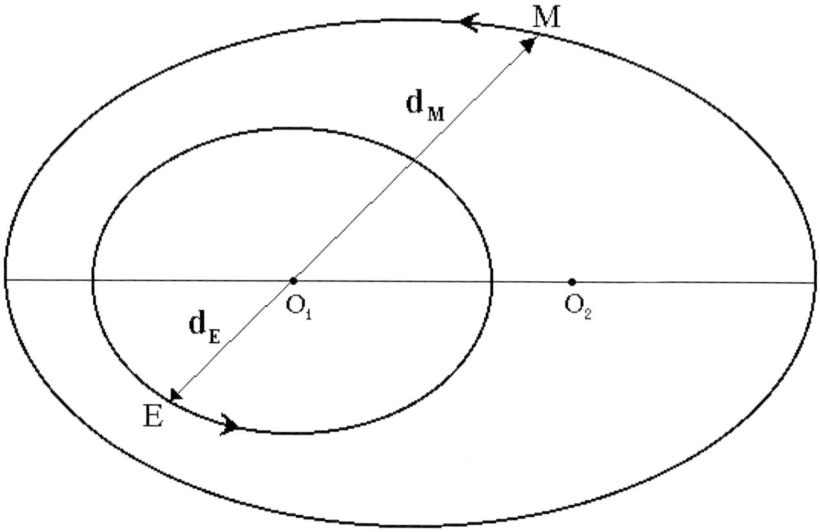

Figure 2.2 Revolution of the Moon and Earth around the center of inertia O_1 of the Earth–Moon system.

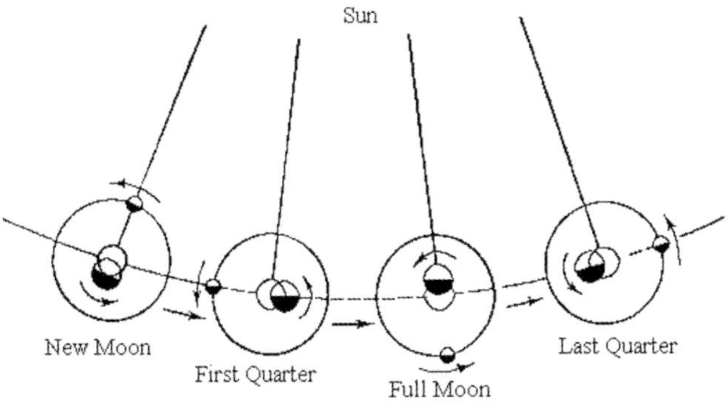

Figure 2.3 Earth's and Moon's motion around the barycentere. Phase of the Moon.

smaller than the Earth's mass M_E. Due to the motion of the Earth and the Moon around the barycenter, the observer from the Sun will see the Earth in front of the barycenter at the Moon first quarter and behind the barycenter at the Moon last quarter (half the synodic month, or 14.76 mean solar days later) (Figure 2.3). This is the so-called lunar inequality in the Earth's motion. The period of this inequality is the synodic month and its value $L = 6''.44$. The same pattern, that is, the same inequality, is observed from the Earth in the motion of the Sun among the stars. But the plane of the lunar orbit in its motion around the Earth does not coincide with the barycenter orbital plane and is inclined to it at an angle of $5°09'$. Hence, the Earth's center occurs above the barycenter orbital plane or below it. The observer from the Earth's center will see the reverse picture: the center of the Sun is situated below the barycenter orbital plane or above it. Due to this inequality (its value is approximately $0''.6$), the Sun's geocentric latitude is not always equal to zero.

In addition to the Earth's complicated motion caused by the Moon, other changes in the Earth's motion occur, namely in the apsis (a straight line connecting the apses) and the orbital elements. The apsis rotates in its plane towards the Earth's motion, due to which the orbital perihelion longitude π, that is, the angle between the directions from the Sun to the perihelion and to the equinox point γ increases by $61''.9$ per year. At present, the perihelion longitude is about $102°8'$. The period of the longitude change π is equal to 20 900 years.

As for the orbital elements of the mass center of the Earth–Moon system moving around the Sun, they may be considered very stable, due to which the orbit is close to an ellipse (Kepler's motion). The semimajor axis a, eccentricity e, and the mean velocity change only slightly and periodically. Instead of the mean velocity astronomers use the mean motion $2\pi/P$, where P is the period of revolution. These elements (they are called the osculation elements) indicate that the orbit gradually changes in its form and dimensions. Notice that if the orbital elements of a body moving along an ellipse do not change or change periodically we may say about a stable motion: the body can move in the same position forever. If the orbital elements change progressively (this, for example, is characteristic of the satellites motion near the Earth), then the motion is unstable. In this case the elliptic orbit of the body can change its orbital elements, and the satellite will fall down onto the Earth.

Frequently, one asks how long the Earth will move around the Sun and whether it will suffer a catastrophe through descending from its orbit? In other words, whether one can say about the stable Earth's motion or, generally speaking, about the stable Solar System?

Laplace, the well-known French scientist, was the first who tried to solve the problem of the Solar System stability. He proved this stability, but not completely (only in the first and second approaches). In any case, the secular perturbations of the first and second orders are absent in the semimajor axes of the major planets of the Solar System. Lagrange, another French scientist, has also approximately proved that the secular perturbations in the eccentricities and the orbital inclinations of the major planets are of the type oscillatory and are rather small. At present, one can tell with confidence that the changes in the orbits of the major planets over a future ten thousand years will be small. In particular, the Earth's and the Moon's orbits around the Sun over the last hundred thousand years have differed little from their present-day orbits.

In the 1960s, A.N. Kolmogorov and A.I. Arnold, the soviet mathematicians, had proved the stability of the bodies system that differed considerably from the Solar System. A.I. Arnold has also obtained the examples of the systems that are unstable in individual cases. However, all these results cannot be extended to the Solar System.

The pattern of motion will change slightly, if the motion of the barycenter Taround the Sun is considered taking into account the attraction of the Earth–Moon system by other planets (that is, the disturbed motion of the center of mass of the Earth–Moon system). Under the influence of the gravitational perturbations, the Earth–Moon system's center of mass moves around the Sun along an orbit close to an elliptic one but somewhat changed (perturbed) due to the Earth's and Sun's attraction by planets. Owing to these perturbations, the motion of the Earth–Moon system's center of mass deviates from the motion following by Kepler's laws. The deviations are insignificant: the change of the longitude l caused by the Moon does not exceed $\pm 7''.377$, by Mercury – $\pm 0''.050$, by Venus – $\pm 17''.57$, by Mars – $\pm 7''.02$, by Jupiter – $\pm 15''.65$, and by Saturn – $\pm 1''.04$ (in total – $\pm 48''.71$). The latitude deviation does not exceed $\pm 0''.8$.

2.2
Motion of the Earth's Spin Axis in Space

2.2.1
Dynamics of the Spinning Top

The Earth can be considered as a huge celestial gyroscope. To understand the motion peculiarity of the Earth's rotation axis, let us remember the laws of dynamics of rigid bodies. Let us consider the quickly rotating top. Let its axis of rotation be deviated from the normal by angle θ (Figure 2.4). The gravity acting on the top is $\boldsymbol{P} = m\boldsymbol{g}$, where m is the top mass, $\boldsymbol{g}$ is the gravity acceleration vector. One would think that under the gravity influence the top should fall. In reality the fall is not observed. The top rotation axis is continuously shifting perpendicularly to gravity rather than in its direction.

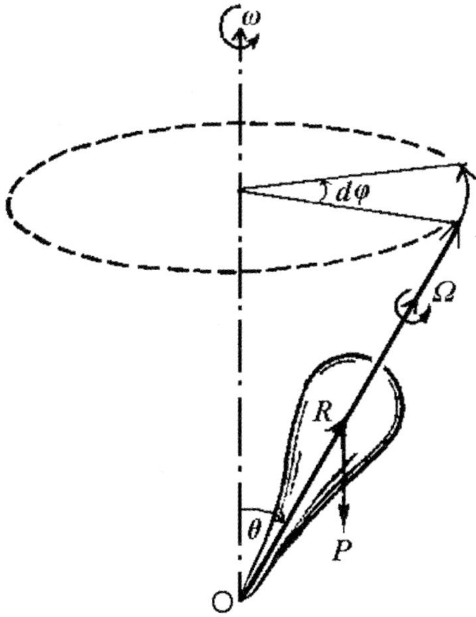

Figure 2.4 Precession of the top.

The axis describes a cone around the normal. This top axis motion is called the *precession*. Why does the top behave in such a way? Let us analyze its dynamics in order to understand this behavior.

The top's vector of angular momentum $\boldsymbol{H} = J\Omega$, where J is the top's moment of inertia with respect to its axis of rotation and Ω is the angular velocity vector. Gravity $\boldsymbol{P}$ creates the force moment $\boldsymbol{L}$ with respect to the fulcrum O: $\boldsymbol{L} = [\boldsymbol{RP}]$, where $\boldsymbol{R}$ is the radius-vector of the center of gravity. Under the effect of the force moment $\boldsymbol{L}$ the top's angular momentum will change with a speed $\frac{\mathrm{d}\boldsymbol{H}}{\mathrm{d}t} = \boldsymbol{L}$. Since the vector of $\boldsymbol{L}$ is perpendicular to $\boldsymbol{R}$ and $\boldsymbol{P}$ and the direction of vector $\boldsymbol{H}$ coincides with that of $\boldsymbol{R}$, the end of vector $\boldsymbol{H}$ and the top rotation axis shift in the direction perpendicular to the direction of gravity $\boldsymbol{P}$. When friction is absent, vector $\boldsymbol{H}$ will change only in the direction: it will rotate, describing a cone with its apex at fulcrum O.

What is the angular velocity ω of the top precession? During the time interval dt vector $\boldsymbol{H}$ receives increment $\mathrm{d}\boldsymbol{H} = \boldsymbol{L}\mathrm{d}t$, perpendicular to itself and lying in the plane of horizon. The ratio $\mathrm{d}H$ to the projection of vector $\boldsymbol{H}$ on the plane of horizon $H\sin\theta$ gives angle dφ through which this projection turns during time interval dt:

$$\mathrm{d}\varphi = \frac{L}{H\sin\theta}\mathrm{d}t \tag{2.5}$$

The derivative $\mathrm{d}\varphi/\mathrm{d}t$ is the required angular velocity of precession:

$$\omega = \frac{L}{H\sin\theta} = \frac{mgR\sin\theta}{J\Omega\sin\theta} = \frac{mgR}{J\Omega} \tag{2.6}$$

The angular velocity of precession is directly proportional to the value of the gravity moment with respect to the top fulcrum and inversely proportional to the top angular momentum. The direction of precession is defined by the rule: the force moment **L** sets segment $R \sin \theta$ into rotation around the point O towards vector **L**.

A more rigorous consideration shows that in addition to the precession the top axis makes fast fluctuations of a small amplitude. These fluctuations (axis tremblings) are called the *nutation* (in Latin *nutatio* is wiggle).

The doubled amplitude $\theta_1 - \theta_0$ and the period τ of the top nutation are approximately equal to:

$$\theta_1 - \theta_0 \approx \frac{2AmgR \sin\theta_0}{(J\Omega)^2}; \quad \tau \approx \frac{2\pi A}{J\Omega} \tag{2.7}$$

where θ_1 and θ_0 are the limits of the angle θ change due to nutation; A is the moment of the top inertia with respect to the axis passing through point O perpendicularly to the axis of rotation.

2.2.2
Precession and Nutation of the Earth's Spin Axis

The Earth rotates around its axis with a velocity of 7.29×10^{-5} rad/s and moves by the elliptic orbit around the Sun S. The inclination angle Ω of the Earth's rotation axis to the ecliptic plane is $66°33'$ (Figure 2.5). The Earth's moment of inertia is equal to 8.04×10^{37} kg m^2. The Earth's figure is close to the ellipsoid of rotation with a bulge along the equator.

Let us consider the action of the solar attraction on the Earth at the moments of the vernal and autumnal equinoxes. The longitude of the Sun during these moments is

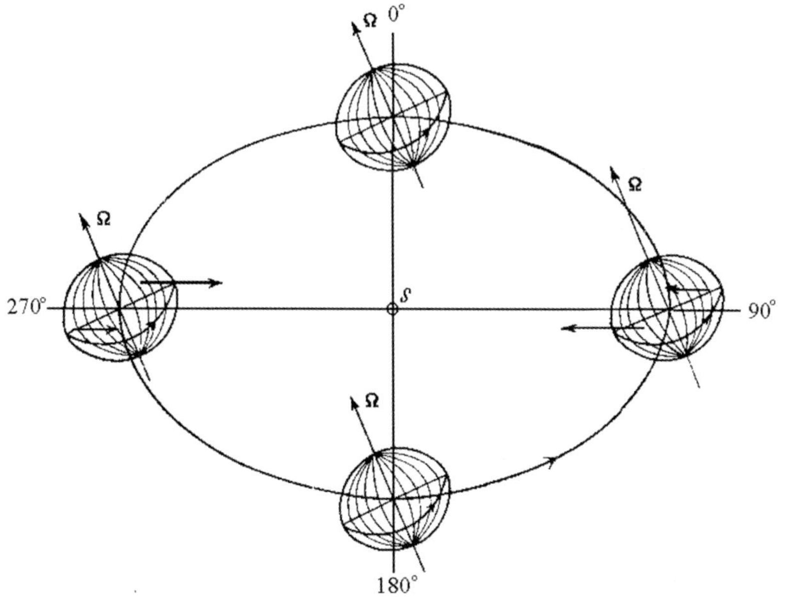

Figure 2.5 Effect of the Sun's gravity on the Earth within different parts of its orbit.

equal to 0° or 180°. The solar attraction force acts on the Earth along the line of intersection of the equator and the ecliptic planes, that is, in the plane of symmetry of the Earth's equatorial bulge. Hence, it passes through the center of the Earth and perturbs only the translation of the Earth around the Sun, not causing any rotation (shift) of the Earth's rotation axis. There is no any pair of forces here.

Let us consider now the action of solar attraction at the moment of the summer or winter solstices, when the Sun's longitude is 90° or 270° (Figure 2.5). In this case the Earth's rotation axis and the straight line connecting the Earth and Sun lie in the plane perpendicular to the ecliptic plane. Since the Sun's attraction force that acts on a part of the Earth equatorial bulge closer to it is stronger than the force acting on its more distant part, their resultant force does not pass through the Earth's center. Here, the attraction force is equivalent to some force applied to the Earth's mass center and to some pair forces striving to straighten the Earth's axis and to make it perpendicular to the ecliptic plane. However, since the rotating Earth has the possibility to turn around its mass center, this pair of forces can cause the precession and nutation of the Earth's rotation axes around the perpendicular to the plane of ecliptic.

Below, we consider the intermediate positions of the Sun on the ecliptic, when the Sun's longitudes differ from those of the equinoxes and solstices.

At the Sun's longitude equal, for example, to 45°, (it is possible to decompose the moment m of the pair of forces lying in the plane of the Earth's meridian perpendicular to this plane, into two moments m_γ and m_β by the parallelogram rule (Figure 2.6). One of them will be directed to the vernal equinox point and the pair corresponding to this moment will lie in the solstitial colure plane. Another moment will be directed to point B, and the pair corresponding to it will lie in the plane of equinoctial colure. This pair will turn the Earth's rotation axis to the point of the autumnal equinox.

At the Sun's longitude equal to 135° the action of the first pair, having moment m_γ, on the Earth will be the same as at a longitude of 45°. The second pair, having moment m_β, will make the reverse action (than at the longitude of 45°), that is, it will turn the Earth's rotation axis to the vernal equinox point.

Thus, in any position of the Sun on the ecliptic, the pair of forces created by the Sun can be decomposed into two pairs. One of them lies in the solstices plane and will always turn the Earth counterclockwise around the straight line directed to the vernal equinox point. Another pair of forces that lies in the equinoxes plane changes its direction of rotation depending on the Sun's position on the ecliptic, shifting the Earth's rotation axis periodically either towards the autumnal or the vernal equinoxes. It is easy to understand that the value of the pair of forces created by the Sun depends on the position of the Sun on the ecliptic (in other words, on its longitude and declination).

The Moon is closer to the Earth than the Sun and exerts a stronger influence on it. The Moon creates another pair of forces, which tends to bring the terrestrial equator plane into coincidence with the lunar orbital plane. But the motion of the Moon is much more complicated than the motion of the Sun, because the Moon moves along the orbit inclined by 5°09′ to the ecliptic. Moreover, the points of intersection of the ecliptic and the lunar orbit (called the lunar nodes) move from east to west (in the

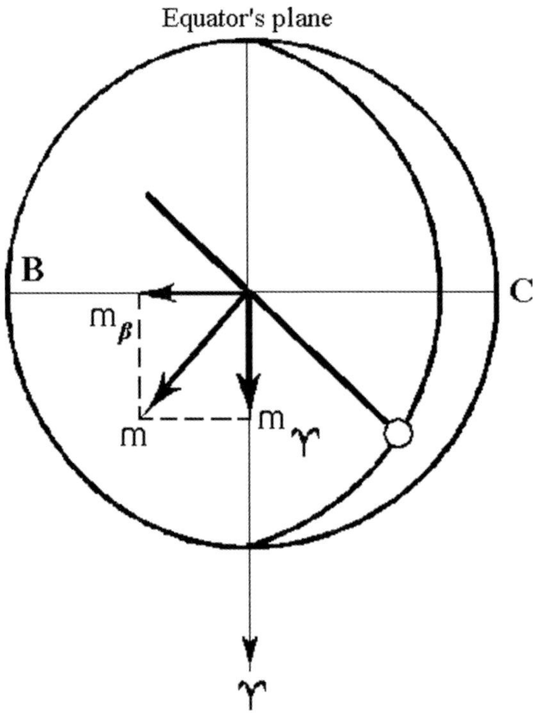

Figure 2.6 Effect of the Sun's gravity on the Earth at the Sun's longitude equal to 45°.

direction opposite to the Moon motion), with a period of about 18.6 years, which markedly complicates the Moon's motion.

Since the pair of forces caused by the Moon acts similarly to the pair caused by the Sun's attraction, it can also be decomposed into two pairs: one situated in the solstitial colure plane, the other in the plane of the equinoctial colure.

Thus, there are two perturbing pairs of forces (from the side of the Sun and the Moon) acting counterclockwise in the plane of solistial colure and rotating the Earth around the axis directed to the vernal equinox point. This rotation, together with the Earth's proper daily rotation, creates a new vector of angular velocity and, consequently, a new axis of rotation, which shifts to the vernal equinox point. Owing to the continuous total action of these two pairs of forces, the Earth's rotation axis slowly shifts (precesses) around an immobile axis, which is directed to the ecliptic pole Π (Figure 2.7) and describes a conic surface with the vertex in the center of the Earth. The described secular shift of the Earth's axis is called the lunisolar precession.

Also, two pairs of forces (from the side of the Moon and the Sun) form in the plane of the equinoctial colure. They change the direction of rotation depending on the position of the Sun in the ecliptic and the Moon in its orbit. As a result, we do not have the translation movement of the Earth's rotation axis in the plane (the plane of the equinoctial colure) but the oscillatory–periodical movement with different periods. For example, the Sun's motion in the ecliptic causes the annual and

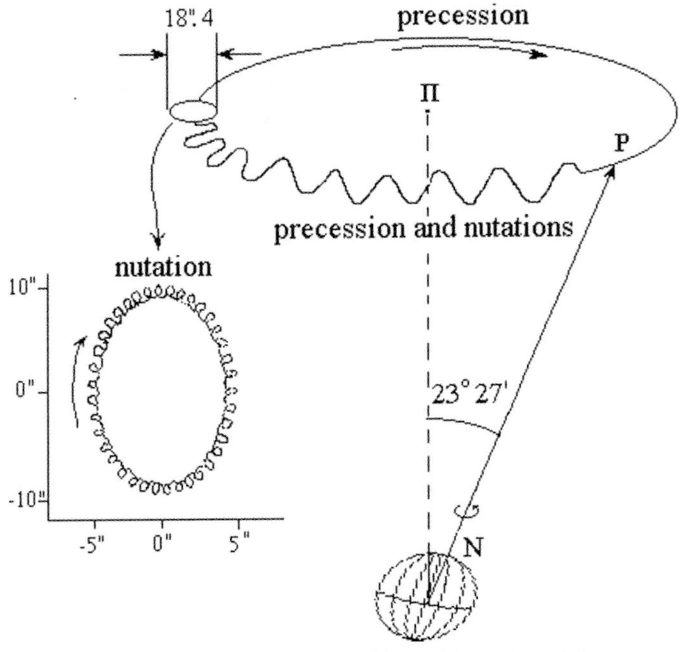

Figure 2.7 Diagram of the spatial motion of the Earth's rotation axis for an extraterrestrial observer.

semiannual periods of the Earth's axis oscillation; the Moon's motion produces the oscillation period of 18.6 years that corresponds to the motion of the lunar nodes by the ecliptic with the monthly and shorter oscillations. This oscillatory–periodical motion of the Earth's rotation axis in the plane of equinoctial colure, which is perpendicular to the precession, is called the nutation of the Earth's axis (Figure 2.7).

The kinematic pattern of the precession and the Earth's proper rotation can be interpreted as follows. One can imagine a narrow circular cone with its vertex in the Earth's center, which is symmetric relatively to its figure axis OC and rigidly connected with the Earth (Figure 2.8). Next, let us imagine the second circular cone that does not rotate with respect to the ecliptic. Its central axis coincides with the ecliptic axis $O\Pi$ and the vortex is in the center of the Earth. When the Earth rotates, the first cone rolls without sliding along the second cone; the Earth, rigidly connected with it, rotates and precesses simultaneously.

The instantaneous axis of rotation is always a tangent, or the cones' common generating line OP. As the Earth rotates together with the small cone, its instantaneous rotation axis (in the planet body) revolves every day along the circular basis of the small cone around the axis of figure OC. Hence, the pole of rotation P shifts over the Earth's surface. This occurs when the initial rotation around the axis of figure OC is disturbed by an external force; as a result, the rotation axis OP deviates from the figure axis OC. In this case, there is an additional shift of the poles, which is superimposed upon the poles' motion caused by the perturbing forces from the side of the Moon and the Sun. The radius of this shift is the hundredth of an arcsec. But it is impossible to disregard this shift when discussing the kinematic pattern of

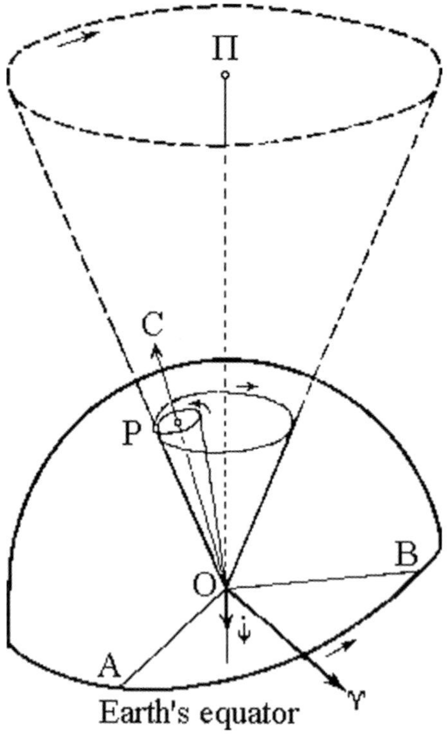

Figure 2.8 Kinematic patterns of the Earth's rotation and of the precession of the Earth's rotation axis.

the Earth's rotation. The motion of the poles described above is mainly caused by the precession of the terrestrial axis, that is, it is a consequence of the Moon and the Sun attraction. Therefore, it would be more correct to call it the lunisolar motion of the poles.

The nutation distorts the above pattern of the motion of the Earth rotation axis, being the cause of changes in the rate of precession ψ and the inclination angle ε of the instantaneous rotation axis to the ecliptic plane. The deviations of the angles of precession $\Delta\psi$ and inclination $\Delta\varepsilon$ are described by the trigonometric series:

$$\Delta\psi = -17''.2064\sin\Omega - 1''.3171\sin 2h - 0''.2276\sin 2s + 0''.2075\sin 2\Omega + \ldots \tag{2.8}$$

$$\Delta\varepsilon = 9''.2052\cos\Omega + 0''.573\cos 2h + 0''.0978\cos 2s - 0''.0897\cos 2\Omega + \ldots \tag{2.9}$$

where Ω is the mean longitude of the ascending node of the lunar orbit, which regularly decreases with time and changes by 2π per 18.613 years; h is the mean longitude of the Sun, regularly increasing by 2π per the tropical year; s is the mean longitude of the Moon, regularly increasing with a speed of 2π per the tropical month (27.32 days). The periods of the nutation harmonics are approximately equal to 18.61 years, 0.5 year, 13.66 days, 9.3 years, respectively. The entire series of the nutation harmonics contain the components with periods from 18.61 years to several hours.

Summarizing the aforesaid, it is possible to make the following conclusion. The total action of the Sun and the Moon attraction onto the flattened Earth leads to shifting the celestial pole around the ecliptic pole along a complicated curve. Methodically, this is a complicated motion; it is divided into two simpler motions: the secular motion (that is, the precession motion) and the periodic motion (that is, the nutation motion) (Figure 2.7).

Due to the precession, the north celestial pole P moves around the ecliptic pole Π that is considered to be immobile among the stars. This pole moves along a small circle with a spherical radius equal to $23^\circ 23'$ and a constant velocity of approximately $20''$/year. The trajectory of the pole P passes through the constellations: Ursa Minor, Cepheus, Cygnus, and Draco. Since the vector of the pole's shift is directed to the vernal equinox point, this point continuously steps back (shifts) along the ecliptic toward the Sun annual motion with a rate of $20''/\sin 23^\circ 26' = 50''.371$ per year ($\approx 1^\circ$ during 72 years). This motion of the vernal equinox point along the ecliptic towards the Sun is the lunisolar precession (the forestalling of equinoxes). The period of precession is equal to $360 \times 60 \times 60''/50''.371 = 25\,729$ years, where the numerator is the number of seconds in the circle ($1\,296\,000''$), the denominator is the distance covered by the vernal equinox point along the ecliptic over a year.

Due to the precession, the time period from one equinox to another (that is, the tropical year) is 20 min shorter than the period of one revolution of the Sun with respect to the zodiac constellations (that is, the sidereal year). The tropical year presents the basis for the calendar. The calendar dates, months and seasons are calculated using this year. The cyclic recurrence of the Sun's motion by the zodiac constellations is determined using the sidereal year. The moments of the Sun's entrance into the zodiac constellations are constantly displaced with respect to the calendar dates and seasons due to the 20-min difference between the sidereal and the tropical years.

Thus, 2100 years ago the spring came when the Sun passed the Aries constellation (the vernal equinox point was in this constellation). At present, the Sun is in the Pisces constellation in spring. After a lapse of 6000 years, the Sagittarius will be the spring constellation, 12 500 years later – the Virgo, 19 000 years – the Gemini constellation; 25 729 years later the cycle will finish, and the Pisces constellation will again be vernal. Over 25 729 years, each sign of the zodiac passes through all calendar dates, months, and seasons.

The designation of the points of equinoxes and solstices and the signs of the zodiac were introduced during the astronomy's blossoming times in ancient Babylon and Greece over 2100 years ago. At that time the Sun passed the Aries and the Libra constellations during the moments of the vernal and the autumnal equinoxes, and the Cancer and the Capricornus constellations – during the summer and winter solstices, respectively. At present, the equinoxes coincides with the Sun's passage through the Pisces and the Virgo constellations, and the solstices – with the position of the Sun in the Gemini and Sagittarius constellations, respectively, (that is, almost a month earlier). All the calendar dates of the zodiac signs, which are used by astrologers, are also a month ahead of the true moments of the Sun's entrance into the corresponding zodiac constellations. Nevertheless, the ancient schedule of dates of the zodiac signs is still used by astrologers for making horoscopes. Therefore, there is no sense in connecting the signs of the zodiac with the respective constellations.

The angle of inclination of the Earth's rotation axis to the ecliptic ($66^\circ 33'$) due to the lunisolar precession does not change, that is, the geographical coordinates of the Earth's stations remain invariable.

The fluctuations with the periods equal to the tropical year (the period of revolution of the Sun along the ecliptic), half a year, 18.6 years (the period of the lunar nodes' shift along the ecliptic), half the lunar month (13.7 days), and with many shorter periods are superimposed on the secular motion of the celestial pole.

The celestial pole that has only the secular – the precession – motion is called the mean celestial pole P_0. The celestial equator corresponding to the mean celestial pole, is the mean equator at the given moment; and the vernal equinox point on the mean equator, is the mean point of the vernal equinox at this moment.

The celestial pole having not only the secular (the precession) but also the periodical (the nutation) motion is called the true celestial pole P. The celestial equator and the vernal equinox point corresponding to the true polar position for the given moment of time is the true equator and the true vernal equinox point at this moment.

In the nutation relative motion, the true celestial pole P moves around the mean pole, following a complicated elliptical curve, whose major axis passes through the pole of ecliptic Π. As it is mentioned above, the mean point of the vernal equinox γ_0 is progressively moving due to the secular motion of the mean celestial pole and is shifting (with a nearly constant rate) along the ecliptic to meet the Sun annual motion. As a result of periodic motion of the true celestial pole, the true vernal equinox point γ oscillates around the mean point of the vernal equinox with different periods, the main of which is 18.6 years. The motion with this period occurs along the ellipse. The major axis of the ellipse is perpendicular to the direction of precession motion and equal to $18''.4$; the minor axis is parallel to this direction and equal to $13''.7$. Thus, the Earth's rotation axis describes a wave-like trajectory on the celestial sphere, the points of the trajectory lying at an average angular distance of about $23^\circ 27'$ from the pole of the ecliptic (Figure 2.7).

The forces acting on the Earth and the Moon from the side of the planets of the Solar System disturb the motion of the mass center of the Earth–Moon system and, hence, displace the ecliptic plane and its pole Π. Approximately, the motion of the north ecliptic pole occurs as follows. The ecliptic pole Π moves with a rate of $0.47''$/year (Figure 2.9) along the arc of the major circle at an angle of 7° to line ΠP passing through the celestial pole P and the pole of ecliptic Π. As the ecliptic shifts (due to the planets' perturbations), the vernal equinox point also shifts along the ecliptic towards the Sun's motion.

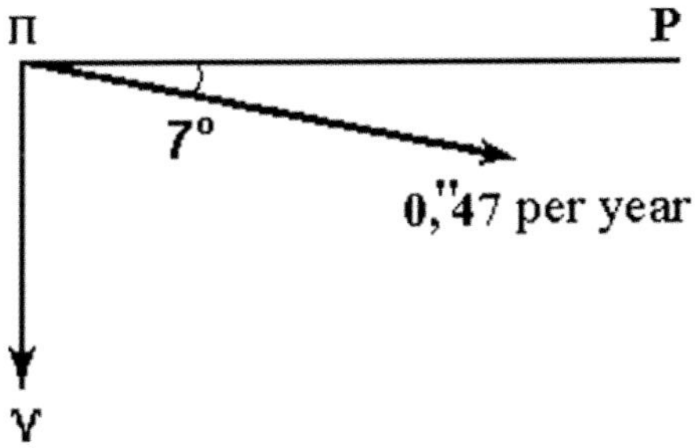

Figure 2.9 Spatial motion of the ecliptic pole Π.

Thus, the full shift of the vernal equinox point along the ecliptic is equal to $p - \lambda' \cos \varepsilon$, where p is the lunisolar precession at longitude and λ' is the planet's precession equal to 0.13″/year.

The orbital revolutions of many planets of the Solar System are resonantly intercorrelated. In addition, the orbital revolutions of the major natural satellites of the planets are in most cases synchronous with their spin rotations, so that the satellite's angular velocities of orbital revolution and spin rotation appear to be equal. Due to this, the satellite always presents the same hemisphere to an attracting planet. In the celestial mechanics, this phenomenon is called "the synchronous rotation" of a satellite.

3
Polar Motion and Irregularities in the Earth's Rotation Rate

3.1
Motion of the Earth's Poles

3.1.1
Motion of the Spin Axis in the Earth's Body

In 1765 Leonard Euler theoretically proved that if the axis of the Earth's rotation does not coincide with the axis of the Earth's figure (that is, the axis of the maximum moment of inertia), then the movement of the geographical poles around the figure's poles should take place with a period of $T_R = \frac{A}{C-A}\frac{2\pi}{\Omega} \approx 305$ sidereal days (where A and C are the equatorial and polar Earth's moments of inertia; Ω is the Earth's rotation angular velocity expressed in radians per day).

Let us explain the kinematics of this process with the help of Figure 3.1. Let the Earth's figure axis OC be deviated by the angle μ from the rotation axis OP along which the vector of the instantaneous angular velocity ω is directed. Then, the direction OH of the vector of the Earth's angular momentum $\boldsymbol{H}$ will not coincide with direction OP of the vector ω but will be situated between OC and OP axes and deviated from axis OP by the angle ν. When the external forces are absent, $\boldsymbol{H} = \text{const}$ and direction OH does not change in space. In this case, the Earth's motion occurs in such a way that axis OP describes two cones: the movable wide cone with the figure axis OC connected with the Earth and the, motionless in space, narrow cone with an axis of the angular momentum OH. The angles at the vortexes of the wide and narrow cones are equal to 2μ and 2ν, respectively.

The wide cone hooks (by its inside surface) the narrow cone and rolls around it, whereas axis OH remains motionless in space all the time, that is, it is always directed to the same point of the sky (Figure 3.2). Hence, the narrow cone as if rotates around the axis of the angular momentum OH. Therefore, owing to the Earth's rotation, the wide cone rolls around the motionless narrow cone; the wide cone center C moves along the circle (dotted line in Figure 3.2). The cones generatrix OP is an instantaneous axis of the Earth's rotation. In the counterclockwise motion, as viewed from the north celestial pole, point P describes two circles – the bases of the cones: the circles with C and H centers. One revolution of the Earth around H (the sidereal day)

The Interaction Between Earth's Rotation and Geophysical Processes. Nikolay S. Sidorenkov

ISBN: 978-3-527-40875-7

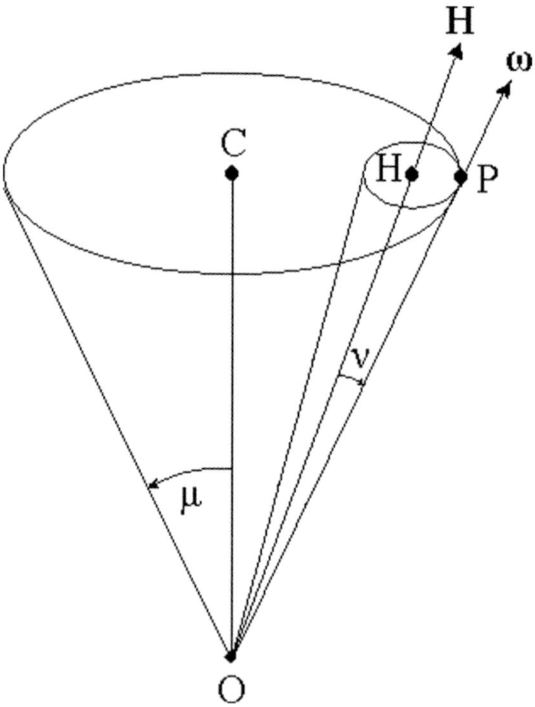

Figure 3.1 Euler's motion of the Earth's rotation axis.

and of point C around H (the dotted circle in Figure 3.2) occurs over the same time interval. Point P will pass the full circle (the base of the wide cone) over as many sidereal days as the circles of the narrow cone are contained in the circle of the wide cone (or however many radii of the narrow cone base are contained in the radius of the wide cone base). Since the wide cone is connected with the Earth, then point C on the Earth's surface and the large circle with center C are motionless; the small circle with center H rolls around the inside surface of the great circle with a center C (Figure 3.3). At this time point P – the point of the circles' inside contact – moves counterclockwise along the circle with center C. Point P is the Earth's north instantaneous pole; point C is the figure's North Pole. Hence, as the Earth rotates, its pole moves constantly around the figure's pole, which is called the instantaneous pole of the Earth's rotation.

According to the theory,

$$\frac{\mu}{\nu} = \frac{C}{C-A} \approx 305.6 \tag{3.1}$$

The angle μ, being the deviation of the rotation axis OP from the figure axis OC of the Earth, is obtained from the astronomical observations. Its value does not exceed $0''.5$. Consequently, angle $\nu \approx 0''.001$. Because of its small value, this angle cannot be obtained from observations. Therefore, it is assumed that the axis of the angular momentum H coincides with the Earth's rotation instantaneous axis OP.

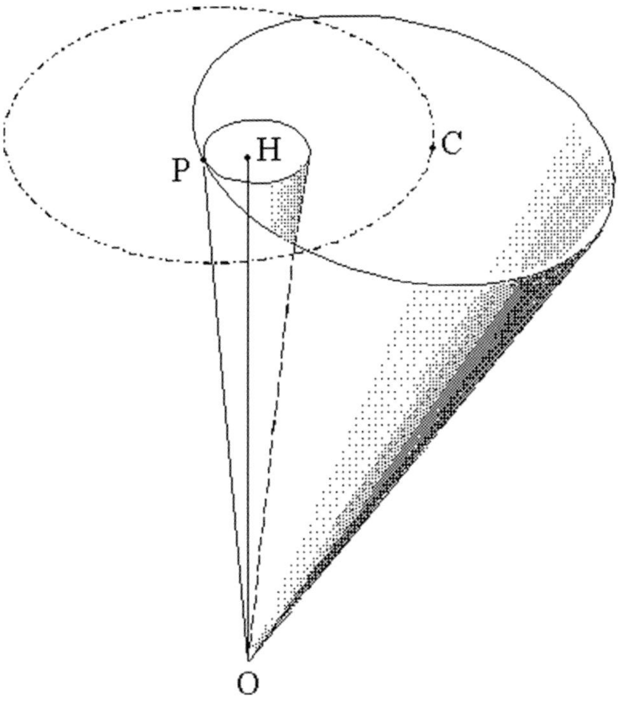

Figure 3.2 Geometric interpretation of the Earth's rotation axis motion.

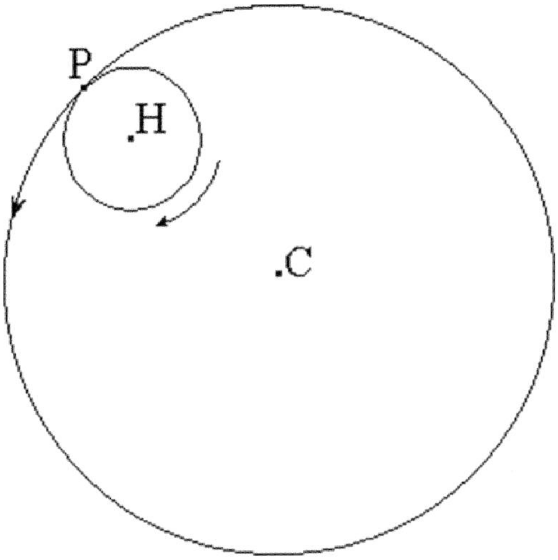

Figure 3.3 A small circle with center *H* rolls along the inside surface of a large circle with center *C*.

3.1.2
International Latitude Service

The conclusions of Euler's theory have stimulated the latitude observations. The analysis of some observations in the nineteenth century has shown that the geographical latitudes of observatories do not remain constant. The geographical latitude of a certain site is the angle between the vertical line at this site and the plane of the Earth's equator. The geographical latitude changes either with the change in the direction of the local vertical line, or with the displacement of the Earth's rotation axis inside the Earth's spheroid.

Let us assume that the latitude of an observational station changes with the changing position of the rotation axis within the Earth's body. Then, the values of the latitude' changes at two stations located at the meridians 180° away from each other should be equal (the same) but of opposite sign: at one site the latitude increases, at another site it decreases by the same value.

To test this conception, the special latitude station has been built in Honolulu on the Sandwich (nowadays the Hawaiian) islands in the Pacific Ocean in 1890. It was away from Berlin at 180° in longitude. Simultaneous observations at two stations, Berlin and Honolulu, had to either confirm or deny the hypothesis about the rotation axis displacement in the Earth's body. It was found that the curve of latitude change for Honolulu is a mirror reflection of the respective curve for Berlin. Hence, it has been proved that the change in the latitude of any point on the terrestrial surface occurs due to the displacement of the Earth's rotation axis in the planet body itself. However, if the Earth's rotation axis moves, the point of intersection of this axis with the Earth's surface (in other words, the Earth's pole) shifts along this surface, describing some curve.

Vain attempts to find the 305-day period in the series of latitude observations at the observatories of Pulkovo, Washington, Berlin, and others were repeatedly undertaken in the nineteenth century. In 1891 S. Chandler (Chandler, 1891) published the results of his analysis of the latitude observations and showed that those series had terms with periods of 428 and 365 days. At first, such unexpected results evoked some doubt. However, soon S. Newcomb pointed out that the period of 305 days is correct if we assume the Earth to be an absolutely rigid body. The Earth's and ocean's elastic deformations can increase this period from 10 to 14 months.

The 11th Conference of the International Geodetic Association (IGA) in Berlin (1895) proposed to organize the International Latitude Service (ILS) for obtaining more accurate data that would allow studying more comprehensively the motion of the Earth's geographic poles. The Central Bureau (which was founded just after the conference) started practical activity – the development of a uniform observational program, choosing the sites for stations, the instrumentation, the design of pavilions identical to all stations, and so forth.

The preliminary results of the Central Bureau activity were considered and confirmed by the 12th IGA Conference in Stuttgart in 1898.

In 1898 six latitude stations were built at the parallel of 39°08′ N, at the sites most suitable in terms of meteorological and seismological conditions. These sites are

Mizusawa (Japan); Tschardjui (Russia); Carloforte (Italy); Gaithersburg, Cincinnati, and Ukiah at the east, middle, and west parts of North America, respectively. All stations had pavilions of similar design and were equipped with zenith telescopes. The systematic observations at the ILS stations have been carried out since 1899. In 1962, the ILS was reorganized into the International Polar Motion Service (IPMS). In the 1980s, the astrooptical observations were replaced by new measurement methods based on using the very long baseline interferometers (VLBI), the satellites laser ranging (SLR), the lunar laser ranging (LLR), and the global positioning systems (GPS). It has become clear that due to new methods the accuracy of observations increased by many times. Therefore in 1988, IPMS and the Earth rotation section of the Bureau International de l'Heure (BIH) were reorganized into the International Earth Rotation Service (IERS). The IERS function was to calculate the universal time corrections and the coordinates of the Earth's pole, using the results obtained by new observational methods.

3.1.3
North Pole Motion

The position of the pole is specified by the rectangular coordinates x and y. The origin of coordinates is taken to be the pole's average position for 1900–1905. It is called the Conventional International Origin (CIO). The x axis is directed from CIO origin along the zero meridian (to Greenwich), and the y axis – along the meridian of 270° E. The measurement units are tenths of an arcsec.

The instantaneous pole moves along the Earth's surface around the mean pole in the direction of the Earth's rotation, that is, from west to east. The trajectory of the polar motion looks like a helix that periodically twists or untwists. The maximum deviation of the instantaneous pole from the helical center does not exceed 15 m.

Figure 3.4 shows the trajectory of the instantaneous North Pole in 1990–1996. In 1990–1993, the pole was twisting, more and more approaching its mean position. In 1993 it most closely approached the helical center; then, since 1994, the pole was untwisting, moving off the mean polar position. Its maximum deviation from the mean position was registered in May–July, 1996. Then, the pole began to twist and continued twisting up to 2000, when the pole approached again at the minimum distance to the helical center (Figure 3.5).

It is shown in Figure 3.5 that the center of the helix is away from the origin of coordinates – the Conventional International Origin. The reason is the so-called secular polar motion. To obtain the coordinates of the mean pole, the annual and Chandler components are eliminated from the polar coordinates. It is found that the mean pole shifts too. The trajectory of the mean pole during 1890–2000 is shown in Figure 3.5 (solid zigzag line). One can see that during the whole observational period the mean pole was shifting along a complex zigzag curve with the predominant direction towards North America (the 290° E meridian) at a velocity of about 10 cm/year.

At present, there is a continuous series of the North Pole coordinates calculated by the International Earth Rotation Service (IERS) since1890 up to the present, with

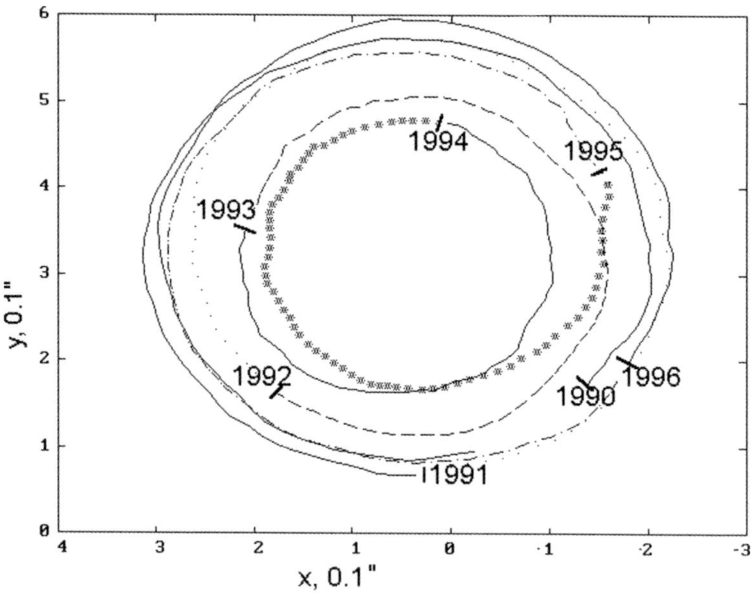

Figure 3.4 Trajectory of movement of the North geographical pole of the Earth in 1990–1996 (the yearly parts of the trajectory are marked by the respective year).

a discrete interval of 0.05 year (IERS Annual Report). Figures 3.6 and 3.7 shows the changes in the x and y polar coordinates over the whole observational period. One can see the secular- and decade-long variations and the six-year beats of the annual component. An interesting peculiarity of the polar motion is its drastic attenuating in 1925–1940 and strong "untwisting" in 1905–1915 and 1950–1960.

The classical spectral analysis of these series allows us to distinguish two dominant harmonics: Chandler's harmonic with a frequency of about $1/1.18 = 0.85$ cycle/year and the annual one – 1.0 cycle/year (Figure 3.8).

Chandler's harmonic is the main mode of the Earth's free nutation. The annual harmonic is the forced oscillation caused by the seasonal movements of the air and water masses over the Earth's surface. An addition of the Chandler and annual harmonics generates the beats, as a result of which the radius of the polar motion trajectory changes from the maximum to the minimum value (with a period of approximately 6 years) (Figures 3.6 and 3.7).

The curves 2 and 3 in Figures 3.6 and 3.7 present the temporal changes of the Chandler and annual harmonics over the whole period of observations. One can see that the Chandler harmonic has the amplitude modulation with a period of about 40 years. The maximum amplitudes were observed about 1915 and 1955, and the deep minimum – about 1930. Due to this, the spectra of the polar coordinates have a side peak at a frequency of 1/1.24 cycle/year near the Chandler frequency peak (Figure 3.8). The parameters of the annual harmonic change with time without pronounced peculiarities. The bottom curve in Figures 3.6 and 3.7 demonstrates the

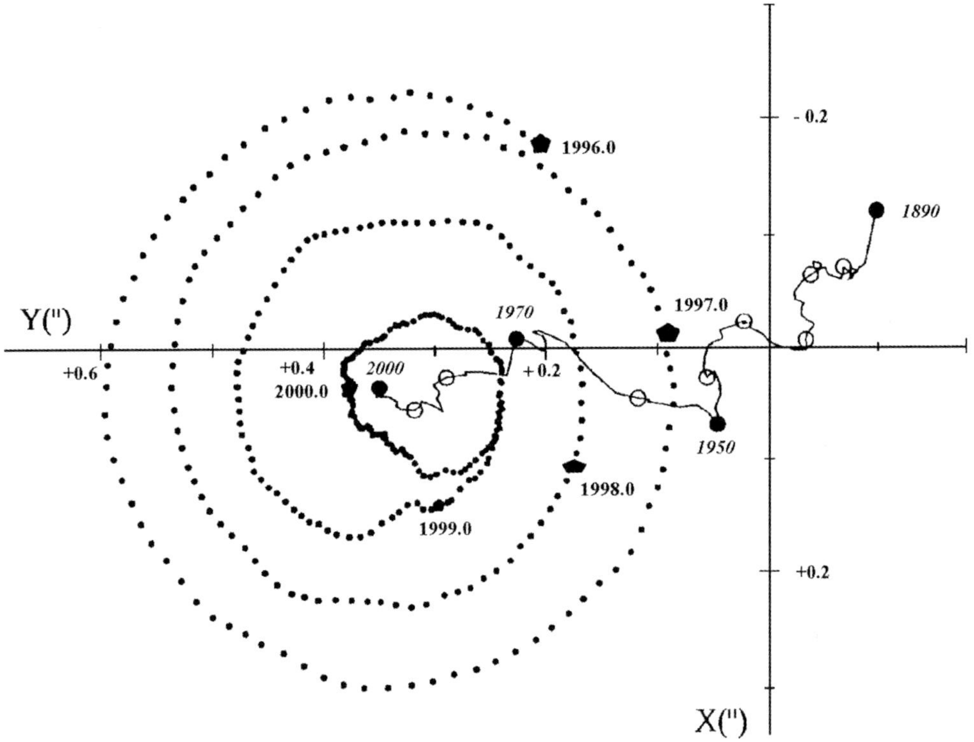

Figure 3.5 Polar motion trajectory in 1996–2000. The continuous polygonal curve denotes the path of the mean polar motion in 1890–2000 (IERS Annual Report . . ., 2002).

temporal course of residuals after the elimination of the trend and the Chandler and annual components.

3.2
Irregularities in the Earth's Rotation Rate

It is known that the Earth rotates around its axis with periods equal to 86 164.099 s and 86 400 s with respect to the motionless stars and the Sun, respectively. The mean solar day is equal to 1.002 738 sidereal days. However, the velocity of the Earth's rotation is variable.

At the end of nineteenth – the beginning of twentieth centuries, the fluctuations in the motions of the Moon, Sun, Mercury, Venus, Mars, and the satellites of Jupiter were found. It appeared that all the obtained residuals are proportional to the rates of the apparent motions of these celestial bodies. This has meant that the deviations of the observed positions of celestial bodies from their calculated positions (ephemerides) are due to the fluctuations of the rate of the Earth's daily rotation. Really, if the Earth rotation slows down, the day becomes longer and the celestial body passes this

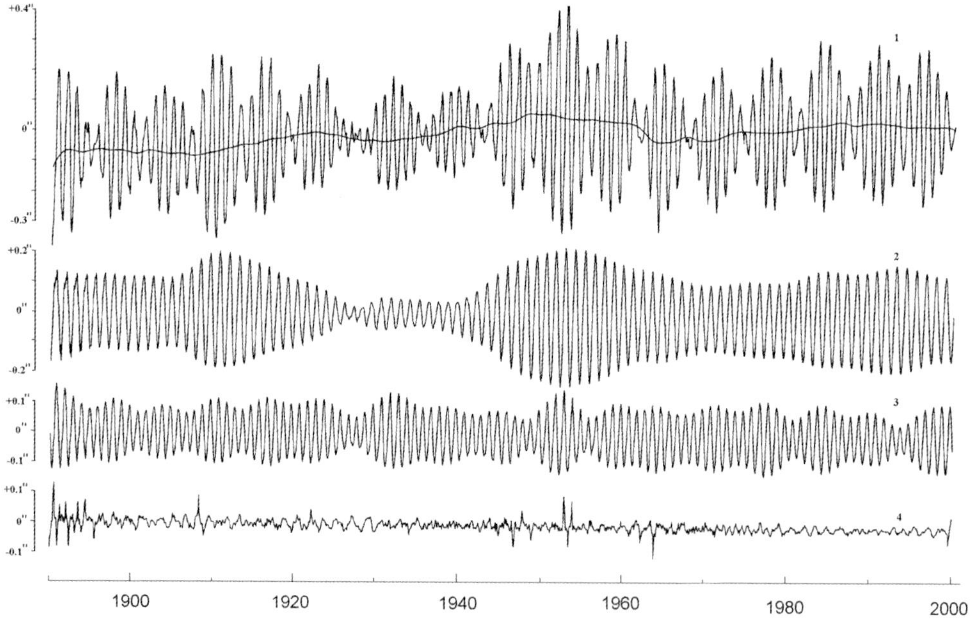

Figure 3.6 Pole's coordinate *x* and its components in 1890–2000 (IERS Annual Report . . ., 2002).

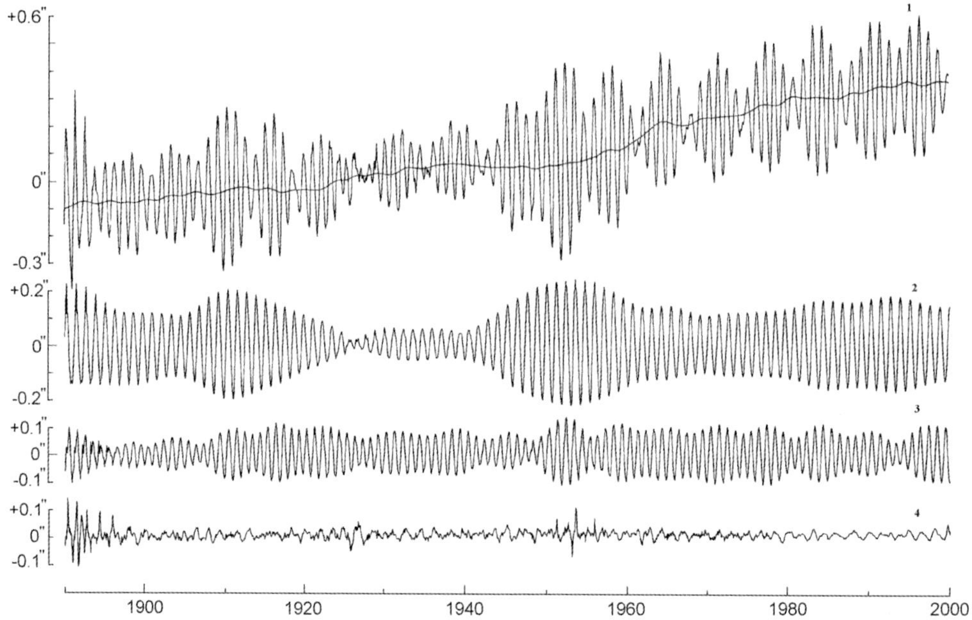

Figure 3.7 Pole's coordinate *y* and its components in 1890–2000 (IERS Annual Report . . ., 2002).

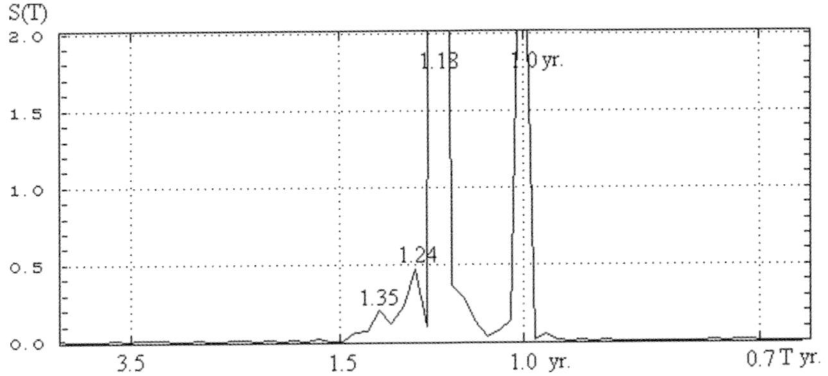

Figure 3.8 Spectrum of variations of the pole's coordinate *x*.

or that route over a shorter time period. When calculating the ephemerides, the day length is considered to be constant. Therefore, the impression is gained that the celestial body movement accelerates. In the case of acceleration of the Earth's rotation, on the contrary the impression is gained that the celestial body movement slows down. Thus, the necessity arose to separate the time scales.

In 1950 the concept of the ephemeris time (ET) – a uniformly running time – was introduced and had been used for calculations of the celestial bodies' ephemerides. The Universal Time (UT) (the local mean solar time on the meridian of Greenwich) is the irregular time, because the Earth's rotational velocity is variable. Analyzing the differences between the UTand ET time scales, it is possible to calculate the variations in the Earth's rotation velocity around the axis, or the changes in the diurnal length. Since the seventeenth century, the instrumental observations of the Moon, the Sun and some of the planets were episodically carried out, the coordinates of the planets being calculated with low precision. It was only in the nineteenth century that the accuracy of observations made it possible to satisfactorily estimate the changes in the Earth's rotational velocity.

A progressive jump in the accuracy of determining the changes in the Earth's rotation velocity occurred in 1955, when the atomic clock was created and the atomic time scale (AT or TA, according to English or French) was introduced. The introduction of the atomic time has opened up a new stage in studying the irregularity of the Earth's rotation. The reliability of determination of changes in the Earth's rotation velocity became dependent only on the accuracy of the Universal Time determination. Over the last 20 years, the classical astronomical methods of the UT determination have been gradually replaced by the new methods based on the Doppler observations of the Earth's artificial satellites, lunar laser ranging, and observations with the very long baseline interferometers. The accuracy of the UT determination has increased by two orders. As a result, there appeared the possibility to study the short periodical oscillations of the Earth's rotation velocity with periods as short as one day and even (during some periods of special series of observations) several hours.

The Earth's rotation velocity can be characterized by the dimensionless value

$$\nu = \frac{(\omega - \Omega)}{\omega} = -\frac{(T-P)}{P} \tag{3.2}$$

where T and P are the length of the terrestrial day and the length of the standard (the ephemeris or atom) day, respectively. The standard day is equal to 86 400 s, $\omega = 2\pi/T$ and $\Omega = 2\pi/P$ are the angular velocities corresponding to the terrestrial and standard days. Since the value of ω changes only in the ninth-eighth decimal digit, then the ν values have an order of 10^{-9}–10^{-8}.

Beginning from the seventeenth century, the velocity of the Earth's rotation has been determined from the observations of the Moon, the Sun and some of the planets. We have the series of the mean semiannual deviations of the day length $(T-P)$ since 1656 to the present time. The accuracy of these data for the seventeenth and eighteenth centuries is very low.

Figure 3.9 represents the curves of changes in the mean annual velocity of the Earth's rotation obtained by the data of (McCarthy and Babcock, 1986) since 1656, (Stephenson and Morrison, 1984) since 1650, and (Brower, 1952; Sidorenkov and Svirenko, 1991; Sidorenkov, 2000c) since 1820. The years and the relative deviations of the angular velocity ν (multiplied by 10^{10}) are plotted on the abscissa and ordinate axes, respectively. The divergence of the above curves allows us to judge the accuracy of determination of the rotation velocity. One can see that with time, as the accuracy increases, the agreement between the curves becomes closer and closer. The Earth's rotation velocity changed considerably in the seventeenth century and since 1860 to

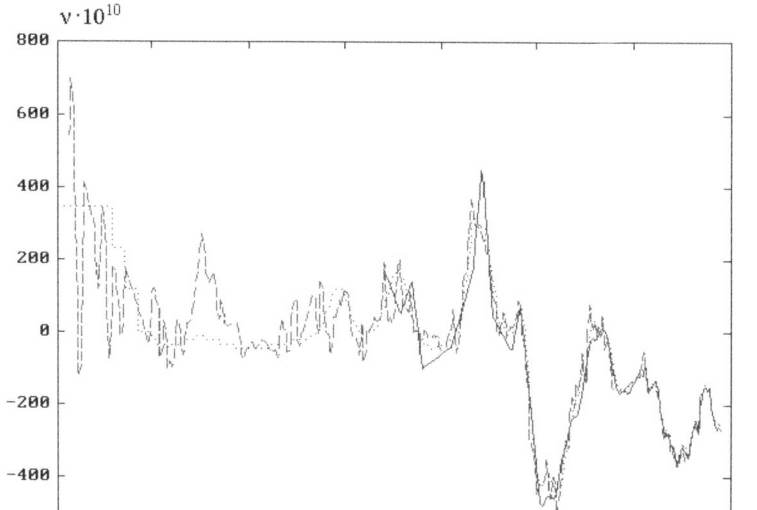

Figure 3.9 Changes in the Earth's rotation angular velocity over the last 350 years.

the present time. In 1700–1860, the fluctuations of the velocity were much smaller. Irregular fluctuations of the angular velocity of the Earth's rotation with characteristic periods of the order of several decades are well pronounced. The Earth rotated most quickly about 1870 (the ν values attained $+340 \times 10^{-10}$) and most slowly about 1903 (the value ν attained -470×10^{-10}). From 1903 to 1934, the acceleration of the Earth's rotation was observed and from the late 1930s to 1972 – the deceleration of rotation (which was sometimes replaced by short periods of slight acceleration). From 1973 to now, the velocity of rotation increases (the deceleration recorded in 1989–1994 was an ordinary fluctuation).

A low resolution of the Moon's longitude discrepancy has generated a problem of "the turning points" (Munk and MacDonald, 1960) in calculating the characteristics of the Earth's rotational velocity. Brown (Brown, 1926) thought that "the changes of the angular velocity occur most probably by jumps rather than gradually". De Sitter (de Sitter, 1927) approximated the curve of correction $\Delta t = UT - ET$ by straight lines, replacing the turning points by the arcs of parabolas. Brower (Brower, 1952) represented the curve Δt in the segments of parabolas arcs, which corresponds to the jump-like changes in accelerations $d\nu/dt$. He assumed that fluctuations $d\nu/dt$ are accidental and there is no correlation between their values in various years. According to Brower, the accumulation of small random fluctuations is responsible for the form of the curve of the decadal fluctuations ν.

In 1955, the International Atomic Time scale (TAI) was introduced. The comparison of TAI with UT makes it possible to calculate the $(T - P)$ or the ν values with a much higher accuracy than was possible earlier. There appeared the possibility to calculate the mean monthly or even five-day values of ν. The International Earth Rotation Service calculates the corrections UT–TAI of the Universal and Atomic Time scales. Using these data, it is easy to calculate the monthly mean velocity ν of the Earth's rotation from 1955 to now (Sidorenkov, 2002a, 2002b). Since 1984 the International Earth Rotation Service calculates daily values of ν.

The temporal course of the monthly means of the Earth's rotation angular velocity ν during 1955–2008 is presented in Figure 3.10. It gives a picture of the variations in the Earth's rotation velocity with the characteristic times of more than two months (that is, about the spectrum in the range of frequencies below 2×10^{-7} Hz). One can see that in 1956–1961 and in 1973–2003 the rotation of the Earth accelerated, and in 1962–1972 and from 2004 to now it decelerates. The deceleration of the Earth's rotation, which ended in 1972, had begun in 1935, that is, outside of the period represented in Figure 3.10. The acceleration that had begun in 1973 was ended in 2003. The acceleration of rotation in 1956–1961 and the deceleration in 1989–1994 are the short-term fluctuations. The seasonal variations of the Earth rotation velocity are well seen against the background of the long-term fluctuations. The rotational velocity is minimum in April and November, and maximum in January and July. One observes the quasi-two-year cyclicity of ν maxima in January and the six-year cyclicity of ν maxima in July.

Figure 3.10 shows that the Earth's rotation velocity changes gradually, without any jumps. The jumps in the data prior to 1900 are due to a large discrete interval of calculated values ν (more than 5 years). The angular accelerations of the Earth's

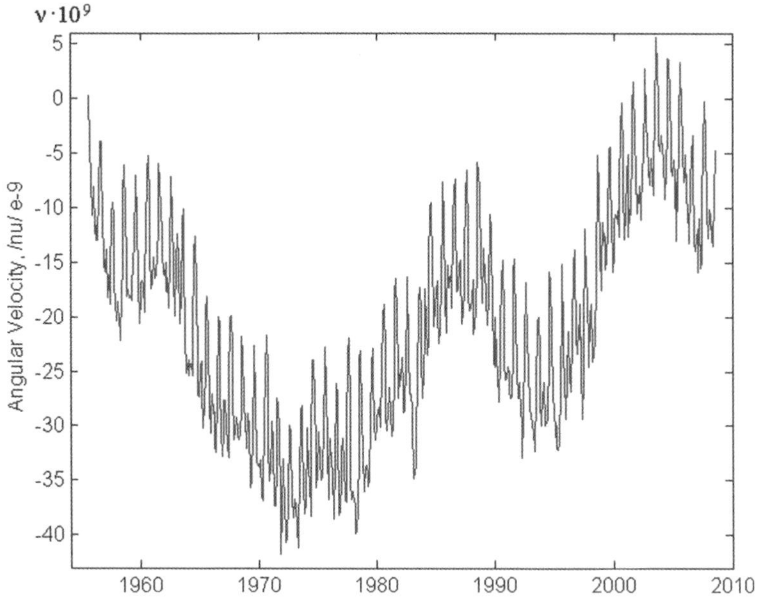

Figure 3.10 Monthly mean angular velocity of the Earth's rotation in 1955–2007.

rotation feature neither of the unexpected jump-like changes. The seasonal variations in $d\nu/dt$ are most regular.

Over the recent twenty years, there is the possibility to calculate the daily values of the Earth's rotation angular velocity ν. The temporal course of the ν daily values in 2007 as reproduced according to the IERS Annual Report data is represented in Figure 3.11 (curve 1). In addition to the seasonal variations caused by hydrometeorological processes, the tidal variations of the Earth's rotation velocity are also seen here. They rank slightly below the seasonal variations in amplitudes, but their periods are shorter by tens of times than the periods of seasonal variations, being close to 14 days.

The components with the annual and semiannual periods, the 13.7, 27.3, 9.1 day-long periods are distinguished in the spectrum of the variations of the Earth's rotation velocity. The spectral analysis of all the available series of the ν mean values reveals the maximum of the spectral density at a period of about 70 years. Fluctuations with this period are most pronounced over the last 150 years. At the beginning of the twentieth century, the amplitude of the 70-year long fluctuations was as long as 2 ms. Slight fluctuations with periods of 33, 22, and 6 years are also distinguished.

It is known that the Earth's rotational velocity decelerates under the influence of the tidal friction. It is very difficult to reveal this deceleration using the observations for the last 350 years. A longer series of observations is needed for this purpose. The tendency of rotation to slow down over the last 100–150 years cannot be indentified with confidence with the secular deceleration, because it can also be considered as a drop in some irregular speed change, having a typical time of some hundreds of years.

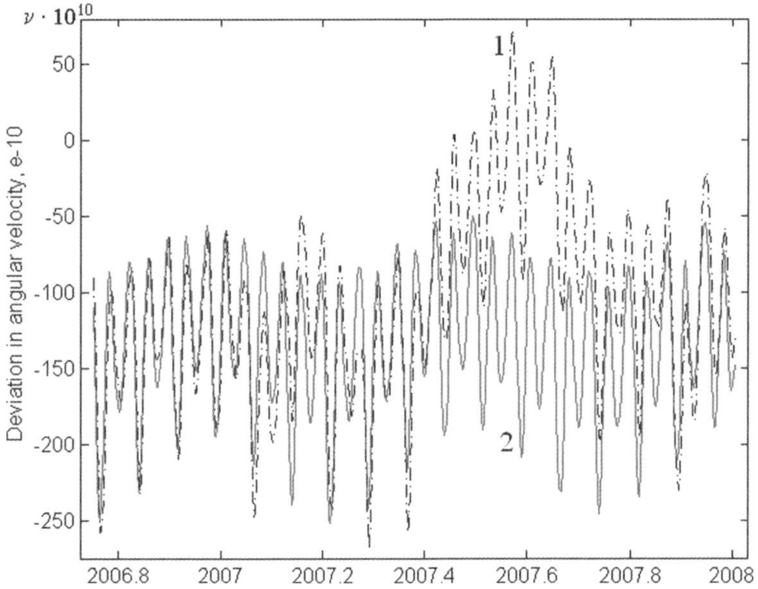

Figure 3.11 Daily angular velocity of the Earth's rotation during the period of October 1, 2006–December 31, 2007: (1) data of observations; (2) tidal oscillations, which were calculated using the theory of Section 5.3.

It is always necessary to take into account that if a random process has a component, the period of which exceeds the length of realization, a systematic change or a trend can be observed. When investigating the tidal deceleration of the Earth's rotation velocity, one often forgets about this fact and, as a result, the estimations of the tidal deceleration are not identical for the various observational periods. For example, Morrison (Morrison, 1973), after having treated the instrumental observations for 1663–1972, has obtained the velocity deceleration equal to 1.5 ms per century. The investigations of the ancient astronomical observations (first of all, the Sun and Moon eclipses) show that for the last 3000 years the deceleration of the velocity was 2.3 ms per century (Newton, 1970; Stephenson, 1997).

4
Estimation Theory of the Effect of Atmospheric Processes on the Earth's Spin

4.1
General Differential Equations of the Rotation of the Earth Around its Mass Center

Let us consider the motion of the Earth around its mass center. The translational motion of the Earth's mass center in space does not concern us in this study. We assume it to be known and completely eliminated. Let us take the system of the moving Cartesian coordinates (Ox_i), which is constantly fixed in the Earth and rotates with it, with angular velocity $\boldsymbol{\omega}$, relative to the inertial coordinates system ($O\boldsymbol{\xi}_i$) fixed in space. Let both coordinate systems have the same datum point – the Earth's mass center O and let axes Ox_i coincide with the principal axes of the ellipsoid of inertia of the undisturbed Earth. The basis of the rotating system $\{\boldsymbol{e}_1, \boldsymbol{e}_2, \boldsymbol{e}_3\}$ is chosen in such a way that the first two unit vectors specify the equatorial plane and the third one is directed along axis Ox_3.

As is known from theoretical mechanics, in the inertial system of coordinate $O\boldsymbol{\xi}_i$ the velocity of changes in the Earth's angular momentum $\boldsymbol{H}$ relative to the mass center is equal to the moment of external forces (torque) $\boldsymbol{M}$ on the Earth relative to the same center O.

$$\frac{\mathrm{D}\boldsymbol{H}}{\mathrm{D}t} = \boldsymbol{M} \tag{4.1}$$

where $\mathrm{D}/\mathrm{D}t$ is the sign of the time derivative in the inertial coordinates system $O\boldsymbol{\xi}_i$.

If we write Equation 4.1 for the inertial coordinate system $O\xi_i$, then, differentiating $\boldsymbol{H}$ with respect to time, we must also find the derivatives of the moments of inertia with respect to time, which is very inconvenient.

It is much more convenient to consider the variations in the Earth's angular momentum $\boldsymbol{H}$ with respect to the moving coordinate system Ox_i. It is known that the absolute derivative of vector $\boldsymbol{H}$ with respect to coordinate system $O\boldsymbol{\xi}_i$ is connected with the local derivative of the same vector relative to coordinate system Ox_i by the following equation:

$$\frac{\mathrm{D}\boldsymbol{H}}{\mathrm{D}t} = \frac{\mathrm{d}\boldsymbol{H}}{\mathrm{d}t} + [\omega \times \boldsymbol{H}] \tag{4.2}$$

The Interaction Between Earth's Rotation and Geophysical Processes. Nikolay S. Sidorenkov

ISBN: 978-3-527-40875-7

The variation of vector $\boldsymbol{H}$ relative to the moving coordinate system Ox_i consists of two components: one is due to the variation of vector $\boldsymbol{H}$ itself, whereas another is due to the motion of axes Ox_i onto which it is projected. For the vector $\boldsymbol{H}$ this variation is equal to $[\boldsymbol{\omega} \times \boldsymbol{H}]$, similar to the way that it was equal to $[\boldsymbol{\omega} \times \boldsymbol{r}]$ for radius vector $\boldsymbol{r}$ for the linear velocity of the point $\boldsymbol{V} = [\boldsymbol{\omega} \times \boldsymbol{r}]$.

Substituting expression (4.2) into (4.1), we obtain the differential equation of the Earth's motion with respect to its mass center in the moving coordinate system Ox_i:

$$\frac{\mathrm{d}\boldsymbol{H}}{\mathrm{d}t} + [\boldsymbol{\omega} \times \boldsymbol{H}] = \boldsymbol{M} \tag{4.3}$$

If expressed in the projections onto axes Ox_1, Ox_2, and Ox_3, this equation, with account for the expansion in terms of the basis:

$$\boldsymbol{H} = \sum_{i=1}^{3} H_i \boldsymbol{e}_i \equiv H_i \boldsymbol{e}_i, \quad \omega = \omega_i \boldsymbol{e}_i, \quad \boldsymbol{M} = M_i \boldsymbol{e}_i,$$

takes the form:

$$\begin{aligned} \frac{\mathrm{d}H_1}{\mathrm{d}t} + \omega_2 H_3 - \omega_3 H_2 &= M_1 \\ \frac{\mathrm{d}H_2}{\mathrm{d}t} + \omega_3 H_1 - \omega_1 H_3 &= M_2 \\ \frac{\mathrm{d}H_3}{\mathrm{d}t} + \omega_1 H_2 - \omega_2 H_1 &= M_3 \end{aligned} \tag{4.4}$$

Equations 4.4 are the differential equations of the Earth's motion that are related to the moving coordinate system Ox_i that is rigidly bound to the Earth.

The vector of angular momentum $\boldsymbol{H}$ is specified by the equation:

$$\boldsymbol{H} = \int_A (\boldsymbol{r} \times \boldsymbol{W}) \rho \, \mathrm{d}V \tag{4.5}$$

where $\boldsymbol{r}$ is the radius-vector of the unit volume $\mathrm{d}V$, ρ is the density, A is the volume of the body, $\boldsymbol{W}$ is the velocity of the volume motion with respect to the inertial system $O\xi_i$ (the absolute velocity):

$$\boldsymbol{W} = (\boldsymbol{\omega} \times \boldsymbol{r}) + \boldsymbol{u} \tag{4.6}$$

where $\boldsymbol{u}$ is the velocity of motion of volume $\mathrm{d}V$ with respect to the moving system Ox_i (the relative velocity). Substituting the expression for $\boldsymbol{W}$ into (4.5), we find:

$$\boldsymbol{H} = \int_A \boldsymbol{r} \times (\omega \times \boldsymbol{r}) \rho \, \mathrm{d}V + \boldsymbol{h} = \int_A [r^2 \omega - \boldsymbol{r}(\omega \, \boldsymbol{r})] \rho \, \mathrm{d}V + \boldsymbol{h} \tag{4.7}$$

where $\boldsymbol{h} = \int_A (\boldsymbol{r} \times \boldsymbol{u}) \rho \mathrm{d}V$ is the relative angular momentum. Introducing the tensor notations, one can rewrite Equation 4.7 in the form:

$$H_i = \int_A (x_l^2 \omega_i - x_i x_k \omega_k) \rho \mathrm{d}V + h_i = \omega_k \int_A (x_l^2 \delta_{ik} - x_i x_k) \rho \mathrm{d}V + h_i$$
$$= N_{ik} \omega_k + h_i \tag{4.8}$$

Here, N_{ik} is the inertia tensor.

Hence, the projections of the angular momentum H_i onto the axes Ox_i are the linear functions of the projections of the angular velocity ω_i, whose coefficients are the components of the inertia tensor N_{ij}:

$$\begin{aligned} H_1 &= N_{11}\omega_1 + N_{12}\omega_2 + N_{13}\omega_3 + h_1 \\ H_2 &= N_{21}\omega_1 + N_{22}\omega_2 + N_{23}\omega_3 + h_2 \\ H_3 &= N_{31}\omega_1 + N_{32}\omega_2 + N_{33}\omega_3 + h_3 \end{aligned} \tag{4.9}$$

The components of the inertia tensor are calculated by the formulas:

$$N_{ij} = \begin{pmatrix} \int_A \rho(x_2^2 + x_3^2)\,\mathrm{d}V & -\int_A \rho x_1 x_2\,\mathrm{d}V & -\int_A \rho x_1 x_3\,\mathrm{d}V \\ -\int_A \rho x_2 x_1\,\mathrm{d}V & \int_A \rho(x_1^2 + x_3^2)\,\mathrm{d}V & -\int_A \rho x_2 x_3\,\mathrm{d}V \\ -\int_A \rho x_3 x_1\,\mathrm{d}V & -\int_A \rho x_3 x_2\,\mathrm{d}V & \int_A \rho(x_1^2 + x_2^2)\,\mathrm{d}V \end{pmatrix} \tag{4.10}$$

The diagonal components of the inertia tensor are called the Earth's axial moments of inertia relative to the axes of Ox_1, Ox_2, and Ox_3; the other components are the centrifugal moments of inertia with respect to the same axes. The axes of the coordinates in which the inertia tensor is brought to the diagonal form are called the principal axes of inertia; and the corresponding axial moments of inertia are called the principal moments of inertia.

The h_i values are the relative angular momentums that are due to the interior motions of substances (air, water, and so on) with respect to the coordinate system Ox_i:

$$\begin{aligned} h_1 &= \int_A \rho(x_2 u_3 - x_3 u_2)\mathrm{d}V \\ h_2 &= \int_A \rho(x_3 u_1 - x_1 u_3)\mathrm{d}V \\ h_3 &= \int_A \rho(x_1 u_2 - x_2 u_1)\mathrm{d}V \end{aligned} \tag{4.11}$$

4.2
Disturbed Motion of the Absolutely Solid Earth

The redistribution of masses and the interior relative motions cause small displacements of the Earth relative to the axis of rotation. To describe these displacements, we use the dimensionless components of the instant angular velocity of the Earth's rotation ν_i:

$$\nu_1 = \frac{\omega_1}{|\omega|}, \quad \nu_2 = \frac{\omega_2}{|\omega|}, \quad \nu_3 = \frac{\omega_3 - |\omega|}{|\omega|}$$

Here, $|\omega| \approx \Omega = \text{const}$ is the modulus of the vector of the angular velocity of the Earth's rotation; ν_1, ν_2, and $(1 + \nu_3)$ are the direction cosines of the instant axis of the Earth's rotation ($\nu_3 \leq 0$). They can also be interpreted as the components of the vector of small additions ν:

$$\nu = \frac{\omega - \Omega e_3}{\Omega} = \nu_1 e_1 + \nu_2 e_2 + \nu_3 e_3 \qquad (4.12)$$

where $\Omega = 7.29 \times 10^{-5}\,\text{s}^{-1}$ is the undisturbed angular velocity of the Earth's rotation, which corresponds to the period $T_0 = 2\pi/\Omega = 86\,174$ s; e_i are the unit vectors of the axes Ox_i.

In the undisturbed state, the axial moments of inertia can be considered constant values and the centrifugal moments are equal to zero. The redistribution of masses in the atmosphere, hydrosphere, and so on slightly changes (disturbs) the components of the inertia tensor. It is natural to assume these disturbances to be small values. They can be presented in the form of small additions to the values $(A, B, C, 0, 0, 0)$ of the components of the inertia tensor of the undisturbed Earth–atmosphere system:

$$N_{11} = A + \tilde{n}_{11};\ N_{22} = B + \tilde{n}_{22};\ N_{33} = C + \tilde{n}_{33};$$
$$N_{12} = N_{21} = \tilde{n}_{12};\ N_{13} = N_{31} = \tilde{n}_{13};\ N_{23} = N_{32} = \tilde{n}_{23}.$$

The relative angular momentums h_i are also assumed to be small values.

Linearizing Equation 4.4 relative to small values ν_i, $\tilde{n}_{ij}$, h_i, that is, neglecting the productions of the form $\tilde{n}_{ij} d\nu_k/dt$, $\nu_j d\tilde{n}_{ik}/dt$, $\Omega \nu_j \tilde{n}_{kl}$, $\nu_j \nu_l C\Omega$, we obtain:

$$\begin{aligned}
&A\Omega \frac{d\nu_1}{dt} + \Omega^2 (C - B)\nu_2 = \Omega^2 \tilde{n}_{23} + \Omega h_2 - \Omega \frac{d\tilde{n}_{13}}{dt} - \frac{dh_1}{dt} + M_1 \\
&B\Omega \frac{d\nu_2}{dt} - \Omega^2 (C - A)\nu_1 = -\Omega^2 \tilde{n}_{13} - \Omega h_1 - \Omega \frac{d\tilde{n}_{23}}{dt} - \frac{dh_2}{dt} + M_2 \\
&C\Omega \frac{d\nu_3}{dt} = -\Omega \frac{d\tilde{n}_{33}}{dt} - \frac{dh_3}{dt} + M_3
\end{aligned} \qquad (4.13)$$

Let us simplify Equations 4.13. For this purpose, we consider the Earth–atmosphere system in the undisturbed state to be dynamically symmetric, that is, its moments of inertia with respect to the equatorial axes to be similar $(A = B)$.

Dividing all the terms of the first two Equations 4.13 by $\Omega^2(C-A)$ and denoting the circular frequency of the free motion of the poles of the absolutely solid Earth (Euler's nutation) as $\sigma_r = \frac{C-A}{A}\Omega$, we found:

$$\frac{1}{\sigma_r}\frac{d\nu_1}{dt} + \nu_2 = \varphi_2$$
$$\frac{1}{\sigma_r}\frac{d\nu_2}{dt} - \nu_1 = -\varphi_1 \qquad (4.14)$$
$$\frac{d\nu_3}{dt} = \frac{d\varphi_3}{dt}$$

Here, φ_i are the dimensionless components of the excitation function:

$$\Omega^2(C-A)\varphi_1 = \Omega^2\tilde{n}_{13} + \Omega h_1 + \Omega\frac{d\tilde{n}_{23}}{dt} + \frac{dh_2}{dt} - M_2$$
$$\Omega^2(C-A)\varphi_2 = \Omega^2\tilde{n}_{23} + \Omega h_2 - \Omega\frac{d\tilde{n}_{13}}{dt} - \frac{dh_1}{dt} + M_1 \qquad (4.15)$$
$$\Omega^2 C\varphi_3 = -\Omega^2\tilde{n}_{33} - \Omega h_3 + \Omega\int_0^t M_3 dt'$$

Equation 4.14 are called the Euler–Liouville equations. They link the changes in the directing cosines of the Earth's rotation axis with the excitation functions that describe all kinds of geophysical phenomena. In particular, the terms of the excitation functions 4.15 that contain the variable components of the inertia tensor $\tilde{n}_{ij}$ and $\Omega\, d\tilde{n}_{ij}/dt$ reflect the effect of the redistribution of masses (air, water, ice, snow, and so on) occurring on the Earth, whereas the terms of the form of Ωh_i and dh_i/dt account for the relative movements (winds and currents). The terms M_i describe the effect of the moments of external forces acting on the Earth from the outer space. The first two equations of system (4.14) describe the poles' motion and the third equation describes changes in the Earth's rotational velocity.

These forms of Equations 4.14 are given in the publications of Pariiski (1954) and Munk and Macdonald (1960), who, in their turn, refer to Young's (1953) publication. However, the excitation function was already being used from the end of the nineteenth century (see, for example, Volterra (1895)).

It is shown by Barnes *et al.* (1983) that when estimating the effect of the atmosphere and oceans on the instability of the Earth's rotation, it is expedient to use the functions of the angular momentum instead of the excitation functions φ_i:

$$\tilde{\chi}_1 = \frac{1}{\Omega(C-A)}(\Omega\tilde{n}_{13} + h_1)$$
$$\tilde{\chi}_2 = \frac{1}{\Omega(C-A)}(\Omega\tilde{n}_{23} + h_2) \qquad (4.16)$$
$$\tilde{\chi}_3 = \frac{1}{C\Omega}(\Omega\tilde{n}_{33} + h_3)$$

These functions do not contain derivatives, which facilitates both the calculation of their magnitudes and the interpretation of various effects.

Let us use complex values $\underline{m} = \nu_1 + i\nu_2$, $\underline{\varphi} = \varphi_1 + i\varphi_2$, $\underline{\tilde{\chi}} = \tilde{\chi}_1 + i\tilde{\chi}_2$, and $\underline{M} = M_1 + iM_2$ (their meaning is the coordinates of the instant poles of rotation, excitation, angular momentum, and moment of external forces, respectively). Multiplying the first equation of system (4.14) by the value i and subtracting from it the second equation, we obtain the equation for describing the motion of the Earth's poles:

$$\frac{i}{\sigma_r}\frac{d\underline{m}}{dt} + \underline{m} = \underline{\varphi} = \underline{\tilde{\chi}} - \frac{i}{\Omega}\frac{d\underline{\tilde{\chi}}}{dt} + \frac{i\underline{M}}{\Omega^2(C-A)} \tag{4.17}$$

If excitation is absent ($\underline{\varphi} = 0$) but at the initial moment $t = t_0$ there is slope of the instant axis of rotation $\underline{m}(t_0) = \underline{m}_0$, then the solution of Equation 4.17 has the form:

$$\underline{m} = \underline{m}_0 e^{i\sigma_r(t-t_0)} \tag{4.18}$$

Hence, at a small deviation of the axis of the absolutely solid Earth from the axis of the maximum moment of inertia Ox_3, the instant axis of rotation makes the circular motion over the cone surface around Ox_3. The radius of this motion is constant and equal to the initial deviation $\underline{m}_0$; its period is $2\pi/\sigma_r = 304.4$ days, because

$$\frac{2\pi}{\sigma_r} = \frac{A}{C-A}\frac{2\pi}{\Omega} = \frac{1-H}{H} \cdot 1\ \text{day}$$

where $H = \frac{C-A}{C} = \frac{1}{305.44}$ is the precession constant, or the dynamical flattening of the Earth.

However, observations are indicative of the free motion of the poles with a period of not 304.4 days but approximately 430 days (Chandler, 1891). First, Newcomb (1891) qualitatively and then Love and Larmore quantitatively (Jeffreys, 1959) have shown that this elongation of the period of the poles' free motion is due to the fact that the Earth is not an absolutely solid body but an elastic deformable body.

4.3 Disturbed Motion of the Elastic Earth

The deformations of the Earth depend on both the value of the mechanical stress and its duration. In the case of the stress short duration (the characteristic time $t \leq 1$ year), the Earth is generally assumed to be an absolutely elastic body, and its deformations are described by way of introducing the respective Love numbers (Love, 1927). The effects associated with more prolonged processes require accounting for the inelastic deformations, the theory of which has not been developed yet.

The Earth's deformations are reflected in the values of those terms in the excitation function (4.15) that describe the pressure exerted to the Earth. Among these are the terms containing the components of the inertia tensor n_{ij}, which reflect the redistribution of masses over the Earth's surface and the components of stresses of external forces M_i, normal to the geoid.

As an example, let us consider the action of the air mass excess on the Earth. Under the pressure of this mass $\Delta P = -mg$, the Earth deflects; at the same time, the air mass attracts the Earth. As a result, the Earth's surface rises. The total action of pressure and attraction shifts the Earth's surface by value $h'U/g$, where U is the internal gravitational potential caused by the excess of air mass; g is the gravity acceleration; m is the anomaly of mass within a unit section of the atmospheric column; h' is Love's number. The redistribution of masses in the Earth's interior entails an additional gravitational potential $k'U$. Since the effect of pressure exceeds the effect of attraction, the Earth's surface deflects consequently, Love's numbers h' and k' are negative. Munk and Macdonald (1960) have shown that

$$k' = -\frac{1}{1+\mu} \approx -0.30 \tag{4.19}$$

where μ is the parameter reflecting the rigidity of the Earth. Notice that the Earth is here assumed homogeneous and incompressible.

Thus, the effect of the excessive air mass over the absolutely solid Earth is expressed by the additions n_{ij} to the components of the inertia tensor; in the case of the elastic Earth, further additions $k'n_{ij}$ should be introduced. The total addition will be $n'_{ij} = (1+k')n_{ij} \approx 0.70n_{ij}$. It is natural that such effects as winds, forces of surface drag friction, and horizontal components of pressure forces do not cause the vertical deformations of the Earth. Therefore, it is not necessary to change the respective terms involved in the excitation function (4.15).

Apart from the deformations under the effect of prolonged loads and pressure, the elastic Earth is subject to deformations due to its rotation. These deformations are similar to the displacements due to the potential centrifugal forces $U_\omega = \frac{1}{2}\omega^2 l^2$, where ω is the instant angular velocity, l is the distance between point Γ and the Earth's instant axis PP' (Figure 4.1).

It is seen from Figure 4.1 that

$$\begin{aligned} l^2 &= r'^2 \sin^2\theta = r'^2 - r'^2\cos^2\theta = (x_1^2+x_2^2+x_3^2) - [\nu_1 x_1 + \nu_2 x_2 + (1+\nu_3)x_3]^2 \\ &\approx x_1^2 + x_2^2 - 2\nu_3 x_3^2 - 2x_3(\nu_1 x_1 + \nu_2 x_2) \end{aligned}$$

Here, ν_1, ν_2, and $(1+\nu_3)$ are the direction cosines of the instant axis of rotation (4.12); θ is the polar angle.

Thus, the potential of centrifugal forces consists of (with an accuracy to the terms of the second order of smallness) the sum of the time-independent terms $\frac{1}{2}\omega^2(x_1^2+x_2^2-2\nu_3 x_3^2)$, which are additions reflecting an increase in the ellipticity, and the terms $\omega^2 x_3(\nu_1 x_1 + \nu_2 x_2)$, which vary with time. The latter terms present the spherical harmonic of degree 2 and describe the Earth's elastic deformation that creates the gravitational potential outside the Earth:

$$V = -k\frac{a^5}{r^5}\omega^2 x_3(\nu_1 x_1 + \nu_2 x_2) \tag{4.20}$$

where a is the radius of the Earth, r is the distance between the Earth's mass center and the point of the circumterrestrial space under consideration, k is Love's number.

At the same time, it is known that the gravitational potential of the deformed Earth is described by MacCulagh's formula (Jeffreys and Swirles, 1966):

$$W = \frac{\gamma M}{r} + \frac{\gamma}{2r^3}(N_{11} + N_{22} + N_{33} - 3I) \tag{4.21}$$

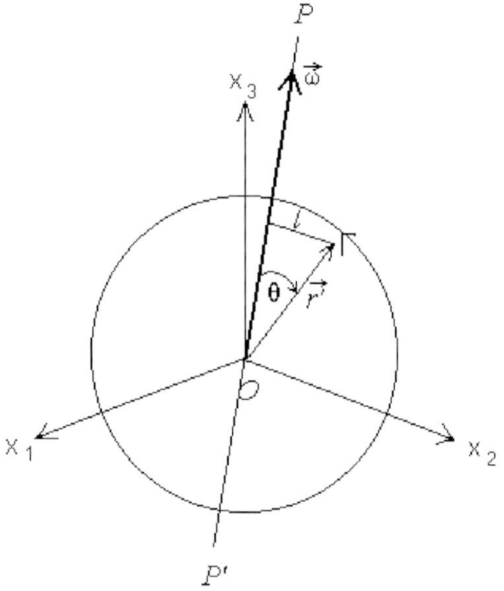

Figure 4.1 A scheme for considering the potential of centrifugal forces.

where M is the mass of the Earth; γ is the gravitational constant; $I = N_{11}\nu_1^2 + N_{22}\nu_2^2 + N_{33}(\nu_3+1)^2 + 2N_{12}\nu_1\nu_2 + 2N_{13}\nu_1(1+\nu_3) + 2N_{23}\nu_2(1+\nu_3)$ is the moment of inertia of the Earth relative to its instant axis of rotation PP′ (Figure 4.1).

Multiplying and dividing the second term in (4.21) by r^2, one can rewrite MacCulagh's formula, with account for equalities $x_1 = r\nu_1, x_2 = r\nu_2$, and, $x_3 = r(1+\nu_3)$ in the form:

$$W = \frac{\gamma M}{r} + \frac{\gamma}{2r^5}\left[\begin{array}{l} N_{11}(x_2^2 + x_3^2 - 2x_1^2) + N_{22}(x_1^2 + x_3^2 - 2x_2^2) \\ + N_{33}(x_1^2 + x_2^2 - 2x_3^2) - 6N_{12}x_1x_2 - 6N_{13}x_1x_3 - 6N_{23}x_2x_3 \end{array}\right] \tag{4.22}$$

Equating the terms that contain the same factors x_ix_j in (4.20) and (4.21), we obtain:

$$\begin{aligned} N_{13} &= \frac{k}{3\gamma}\omega^2 a^5 \nu_1 = \frac{kq}{3} Ma^2 \nu_1 \equiv \Re \nu_1 \\ N_{23} &= \frac{k}{3\gamma}\omega^2 a^5 \nu_2 = \frac{kq}{3} Ma^2 \nu_2 \equiv \Re \nu_2 \end{aligned} \tag{4.23}$$

Here, we use parameter $q = \omega^2 a / \frac{\gamma M}{a^2} = 0.003\,4678$, which is equal to the ratio between the centrifugal acceleration and the gravity acceleration, and introduce the designation $\Re = \frac{kq}{3} Ma^2$.

Now let us deduce the equations of the elastic Earth-disturbed motion. For this purpose we use the same components of the instant angular velocity of the Earth's rotation ν_i that were used before. The components of the inertia tensor of the

Earth–atmosphere system can be represented in the following form:

$$\begin{aligned} &N_{11} = A+\tilde{n}'_{11};\ N_{22} = B+\tilde{n}'_{22};\ N_{33} = C+\tilde{n}'_{33};\ N_{12} = N_{21} = \tilde{n}'_{12} \\ &N_{13} = N_{31} = \Re v_1 + \tilde{n}'_{13};\ N_{23} = N_{32} = \Re v_2 + \tilde{n}'_{23} \end{aligned} \tag{4.24}$$

where $\tilde{n}'_{ij} = (1+k')\tilde{n}_{ij}$; A, B, and C are the principal moments of inertia of the undisturbed Earth; $\tilde{n}_{ij}$ is the variable components of the inertia tensor (for example, those of the atmosphere).

Linearizing Equation 4.4 relative to small values ν_i, $\tilde{n}'_{ij}$, and h_i, we obtain:

$$\begin{aligned} &(A+\Re)\Omega\frac{d\nu_1}{dt} + \Omega^2(C-B-\Re)\nu_2 = \Omega^2\tilde{n}'_{23} + \Omega h_2 - \Omega\frac{d\tilde{n}_{13}}{dt} - \frac{dh_1}{dt} + M_1 \\ &(B+\Re)\Omega\frac{d\nu_2}{dt} - \Omega^2(C-A-\Re)\nu_1 = -\Omega^2\tilde{n}'_{13} - \Omega h_1 - \Omega\frac{d\tilde{n}'_{23}}{dt} - \frac{dh_2}{dt} + M_2 \\ &C\Omega\frac{d\nu_3}{dt} = -\Omega\frac{d\tilde{n}'_{33}}{dt} - \frac{dh_3}{dt} + M_3 \end{aligned} \tag{4.25}$$

With a sufficient degree of accuracy, we may, as previously, assume that $A = B$.

Dividing all the terms of the first two equations of system (4.25) by $\Omega^2(C-A-\Re)$ and designating $\sigma_0 = \frac{C-A-\Re}{A+\Re}\Omega$, we found:

$$\begin{aligned} &\frac{1}{\sigma_0}\frac{dv_1}{dt} + v_2 = \psi_2 \\ &\frac{1}{\sigma_0}\frac{dv_2}{dt} - v_1 = -\psi_1 \\ &\frac{dv_3}{dt} = \frac{d\psi_3}{dt} \end{aligned} \tag{4.26}$$

where ψ_i are the dimensionless components of the excitation function:

$$\begin{aligned} \Omega^2(C-A)\psi_1 &= \kappa\left(\Omega^2\tilde{n}'_{13} + \Omega h_1 + \Omega\frac{d\tilde{n}'_{23}}{dt} + \frac{dh_2}{dt} - M_2\right) \\ &= \Omega^2 n_{13} + \Omega\kappa h_1 + \Omega\frac{dn_{23}}{dt} + \kappa\frac{dh_2}{dt} - \kappa M_2 = \Omega^2(C-A)\left(\chi_1 + \frac{1}{\Omega}\frac{d\chi_2}{dt}\right) - \kappa M_2 \\ \Omega^2(C-A)\psi_2 &= \kappa\left(\Omega^2\tilde{n}'_{23} + \Omega h_2 - \Omega\frac{d\tilde{n}'_{13}}{dt} - \frac{dh_1}{dt} + M_1\right) \\ &= \Omega^2 n_{23} + \Omega\kappa h_2 - \Omega\frac{dn_{13}}{dt} - \kappa\frac{dh_1}{dt} + \kappa M_1 = \Omega^2(C-A)\left(\chi_2 - \frac{1}{\Omega}\frac{d\chi_1}{dt}\right) + \kappa M_1 \\ \Omega^2 C\psi_3 &= -\Omega^2\tilde{n}'_{33} - \Omega h_3 + \Omega\int_0^t M_3 dt' \\ &= -\Omega^2(1+k')n_{33} - \Omega h_3 + \Omega\int_0^t M_3 dt' = -\Omega^2 C\chi_3 + \Omega\int_0^t M_3 dt' \end{aligned} \tag{4.27}$$

Here, we have introduced the designation $\frac{C-A}{C-A-\Re} \equiv \kappa$.

It is known from the theory of the Earth's figure that $C-A=J_2Ma^2$, where $J_2=1.08264\times10^{-3}$ is the gravitational moment (Zharkov and Trubitsyn, 1978). Consequently,

$$\kappa=\frac{J_2}{J_2-kq/3}\approx 1.43 \tag{4.28}$$

In (4.28) Love's number k was assumed equal to 0.28. It is easy to see that $\kappa(1+k')\approx 1$, that is, the increase in the disturbance resulting from deformation due to rotation is compensated by the decrease in the disturbance resulting from deflection due to load. Therefore, the first terms of expressions for ψ_1 and ψ_2 (and also for χ_1 and χ_2), which describe the effect of redistribution of masses, become eventually the same as those for the absolutely solid Earth (4.15). The rest terms of the excitation functions ψ_1 and ψ_1 (and also of the functions of the equatorial angular momentum χ_1 and χ_2), which are irrelevant to deformations due to deflection, increase by κ times because of the effect of deformations due to rotation.

The terms representing the excitation function ψ_3 and the function of the axial angular momentum χ_3 are virtually irrelevant to the effect of deformations due to rotation but are associated with deformations due to load. Therefore, only the term $\Omega^2\tilde{n}_{33}$ that describes the redistribution of masses should be reduced (multiplied by $1+k'\approx 0.7$).

Let us use complex values $\underline{m}=\nu_1+i\nu_2$, $\underline{\psi}=\psi_1+i\psi_2$, and $\underline{\chi}=\chi_1+i\chi_2$. Then the first two Equation 4.26 that describe the poles' motion for the elastic Earth produce:

$$\frac{i}{\sigma_0}\frac{d\underline{m}}{dt}+\underline{m}=\underline{\psi}=\underline{\chi}-\frac{i}{\Omega}\frac{d\underline{\chi}}{dt}+\frac{i\kappa\underline{M}}{\Omega^2(C-A)} \tag{4.29}$$

If excitation is absent ($\underline{\psi}=0$) but there is slope ($\underline{m}(t_0)=\underline{m}_0\neq 0$), then integrating (4.29), we have:

$$\underline{m}=\underline{m}_0\exp\{i\sigma_0(t-t_0)\} \tag{4.30}$$

Thus, in the case of the elastic Earth the period of the poles' free motion T is not equal to 304 days but is described by the following expression:

$$T=\frac{2\pi}{\sigma_0}=\frac{A+\Re}{C-A-\Re}\frac{2\pi}{\Omega}=\frac{J_2\left(\frac{1+H}{H}\right)+\frac{1}{3}kq}{J_2-\frac{1}{3}kq}\cdot 1\ \text{day} \tag{4.31}$$

It is here taken into account that $A=C-HC=J_2\frac{1-H}{H}Ma^2$. Substituting the above values of geophysical constants J_2, H, and q in Equation 4.31 and assuming Love's number to be $k=0.3$, we obtain: $T=448$ days.

Expression (4.31) allows one to also solve the reverse problem – to estimate Love's number k by the observed period of the poles' free motion (Chandler's period T). The statistical analysis of a long series of observations of the poles' coordinates yielded the following values for Chandler's period (in the mean solar days): 433.15 ± 2.23 (for 1899–1967, Jeffreys (1968)), 441.21 ± 2.56 (1846–1965, Rykhlova (1971)), and 433.54 ± 2.12 (1846–1971, Yatskiv *et al.* (1976)). Substituting these T values into (4.31), we find that Love's number k is equal to 0.28, 0.29, and 0.28, respectively.

It is worth mentioning that the above values of κ, k', and k are only valid for the disturbances that are described by the spherical harmonics of the second order and whose characteristic time is less than one year. They are not applicable for higher harmonics and low-frequency disturbances.

Thus, the theory of motion of the elastic Earth adequately describes the extension of the period of the poles' free motion. This theory allows one to calculate the poles' coordinates ν_1 and ν_2 and the Earth's rotational velocity ν_3 by the components of excitation function ψ_i determined from geophysical data. It is also possible to solve the reverse problem – to calculate the components of excitation function ψ_i, using the values of ν_i determined from astronomical observations, and thus to assess the respective geophysical processes (Sidorenkov, 2002a, 2002b).

4.4
Interpretation of Excitation Functions

The dimensionless excitation functions ψ_i are applicable for describing all kinds of geophysical phenomena governing the motion of the Earth around its center of masses. The investigation and calculation of functions ψ_i are among the fundamental issues of studying the instabilities of the Earth's rotation. It should be kept in mind that the meaning of excitation functions ψ_i depends on the particular problems and methods of investigation. Let us consider functions ψ_i as applied to estimating the effect of the atmosphere on the diurnal rotation of the Earth. Two approaches are possible here (Sidorenkov, 1968a).

The first approach is based on the assumption that the Earth–atmosphere system is closed; the problem is to study the balance of angular momentum in this system. The variations in the absolute angular momentum of the atmosphere are attended with the equal in magnitude but opposite in sign variations in the angular momentum of the Earth and, as a consequence of it, with small instabilities of the Earth's rotation. The initial causes of these changes are the variations in the intensity of atmospheric circulation and the redistribution of air masses. The effect of atmospheric circulation (of winds) is accounted for through the components of the atmosphere relative angular momentum h_i and their derivatives $\frac{dh_i}{dt}$. The effect of air mass redistribution is estimated by way of calculating the variable parts of the components of the atmosphere's inertia tensor $\tilde{n}_{ij}$ and their derivatives $d\tilde{n}_{ij}/dt$. In this case, the moments of forces of mechanical interaction between the atmosphere and the Earth cancel out (by virtue of Newton's third law) and are completely eliminated from consideration; that is, the components of the excitation function have the form:

$$
\begin{aligned}
\Omega^2(C-A)\psi_1 &= \Omega^2 n_{13} + \Omega\kappa h_1 + \Omega\frac{dn_{23}}{dt} + \kappa\frac{dh_2}{dt} = \Omega^2(C-A)\left(\chi_1 + \frac{1}{\Omega}\frac{d\chi_2}{dt}\right)\\
\Omega^2(C-A)\psi_2 &= \Omega^2 n_{23} + \Omega\kappa h_2 - \Omega\frac{dn_{13}}{dt} - \kappa\frac{dh_1}{dt} = \Omega^2(C-A)\left(\chi_2 - \frac{1}{\Omega}\frac{d\chi_1}{dt}\right)\\
\Omega^2 C\psi_3 &= -\Omega^2(1+k')n_{33} - \Omega h_3 = -\Omega^2 C\chi_3
\end{aligned}
\tag{4.32}
$$

where n_{ij} and h_i are the components of the atmosphere inertia tensor and relative angular momentum.

Equation 4.32 involve the effective angular momentum functions χ_i, which are specified by formulas (4.16) with account for corrections for the elastic deformations of the Earth. One can see that functions ψ_i and χ_i are linked by the following equations:

$$\psi_1 = \chi_1 + \frac{1}{\Omega}\frac{d\chi_2}{dt}$$
$$\psi_2 = \chi_2 - \frac{1}{\Omega}\frac{d\chi_1}{dt} \qquad (4.33)$$
$$\psi_3 = -\chi_3$$

The above method of estimating the instabilities of the Earth's rotation is called the angular momentum approach.

The second approach is based on the consideration of the Earth without the atmosphere but with account for the angular momentum flux through the Earth's surface due to mechanical interaction between the atmosphere and the Earth. In this case, the absolute angular momentum of the atmosphere in not involved into the equation of motion; instead, one estimates the moments of forces M_i acting on the Earth from the side of the atmosphere. Thus, the excitation functions have the following form:

$$\Omega^2(C-A)\psi_1 = -\kappa M_2$$
$$\Omega^2(C-A)\psi_2 = \kappa M_1 \qquad (4.34)$$
$$\Omega C\psi_3 = \int_0^t M_3 dt'$$

The method described above is called the torque approach. It is fully valid for depicting the changes in the Earth's angular momentum due to atmospheric effects. In particular, this method accounts for the effects of both the air mass circulation (which is eventually due to isolation) and the circulation caused by the gravity field of the closest to the Earth celestial bodies (by tides) or by the electromagnetic interaction between the atmosphere and the interplanetary magnetic field, or solar wind. The effects of redistribution of air and moisture masses are also accounted for in this method, because the replacing masses never come into the state of rest relative to the Earth's surface until an excessive (insufficient) angular momentum is transmitted through drag friction.

If the Earth–atmosphere system is closed, then both methods (the balance one and that of the moments of forces) should in principle give similar results. Thus, the choice of a method depends only on the available meteorological data. The balance method requires at least the data series for the distribution of wind over the entire globe and at all heights. The method of the moments of forces is based on the data on the distribution of wind over the entire globe but only at one level – the surface layer of the atmosphere.

However, along with this advantage the latter method has significant disadvantages, the main of which are the inadequacy of the theory of the angular momentum exchange between the atmosphere and the Earth's surface and the uncertainty of the drag coefficients. Also, an essential disadvantage of this method is that it yields not an integral value (the increment of the Earth's rotational velocity) but a differential one (the angular accelerations). The errors, inevitable when calculating the Earth's rotation angular accelerations by the torque approach, accumulate when integrating and thereby misrepresent the progressive changes (trends) in the velocity of the Earth's rotation. If we take into account that the errors of calculated angular accelerations are large whereas the progressive changes in the Earth's rotational velocity are relatively small, then it becomes clear that the torque approach gives no way of estimating the potential contribution of atmospheric circulation to the generation of these (progressive) changes in the Earth's rotational velocity.

The moments of forces of mechanical interaction between the atmosphere and the Earth reflect all irregular and quasiperiodical variations in the atmospheric circulation. Therefore, it is hoped that the respective irregular and quasiperiodical variations in the velocity of the Earth's rotation (of any periods) can be determined by the torque approach with an accuracy to the constant multiplier.

The torque approach has one fundamental advantage over the angular momentum approach: the latter is based on the assumption that the atmosphere's absolute angular momentum can solely change due to the mechanical interaction between the atmosphere and the Earth (the surface drag friction and pressure forces applied to mountain ridges). Only in this case are the moments of forces applied to the atmosphere canceled out by the, equal in magnitude and opposite in direction, moments of forces acting (by virtue of Newton's third law) onto the Earth. If there are any other phenomena or presently unknown forces that are responsible for changes in the absolute angular momentum of the atmosphere, then the balance may fail. In this case, the balance method in the form (4.32) can only be used for the time intervals that are sufficiently small for these unknown forces to exhibit the distorting effects. At the same time, the torque approach can be used no matter what forces and phenomena are responsible for changes in the atmosphere's absolute angular momentum, because this momentum is likely to be transmitted from the atmosphere to the Earth only due to the moments of the drag and pressure forces. Other moments of forces that are capable of transmitting the angular momentum from the atmosphere to the Earth are unknown at present.

The effect of redistribution of moisture over the globe deserves special consideration. The evaporation of moisture entails a decrease in the Earth's inertia tensor, whereas precipitation entails its increase. If precipitation and evaporation had occurred within the same area, the components of the Earth's inertia tensor would not have changed. However, due to the atmospheric circulation, the water vapor entering the atmosphere through evaporation is carried away for distances of thousands of kilometers before it precipitates. This redistribution of moisture alters the Earth's inertia tensor and, as a consequence, causes the instabilities of the Earth's rotation. It should be mentioned that using the balance method in the above interpretation (see (4.32)), we do not obtain these instabilities of rotation, because the atmosphere is solely a mechanism of moisture redistribution, it does not virtually change components n_{ij}, and h_i. To adequately describe the effect of moisture redistribution, it is necessary to introduce the terms that account for changes in the components of the Earth's inertia tensor N_{ij} into the excitation functions (4.32). Then, the excitation functions take the form:

$$\begin{aligned}
\Omega^2(C-A)\psi_1 &= \Omega^2(n_{13}+N_{13})+\Omega\kappa h_1+\Omega\frac{d(n_{23}+N_{23})}{dt}+\kappa\frac{dh_2}{dt} \\
&= \Omega^2(C-A)\left(\chi_1+\frac{1}{\Omega}\frac{d\chi_2}{dt}\right)+\Omega^2 N_{13}+\Omega\frac{dN_{23}}{dt} \\
\Omega^2(C-A)\psi_2 &= \Omega^2(n_{23}+N_{23})+\Omega\kappa h_2-\Omega\frac{d(n_{13}+N_{13})}{dt}-\kappa\frac{dh_1}{dt} \\
&= \Omega^2(C-A)\left(\chi_2-\frac{1}{\Omega}\frac{d\chi_1}{dt}\right)+\Omega^2 N_{23}-\Omega\frac{dN_{13}}{dt} \\
\Omega^2 C\psi_3 &= -\Omega^2(1+k')(n_{33}+\delta C)-\Omega h_3 = -\Omega^2 C\chi_3-\Omega^2(1+k')\delta C
\end{aligned} \tag{4.35}$$

When considering the interannual variations with the characteristic time intervals of the order of a decade or longer, one may neglect all the terms on the right-hand side of each expression from (4.35) except for the first one. Indeed, considerable multiyear variations in the relative angular momentum of the atmosphere are not observed. As has already been mentioned, the values of derivatives dn_{ij}/dt, and dh_i/dt are inversely proportional to the time interval (in days) under consideration.

In this case the torque approach (4.33), unlike the angular momentum approach (4.32), yields correct results and does not require any modifications. The point is that the moisture transferred by the atmospheric circulation does not retain its angular momentum. The moisture exchanges its angular momentum with the surrounding air and, eventually, with the Earth's surface, thereby quenching the velocity of relative motion. This exchange by the angular momentum occurs at the expense of the drag friction and pressure forces. The torque approach, accounts for the total flow of the angular momentum from the atmosphere to the Earth, including the forces that arise at the Earth/atmosphere interface due to moisture redistribution.

Thus, if the torque approach accounts only for the mechanical interaction between the atmosphere and the Earth, then the angular momentum approach allows us to investigate the role of the atmosphere in the instability of the Earth's rotation as well, thus producing more reliable results. However, as is mentioned above, this approach is applicable to relatively short time intervals, insufficient to manifest the forces of distortion. For longer periods only the torque approach is applicable, this approach being very rough. Apart from this method, the modified balance method (4.35) can be used for estimating the effect of moisture redistribution.

When studying the seasonal instabilities of the Earth's rotation, the expressions for the excitation functions ψ_i can be simplified in many cases. In order to show this, let us compare the terms containing time derivatives with the terms without time derivatives. Components n_{ij}, and h_i vary with characteristic time intervals T_k. They can be approximated by expressions:

$$n_{ij} = \bar{n}_{ij} + a_{ij}\cos\frac{2\pi}{T_k}(t+t_0)$$

$$h_i = \bar{h}_i + b_i\cos\frac{2\pi}{T_k}(t+t'_0)$$

where $\bar{n}_{ij}$ and $\bar{h}_i$ are the values of n_{ij}, and h_i averaged over period T_k; a_{ij} and b_i are their amplitudes, and t_0 and t'_0 are the initial phases.

Let us assess the orders of ratios of the following pairs of terms (in days):

$$0\left(\frac{\Omega^2\tilde{n}_{ij}}{\Omega\kappa\frac{dn_{ij}}{dt}}\right) = \frac{\Omega a_{ij}\cos\frac{2\pi}{T_k}(t-t_0)}{ka_{ij}\frac{2\pi}{T_k}\sin\frac{2\pi}{T_k}(t-t_0)} \approx \frac{\Omega}{\frac{2\pi}{T_k}} = T_k$$

$$0\left(\frac{\Omega kh_i}{\kappa\frac{dh_i}{dt}}\right) = \frac{\Omega b_i\cos\frac{2\pi}{T_k}(t+t'_0)}{b_i\frac{2\pi}{T_k}\sin\frac{2\pi}{T_k}(t+t'_0)} \approx \frac{\Omega}{\frac{2\pi}{T_k}} = T_k$$

Here, it is taken into account that the parameters of variations in the components under consideration are approximately equal; that is, $a_{13} \approx a_{23}$, and $b_1 \approx b_2$.

Hence, when considering the disturbances with characteristic time intervals of an order of one year ($T_k = 365$ days), the terms containing derivatives dn_{ij}/dt, and dh_i/dt may be neglected, because they are approximately by 365 times smaller than terms $\Omega^2\tilde{n}_{ij}$ and $\Omega\kappa h_i$, respectively.

4.5
Harmonic Excitation Function and Motion of the Earth's Poles

The instabilities of the Earth's rotation cannot be theoretically determined, because the excitation functions (4.27) are not calculated analytically but are found from empirical data on the processes occurring in the atmosphere, oceans, and solid Earth, these processes change very irregularly. Only some particular cases of the analytical presentation of the excitation functions can be considered. One such case is the excitations' seasonal variations, which are adequately approximated by the sum of harmonics. The relationship between ν_3 and ψ_3 is trivial; therefore we only consider the relationship between complex values $\underline{m}$ and $\underline{\psi}$.

The excitation function $\underline{\psi}$ that changes with time by the harmonic law with an annual period can be represented in the form:

$$\begin{aligned}\underline{\psi} &= \psi_1 + i\psi_2 = (a_1\cos\oplus + b_1\sin\oplus) + i(a_2\cos\oplus + b_2\sin\oplus) \\ &= A\cos(\oplus-\varphi_1) + iB\cos(\oplus-\varphi_2) = \underline{\psi}^C\cos\oplus + \underline{\psi}^S\sin\oplus \\ &= (a_1 + ia_2)\cos\oplus + (b_1 + ib_2)\sin\oplus \end{aligned} \qquad (4.36)$$

where $\oplus = 2\pi t$ is the longitude of the mean Sun, t is the time in fractions of a year. The values a_1, a_2, b_1, and b_2 are expressed in terms of the amplitudes (A and B) and phases (φ_1 and φ_2) of the excitation function $\underline{\psi}$:

$$A = \sqrt{a_1^2 + b_1^2},\ B = \sqrt{a_2^2 + b_2^2} \tag{4.37}$$

$$\varphi_1 = \operatorname{arctg}\frac{b_1}{a_1},\ \varphi_2 = \operatorname{arctg}\frac{b_2}{a_2} \tag{4.38}$$

Alternatively, the same excitation function $\underline{\psi}$ can be expressed as the circular movements in the positive (anticlockwise) and negative (clockwise) directions:

$$\underline{\psi} = \underline{\psi}^+ e^{i\oplus} + \underline{\psi}^- e^{-i\oplus} = \underline{\psi}^+ \cos\oplus + i\underline{\psi}^+ \sin\oplus + \underline{\psi}^- \cos\oplus - i\underline{\psi}^- \sin\oplus \tag{4.39}$$

Comparing (4.39) with (4.36), we find the expressions that connect the linear and circular parameters of the excitation function:

$$\underline{\psi}^C = \underline{\psi}^+ + \underline{\psi}^- = a_1 + ia_2 \tag{4.40}$$

$$\underline{\psi}^S = i(\underline{\psi}^+ - \underline{\psi}^-) = b_1 + ib_2 \tag{4.41}$$

Hence,

$$\underline{\psi}^+ = \frac{a_1 + b_2}{2} + i\frac{a_2 - b_1}{2} = |\underline{\psi}^+| e^{i\lambda^+} \tag{4.42}$$

$$\underline{\psi}^- = \frac{a_1 - b_2}{2} + i\frac{a_2 + b_1}{2} = |\underline{\psi}^-| e^{-i\lambda^-} \tag{4.43}$$

The radii and initial phases of these circular movements are obviously equal to:

$$|\underline{\psi}^+| = \frac{1}{2}\sqrt{(a_1 + b_2)^2 + (a_2 - b_1)^2} \tag{4.44}$$

$$|\underline{\psi}^-| = \frac{1}{2}\sqrt{(a_1 - b_2)^2 + (a_2 + b_1)^2} \tag{4.45}$$

$$\lambda^+ = \operatorname{arctg}\frac{a_2 - b_1}{a_1 + b_2} \tag{4.46}$$

$$\lambda^- = \operatorname{arctg}\frac{-a_2 - b_1}{a_1 - b_2} \tag{4.47}$$

Summing up, the positive and negative circular movements produce the elliptical trajectory. The major axis of the ellipse is equal to the sum of radii ($|\underline{\psi}^+| + |\underline{\psi}^-|$) and the minor axis – to their difference ($|\underline{\psi}^+| - |\underline{\psi}^-|$).

The eastern longitude of the major semiaxis of the ellipse λ^E can be found from the following considerations. When the pole passes one of the ends of the ellipse major axis, the phases of circular movements coincide: $\oplus + \lambda^+ = -(\oplus + \lambda^-)$.

Hence:

$$\oplus = -\frac{1}{2}(\lambda^+ + \lambda^-) \text{ and } \lambda^E = \oplus + \lambda^+ = \frac{1}{2}(\lambda^+ - \lambda^-) \tag{4.48}$$

The longitude of another end of the major axis is $\lambda^E = 180^\circ + \frac{1}{2}(\lambda^+ - \lambda^-)$. The longitude of the minor axis is obviously equal to $\pm 90^\circ + \frac{1}{2}(\lambda^+ - \lambda^-)$.

From Equations 4.42 and 4.43, it is easy to deduce the relationships for calculating the coefficients of trigonometric terms, using the known parameters of the circular movements:

$$\begin{aligned} a_1 &= |\underline{\psi}^+|\cos\lambda^+ + |\underline{\psi}^-|\cos\lambda^- \\ b_1 &= -|\underline{\psi}^+|\sin\lambda^+ - |\underline{\psi}^-|\sin\lambda^- \\ a_2 &= |\underline{\psi}^+|\sin\lambda^+ - |\underline{\psi}^-|\sin\lambda^- \\ b_2 &= |\underline{\psi}^+|\cos\lambda^+ - |\underline{\psi}^-|\cos\lambda^- \end{aligned} \tag{4.49}$$

Also, the formulas similar to (4.36–4.49) can be written for the pole's complex coordinate $\underline{m}$.

The changes in the pole's coordinate $\underline{m} = \nu_1 + i\nu_2$, which are due to the effect of excitation $\underline{\psi} = \psi_1 + i\psi_2$, are described by Equation 4.29 (as it is shown above). With account for expressions $\sigma_0 = 2\pi/T$ and $dt = d\oplus/(2\pi)$, it takes the form:

$$iT\frac{d\underline{m}}{d\oplus} + \underline{m} = \underline{\psi} \tag{4.50}$$

The solution of this linear inhomogeneous differential equation of the first order is easily found (for example, with the help of Bernoulli's substitution or by the method of variation of constants):

$$\underline{m} = \underline{m}_0 e^{\frac{i\oplus}{T}} + \frac{e^{\frac{i\oplus}{T}}}{iT}\int_{-\infty}^{\oplus} \underline{\psi} e^{\frac{\oplus'}{iT}} d\oplus' \tag{4.51}$$

The first item in the right side of Equation 4.51 describes the free circular movement of the pole (with Chandler's period T and radius $\underline{m}_0$) and the second item – the forced oscillation. Substituting the excitation function of the form (4.36) in (4.51) and integrating the received expression, we determine the forced oscillation in the explicit form:

$$\underline{m}_b = \frac{1}{T^2 - 1}\{[-(a_1 + Tb_2) - i(a_2 - Tb_1)]\cos\oplus + [(Ta_2 - b_1) - i(Ta_1 + b_2)]\sin\oplus\} \tag{4.52}$$

And if we use the excitation function of the form (4.38), then the poles' forced motion is described by the expression:

$$\underline{m}_b = -\frac{1}{T-1}\underline{\psi}^+ e^{i\oplus} + \frac{1}{T+1}\underline{\psi}^- e^{-i\oplus} \tag{4.53}$$

To characterize the pole's motion in the form similar to expressions (4.40) and (4.41), we can write the equations:

$$\underline{m}^c = \underline{m}^+ + \underline{m}^- = \nu_1^c + i\nu_2^c \qquad \underline{m}^s = i(\underline{m}^+ - \underline{m}^-) = \nu_1^s + i\nu_2^s \tag{4.54}$$

These equations give:

$$\nu_1^c = \mathrm{Re}(\underline{m}^+ + \underline{m}^-) = -\frac{1}{T-1}|\underline{\psi}^+|\cos\lambda^+ + \frac{1}{T+1}|\underline{\psi}^-|\cos\lambda^-$$

$$\nu_1^s = \mathrm{Im}(-\underline{m}^+ + \underline{m}^-) = \frac{1}{T-1}|\underline{\psi}^+|\sin\lambda^+ - \frac{1}{T+1}|\underline{\psi}^-|\sin\lambda^-$$

$$\nu_2^c = \mathrm{Im}(\underline{m}^+ + \underline{m}^-) = -\frac{1}{T+1}|\underline{\psi}^+|\sin\lambda^+ - \frac{1}{T+1}|\underline{\psi}^-|\sin\lambda^-$$

$$\nu_2^s = \mathrm{Re}(\underline{m}^+ - \underline{m}^-) = -\frac{1}{T-1}|\underline{\psi}^+|\cos\lambda^+ - \frac{1}{T+1}|\underline{\psi}^-|\cos\lambda^-. \quad (4.55)$$

Hence, if the Earth is subject to the action of the factor that is described by the harmonic excitation function with an annual period, the poles move with the same period but different amplitude. The Earth's dynamic system transforms the amplitudes in such a way that the amplitude of the initial excitation grows by $1/(T-1)$ times for the positive circular movement and diminishes by $1/(T+1)$ times for the negative circular movement ($T \approx 1.2$).

Also, Equation 4.50 allows one to solve the reverse problem: to find the parameters of the excitation function by the known coordinates of the pole. In order to derive the calculation formulas, we substitute expressions

$$\underline{m} = \nu_1^c\cos\oplus + \nu_1^s\sin\oplus + i(\nu_2^c\cos\oplus + \nu_2^s\sin\oplus) \quad (4.56)$$

and (4.36): $\underline{\psi} = (a_1 + ia_2)\cos\oplus + (b_1 + ib_2)\sin\oplus$ into Equation 4.50, differentiate $\underline{m}$, and equate the similar terms. As a result we obtain:

$$\begin{aligned} a_1 &= \nu_1^c - T\nu_2^s, \quad b_1 = \nu_1^s + T\nu_2^c \\ a_2 &= \nu_2^c + T\nu_1^s, \quad b_2 = \nu_2^s - T\nu_1^c \end{aligned} \quad (4.57)$$

If we substantiate expressions

$$\underline{m} = \underline{m}^+ e^{i\oplus} + \underline{m}^- e^{-i\oplus} \text{ and } \underline{\psi} = \underline{\psi}^+ e^{i\oplus} + \underline{\psi}^- e^{-i\oplus} \quad (4.58)$$

into (4.50), then in a similar manner we obtain:

$$\underline{\psi}^+ = (1-T)\underline{m}^+ \text{ and } \underline{\psi}^- = (1-T)\underline{m}^- \quad (4.59)$$

Relationships (4.57)–(4.59) make it possible to calculate the needed parameters of the excitation functions with the use of the observed characteristics of the Earth's poles' motion with an annual period.

When using functions χ_i of the atmospheric angular momentum, Equation 4.29 for the poles' motion has the form:

$$\frac{i}{\sigma_0}\frac{d\underline{m}}{dt} + \underline{m} = -\frac{i}{\Omega}\frac{d\underline{\chi}}{dt} - \underline{\chi} \quad (4.60)$$

Here, the equatorial components of the angular momentums of the Earth and atmosphere are connected by the oscillation differential equation. The periods of the free oscillations of the Earth and atmosphere are essentially different: for the Earth $T = 2\pi/\sigma_0 = 430$ days and for the atmosphere $T_0 = 2\pi/\omega = 1$ day.

Let at $t=0$ be $\underline{m} = \underline{m}_0$, and $\underline{\chi} = \underline{\chi}_0$. Then, solving Equation 4.60 by standard methods and using the integration by parts, we find:

$$\underline{m} = e^{i\sigma t}\left[\underline{m}_0 - i\sigma\left(1+\frac{\sigma}{\omega}\right)\int_0^t \underline{\chi} e^{-i\sigma\tau} d\tau\right] - \frac{\sigma}{\omega}\left[\underline{\chi} - \underline{\chi}_0 e^{i\sigma t}\right] \tag{4.61}$$

4.6 Equation of Motion of the Earth's Spin Axis in Space

The motion of the Earth with respect to its center of masses is most convenient to be described with the help of Euler's angles ψ, θ, and φ. They characterize the orientation of the moving coordinate system Ox_i with respect to the inertial space coordinate system $O\xi_i$ and are shown in Figure 4.2. As is seen from this figure, the moving system Ox_i can be transformed from some position of the inertial system $O\xi_i$ into the position shown in Figure 4.2, by way of three consecutive turnings:

1. the turning through angle ψ around axis $O\xi_3$, after which axis $O\xi_2$ takes up the position of the line of knots ON and axis $O\xi_1$–the position of OA;
2. the turning through angle θ around line ON, after which axis $O\xi_3$ matches with axis Ox_3 and line OA–with line OB;
3. the turning through angle φ around axis Ox_3, after which line OB matches with axis Ox_1 and line ON with axis Ox_2.

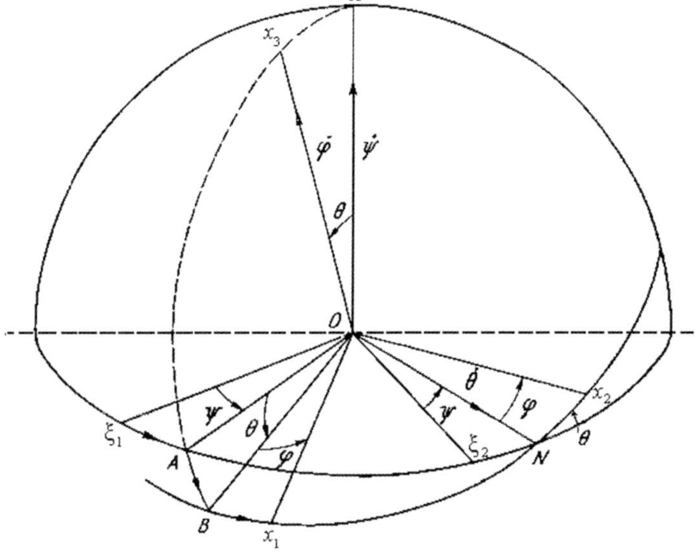

Figure 4.2 Euler's angles.

Resulting from these turnings, axes Ox_i take up the position shown in Figure 4.2. Angles ψ, θ, and φ are called the angles of precession, nutation, and proper rotation, respectively.

Thus, the transition from coordinate system $O\xi_i$ to coordinate system Ox_i may be represented as three successive turnings of $\mathbf{T}_i$. The formula for transforming the coordinates is described by equation $\mathbf{x}=\mathbf{TX}$, where matrix $\mathbf{T}$ may be obtained:

$$\mathbf{T} = \mathbf{T}_3(\varphi)\mathbf{T}_2(\theta)\mathbf{T}_1(\psi)$$
$$= \left\{ \begin{pmatrix} \cos\varphi & \sin\varphi & 0 \\ -\sin\varphi & \cos\varphi & 0 \\ 0 & 0 & 1 \end{pmatrix} \cdot \left[\begin{pmatrix} \cos\theta & 0 & -\sin\theta \\ 0 & 1 & 0 \\ \sin\theta & 0 & \cos\theta \end{pmatrix} \cdot \begin{pmatrix} \cos\psi & \sin\psi & 0 \\ -\sin\psi & \cos\psi & 0 \\ 0 & 0 & 1 \end{pmatrix} \right] \right\}$$
$$= \begin{pmatrix} \cos\varphi\cos\theta\cos\psi - \sin\varphi\sin\psi & \cos\varphi\cos\theta\sin\psi + \sin\varphi\cos\psi & -\cos\varphi\sin\theta \\ -\sin\varphi\cos\theta\cos\psi - \cos\varphi\sin\psi & -\sin\varphi\cos\theta\sin\psi + \cos\varphi\cos\psi & \sin\varphi\sin\theta \\ \sin\theta\cos\psi & \sin\theta\sin\psi & \cos\theta \end{pmatrix} \tag{4.62}$$

The derivatives of Euler's angles are the angular velocities that form the vector of angular velocity ω of the Earth's rotation:

$$\omega = \dot{\varphi}\boldsymbol{e}_\varphi + \dot{\theta}\boldsymbol{e}_\theta + \dot{\psi}\boldsymbol{e}_\psi \tag{4.63}$$

where $\boldsymbol{e}_\varphi$, $\boldsymbol{e}_\theta$, and $\boldsymbol{e}_\psi$ are the unit vectors of axes Ox_3, ON, and $O\xi_3$. Let us express the projections of vector $\boldsymbol{\omega}\{\boldsymbol{\omega}_{x1},\boldsymbol{\omega}_{x2},\boldsymbol{\omega}_{x3}\}$ as Euler's angles and their time derivatives. For this purpose we project expression (4.63) onto axes Ox_i. Then we obtain:

$$\begin{aligned} \omega_1 &= \dot{\theta}\sin\varphi - \dot{\psi}\sin\theta\cos\varphi \\ \omega_2 &= \dot{\theta}\cos\varphi + \dot{\psi}\sin\theta\sin\varphi \\ \omega_3 &= \dot{\varphi} + \dot{\psi}\cos\theta \end{aligned} \tag{4.64}$$

Equations 4.64 are called Euler's kinematics equations.

Also, one can express the derivatives of Euler's angles as projections of ω_i:

$$\begin{aligned} \dot{\psi} &= \frac{1}{\sin\theta}(\omega_2\sin\varphi - \omega_1\cos\varphi) \\ \dot{\theta} &= \omega_1\sin\varphi + \omega_2\cos\varphi \\ \dot{\varphi} &= \omega_3 + (\omega_1\cos\varphi - \omega_2\sin\varphi)\mathrm{ctg}\theta \end{aligned} \tag{4.65}$$

Let us assume Euler's angles ψ, θ, and φ as the generalized coordinates and express the kinetic energy E of the rotation of the dynamically symmetric Earth in terms of these angles and their derivatives:

$$E = \frac{1}{2}\left[A(\dot{\theta}^2 + \dot{\psi}^2\sin^2\theta) + C(\dot{\varphi} + \dot{\psi}\cos\theta)^2\right] \tag{4.66}$$

Knowing the Lagrangian function $\Re = E - U$, we can set Lagrange's equation:

$$\frac{\mathrm{d}}{\mathrm{d}t}\frac{\partial\Re}{\partial\dot{q}_i} - \frac{\partial\Re}{\partial q_i} = 0 \tag{4.67}$$

where U is the potential energy; $q_1 = \psi$, $q_2 = \theta$, and $q_3 = \varphi$ are Euler's angles. Calculating the partial derivatives of E and U and substantiating them in (4.67), we obtain:

$$\begin{aligned}
&\frac{d}{dt}\left[A\dot{\psi}\sin^2\theta + C(\dot{\varphi} + \dot{\psi}\cos\theta)\cos\theta\right] = -\frac{\partial U}{\partial \psi}\\
&A\ddot{\theta} - A\dot{\psi}^2\sin\theta\cos\theta + C(\dot{\varphi} + \dot{\psi}\cos\theta)\dot{\psi}\sin\theta = -\frac{\partial U}{\partial \theta}\\
&\frac{d}{dt}\left[C(\dot{\varphi} + \dot{\psi}\cos\theta)\right] = -\frac{\partial U}{\partial \varphi} = 0
\end{aligned} \tag{4.68}$$

The potential energy does not depend on φ, because the Earth is a body of revolution and its gravity field is symmetrical with respect to the axis of rotation. From the third Equation 4.68 it follows:

$$\dot{\varphi} + \dot{\psi}\cos\theta = \text{const} \approx \Omega \tag{4.69}$$

With account for (4.69), Equation 4.68 can be rewritten in the following form:

$$\begin{aligned}
&A\ddot{\psi}\sin^2\theta + 2A\dot{\psi}\dot{\theta}\sin\theta\cos\theta - C\Omega\dot{\theta}\sin\theta = -\frac{\partial U}{\partial \psi}\\
&A\ddot{\theta} - A\dot{\psi}^2\sin\theta\cos\theta + C\Omega\dot{\psi}\sin\theta = -\frac{\partial U}{\partial \theta}
\end{aligned} \tag{4.70}$$

The velocities of precession $\dot{\psi}$ and nutation $\dot{\theta}$ of the Earth's spin axis are by many orders of magnitude lower than the angular velocity of the Earth's diurnal rotation $\dot{\varphi} = \Omega$. Therefore, one may neglect the first two terms in the left part of (4.70). As a result, we have:

$$\begin{aligned}
&\frac{d\theta}{dt} = \frac{1}{C\Omega\sin\theta}\frac{\partial U}{\partial \psi}\\
&\sin\theta\frac{d\psi}{dt} = -\frac{1}{C\Omega}\frac{\partial U}{\partial \theta}
\end{aligned} \tag{4.71}$$

Complex variables are generally introduced into (4.71). In this case we obtain:

$$\frac{d\theta}{dt} + i\sin\theta\frac{d\psi}{dt} = \frac{1}{C\Omega}\left(\frac{1}{\sin\theta}\frac{\partial U}{\partial \psi} - i\frac{\partial U}{\partial \theta}\right) \tag{4.72}$$

Equations 4.71 and 4.72 are called the Poisson equations. They describe changes in the direction of the Earth's spin axis with respect to the inertial space coordinate system $O\xi_i$, which occur under the effect of gravity of the Moon, Sun, and planets. These equations are the fundamental equations of the theory of precession and nutation, which will be more thoroughly discussed in Chapter 5.

Thus, the theory of instabilities of the Earth's rotation, which is presented in this chapter, shows that these instabilities can be expressed with the help of the excitation functions, which in most cases cannot be analytically obtained. They should be determined from empirical data on the processes occurring on the Earth. In the case of the effect of the atmosphere, the excitation functions are determined either from the data on changes in the components of the moment of inertia (the effect of redistribution of the air and water masses) and the variations in the relative angular momentum of the atmosphere (the effect of winds) or by calculating the torques of the atmosphere acting onto the Earth's surface. In the theory of precession and nutation, the excitations are the moments of forces acting onto the Earth through the attraction of celestial bodies surrounding the Earth. There are no principal difficulties hindering the determination of these forces; the difficulties are associated with the account for the complicated structure and physical properties of all the envelopes of the Earth.

5
Tides and the Earth's Rotation

5.1
Tide-Generating Potential

Every particle of any geosphere (either atmosphere or ocean or lithosphere) is affected by the gravitational forces of the Earth and the surrounding celestial bodies, and also the inertial and electromagnetic forces. The latter two are usually ignored in the tide-generating theories as they are much less than the former two. The surface of the geosphere corresponds to a gravitational equipotential surface of all gravitational and inertial forces. The Earth's gravitational potential and the potential of the centrifugal forces arising from the diurnal Earth's rotation together form an unperturbed level, while the gravitational potentials of the Moon and the Sun and the centrifugal potentials arising from the orbital motion of the Earth vary with time and perturb the geosphere level.

Let us first deduce an expression for the tide-generating potential of the Moon alone. Following Lamb (1932), we take a unit-mass particle of a geosphere (for example, ocean) at point A (Figure 5.1). Let $\mathbf{R}$ and $\mathbf{d_M}$ be the geocentric radii-vectors of points A and Moon M, respectively; the angle z_M is the zenith distance of the Moon at point A, reduced to the Earth's center. Then the gravitational potential of the Moon W_M at point A is as follows:

$$W_M = -\frac{\gamma M_M}{|\boldsymbol{d}_M - \boldsymbol{R}|} = -\frac{\gamma M_M}{d_M}\left(1 - 2\frac{R}{d_M}\cos z_M + \frac{R^2}{d_M^2}\right)^{-\frac{1}{2}} \tag{5.1}$$

Here, γ is the gravitational constant; M_M is the Moon's mass.

$$|\boldsymbol{d}_M - \boldsymbol{R}| = (d_M^2 + R^2 - 2d_M R\cos z_M)^{1/2}, \quad (\mathbf{R}\cdot\mathbf{d}_M) = R d_M \cos z_M \tag{5.2}$$

It is known that the Earth and the Moon revolve translating around the Earth–Moon system's mass center, the Earth and the Moon moving along their trajectories, which are geometrically similar and represent ellipses with a common focus in the center of mass O_1 (Figure 2.2). The Earth's ellipse E is as many times less than the Moon's ellipse of M as the mass of the Moon M_M is less than the mass of the Earth M_E.

Let us take a coordinate system whose origin is in the Earth's center of mass and whose axes are constantly directed at fixed celestial objects. In this coordinate system, movement of substance is defined not only by the gravitational potential of the

The Interaction Between Earth's Rotation and Geophysical Processes. Nikolay S. Sidorenkov

ISBN: 978-3-527-40875-7

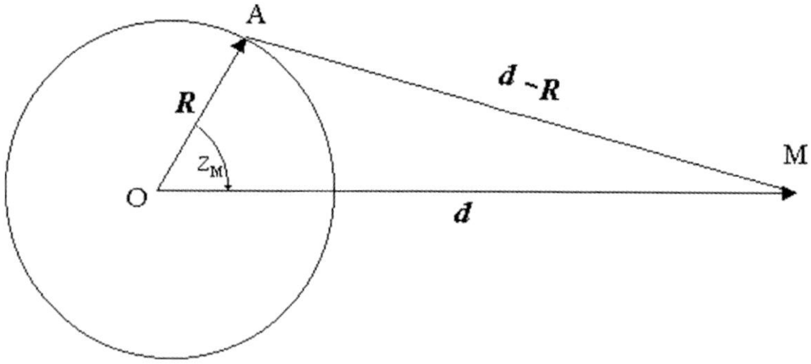

Figure 5.1 A scheme for considering the tide-generating force of the Moon.

Moon (5.1) but also by the potential of centrifugal forces Φ_M arising from the translational motion of the coordinate system revolving, together with the Earth, around the center of mass O_1 of the Earth–Moon system. The centrifugal forces are identical for all particles of the geosphere and parallel to vector $\mathbf{d}_E$ (in the opposite direction of the Moon). This force compensates the lunar gravity force averaged over the total mass of the Earth $-\gamma M_M \frac{\mathbf{d}_M}{d_M^3}$. Hence, we find an equivalent of the potential of the centrifugal force by means of integration by $\mathbf{dR}$:

$$\Phi_M = \frac{\gamma M_M}{d_M^2} R\cos z_M + C \tag{5.3}$$

Summing expressions (5.1) and (5.3), we obtain:

$$U_M = W_M + \Phi_M = -\frac{\gamma M_M}{d_M}\left(1 - 2\frac{R}{d_M}\cos z_M + \frac{R^2}{d_M^2}\right)^{-\frac{1}{2}} + \frac{\gamma M_M}{d_M^2} R\cos z_M + C \tag{5.4}$$

In the center of the Earth ($R = 0$), potential U_M is equal to zero, consequently $C = \frac{\gamma M_M}{d_M}$. If the first term of the right part (5.4) is expanded into a series in powers of equatorial parallax R/d_M, we have:

$$\begin{aligned} U_M = & -\frac{\gamma M_M}{d_M} - \frac{\gamma M_M}{d_M^2} R\cos z_M + \gamma M_M \sum_{n=2}^{\infty} \frac{R^n}{d_M^{n+1}} P_n(\cos z_M) \\ & + \frac{\gamma M_M}{d_M^2} R\cos z_M + \frac{\gamma M_M}{d_M} = \gamma M_M \sum_{n=2}^{\infty} \frac{R^n}{d_M^{n+1}} P_n(\cos z_M) \end{aligned} \tag{5.5}$$

where $P_n(\cos z_m)$ is the Legendre polynomial of degree n.

Function U_M is referred to as the tide-generating potential of the Moon. It represents a series with the rate of convergence R/d_M and the Legendre polynomials $P_n(\cos z_m)$, beginning with the second degree. Since R/d_M is equal to 0.016, all terms of the series, except for the first, can be neglected in the first approximation. Then we have:

$$U_{2M} = \gamma M_M \frac{R^2}{d_M^3}\left(\frac{3}{2}\cos^2 z_M - \frac{1}{2}\right) \tag{5.6}$$

If U_{2M} is considered a function of the position of point A, it is a zonal spherical harmonic of the second degree with an axis parallel to $\mathbf{d}_M$. It sufficiently accurately describes the tide-generating potential of the Moon.

Similarly, expressions for the tide-generating potentials of the Sun, planets and other celestial bodies are derived. Replacing M_M, $\mathbf{d}_M$ and z_M in (5.5) and (5.6) by the mass of the Sun M_S, its geocentric distance $\mathbf{d}_S$ and geocentric zenith distance z_S, respectively, we find the tide-generating potential of the Sun U_S:

$$U_S = \gamma M_S \sum_{n=2}^{\infty} \frac{R^n}{d_S^{n+1}} P_n(\cos z_S) \approx \gamma M_S \frac{R^2}{d_S^3}\left(\frac{3}{2}\cos 2z_S - \frac{1}{2}\right) \tag{5.7}$$

Let the Doodson constant G be introduced. Let us denote

$$G(R) = \frac{3}{4}\gamma'\mu\frac{R^2}{c^3} \tag{5.8}$$

where γ' is the gravitational constant multiplied by the Earth's mass; μ is the mass of a perturbing body (in units of the Earth's mass); c is the semimajor axis of the orbit of the perturbing body; R is the distance from observer A to the Earth's center. The average radius of the Earth is denoted as a, and gravitational acceleration as $g = \gamma'/a^2$. Hence:

$$G(R) = \frac{3}{4}\mu\frac{ga^2}{c^3}R^2 = \left(\frac{R}{a}\right)^2 G \tag{5.9}$$

where $G = G(a) = \frac{3}{4}\mu\frac{ga^4}{c^3} = \frac{3}{4}\gamma'\mu\frac{a^2}{c^3}$ is referred to as the Doodson constant. Its value depends on the perturbing body. For tides caused by the Moon ($\mu = 1/81.30$; $c = 60.27a$; $a = 6\,371\,012$ m; $g = 9.8204\ \mathrm{m/s^2}$), the Doodson constant is equal to:

$$G_M = 2.6364\ \mathrm{m/s^2} \tag{5.10}$$

For solar tides when $\mu = 332\,946$ and $c = 23\,466a$:

$$G_S = 1.2091\ \mathrm{m^2 s^{-2}} = 0.46\ G_M \tag{5.11}$$

The ratio of tide-generating potentials of the Moon and the Sun is equal to:

$$\frac{G_M}{G_S} = \frac{M_M}{M_S} \cdot \frac{d_S^3}{d_M^3} = 2.17$$

that is, the influence of the Moon is 2.17 times as great as that of the Sun. Similarly, it can be shown that the influence of the Moon is by several orders of magnitude greater than that of Jupiter, Venus, Saturn, and other planets.

The total influence of the lunar and solar tides gives height ζ(m) of a static tide:

$$\xi = \frac{U}{g} \approx \frac{G_M + G_S}{g}\left(\cos 2z + \frac{1}{3}\right) = 0.39\left(\cos 2z + \frac{1}{3}\right) \tag{5.12}$$

that is the reference surface of the Earth may range from +52 cm to −26 cm. The Earth's figure deformed by the tides of the type described by (5.12) is a harmonious zonal-type spheroid of the second degree whose axis goes through the perturbing body.

Expression for tide-generating potential U_2 contains a coordinate of horizontal system of celestial coordinates, namely geocentric zenith distance z of a perturbing body. It is inconvenient to use local coordinates. Therefore, the equatorial system of celestial coordinates is used instead of the horizontal system (instead of z, declination δ, hour angle H and astronomical coordinates of an observation point (latitude φ and longitude λ) are used (Melchior, 1971, 1983). The following basic formula of a parallactical triangle is used:

$$\cos z = \sin\varphi \sin\delta + \cos\varphi \cos\delta \cos H \tag{5.13}$$

Let (5.13) be substituted in (5.6), performing the following transformations before:

$$\cos^2 z = \sin^2\varphi \sin^2\delta + \cos^2\varphi \cos^2\delta \cos^2 H$$

$$+ 2\sin\varphi \cos\varphi \sin\delta \cos\delta \cos H$$

$$\cos^2 H = \frac{\cos 2H + 1}{2}$$

$$3\cos^2 z - 1 = 3\sin^2\varphi \sin^2\delta - 1 + \frac{3}{2}\cos^2\varphi \cos^2\delta[\cos 2H + 1]$$

$$+ \frac{3}{2}\sin 2\varphi \sin 2\delta \cos H = 3\sin^2\varphi \sin^2\delta - 1 + \frac{3}{2}\cos^2\varphi \cos^2\delta$$

$$+ \frac{3}{2}\cos^2\varphi \cos^2\delta \cos 2H + \frac{3}{2}\sin 2\varphi \sin 2\delta \cos H$$

The first of these terms can be written as:

$$3\sin^2\varphi \sin^2\delta - 1 + \frac{3}{2}(1 - \sin^2\varphi)(1 - \sin^2\delta)$$

$$= 3\sin^2\varphi \sin^2\delta + \frac{3}{2}\sin^2\varphi \sin^2\delta - \frac{3}{2}\sin^2\varphi - \frac{3}{2}\sin^2\delta - 1 + \frac{3}{2}$$

$$= \frac{3}{2}\left[3\sin^2\varphi \sin^2\delta - \sin^2\varphi - \sin^2\delta + \frac{1}{3}\right]$$

$$= \frac{3}{2}\left[3\left(\sin^2\varphi - \frac{1}{3}\right)\left(\sin\delta - \frac{1}{3}\right)\right]$$

Finally (5.6) and (5.13) give:

$$U_2 = G\left(\frac{R}{a}\right)^2\left(\frac{c}{d}\right)^3$$

$$\times\left[\cos^2\varphi \cos^2\delta \cos 2H + \sin 2\varphi \sin 2\delta \cos H + 3\left(\sin^2\varphi - \frac{1}{3}\right)\left(\sin^2\delta - \frac{1}{3}\right)\right] \tag{5.14}$$

where G is the Doodson constant; c is the major semiaxis of the lunar orbit and d is the geocentric distance of the Moon; a is the average radius of the Earth.

Such separation of the tide-generating potential into three terms was introduced by Laplace, who first gave physical and geometrical interpretation of the terms. Each of the terms describes the corresponding type of tides and the spherical function of the second degree.

The first spherical function (sectorial harmonic) has nodal lines (zero lines) on the meridians located in 45° on every side of the meridian of a perturbing body

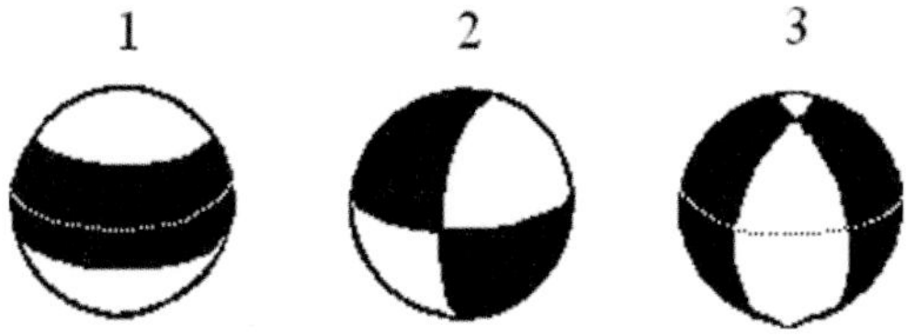

Figure 5.2 Types of the tides: 1 – zonal; 2 – diurnal; 3 – semidiurnal.

(Figure 5.2 (3). These lines divide the Earth into four sectors where the function is alternately positive and negative. One of the function multipliers is cos $2H$, that is the described tides have a semidiurnal period. The maximal amplitudes are observed on the equator provided that the declination of the perturbing body is zero. Tides are zero at the poles. Laplace named these tides of the third type. They affect neither the short-period variations in the rate of the Earth's rotation nor the movement of the poles but cause the Earth's secular spin-down due to the tide-generating friction.

The second spherical function is referred to as a tesseral one. As the nodal lines, it has the equator and the meridian located in 90° from the meridian of the perturbing body (Figure 5.2 (2)). This function divides the sphere into the areas that have the sign changed along with the change of the sign of declination δ. The tesseral function contains factor cos H, therefore it describes the tides with the diurnal period. The maximal amplitudes of the tides are observed at latitudes of $\pm 45^\circ$ provided the maximal declination of the perturbing body. Zero tides are at the equator and the poles. Laplace named these tides of the second type. They influence the position of the pole of inertia, causing precession, nutation and movement of the Earth's poles. Tides of the second type do not affect the rate of the Earth's rotation.

The third spherical function (zonal harmonic) depends only on the latitude of a site. Its nodal lines are parallels $\pm 35^\circ 16'$ (Figure 5.2 (1). This spherical function does not explicitly depend on time (hour angle H) and its periods are defined by changes of declination δ and parallax a/c of a perturbing body. 14-day and 27.3-day periods of lunar tides and 0.5-year and 1-year periods of solar tides prevail here. These tides are referred to as long-period tides or first-type tides by Laplace's classification. They do not displace the pole of inertia but change the polar moment of inertia of the Earth. Therefore, the zonal tides cause significant variations of the rate of the Earth's rotation, the variations having the above-mentioned periods. Due to zonal tides, the reference surface is constantly displaced downwards at the pole (approximately by 28 cm) and raised at the equator (approximately by 14 cm), which somewhat increases the compression of the Earth.

5.2
Expansion of Tide-Generating Potential

Tidal analysis and forecast should be based on solar and lunar motion theories. For a long time the calculation of tides was based on Newcomb's theory of the motion of the Sun and Brown's theory of the Moon's motion (Brown, 1905, 1919). The best known is Doodson's research (Doodson, 1922), who used Newcomb's and Brown's theories in order to perform the harmonic expansion of the tide-generating potential containing 396 harmonic terms. Subsequently, the expansion was performed by Cartwright and Tayler (1971), Cartwright and Edden (1973), Büllesfeld (1985), Xi (1987, 1989), and Tamura (1987, 1995). Until recently the most accurate expansions of tide-generating potential were performed by Hartmann and Wenzel (1994, 1995) and Roosbeek (1996) who used different techniques.

Roosbeek (1996) applied an analytical method. He took ephemerides of attractive bodies from the analytical series of the ELP2000–85 theory (Chapront-Touzé and Chapront, 1988) for the spherical coordinates of the Moon and the VSOP87 theory (Bretagnon and Francou, 1988) for the orbital elements of the major planets of the Solar System. Roosbeek obtained the expansion of the tide-generating potential by consecutive multiplications of the initial analytical series for coordinates of the Moon, the Sun and the planets. The accuracy of the final expansion is obviously limited by the accuracy of the coordinates of attractive bodies in the initial analytical series (at the present time, the accuracy of the coordinates is less than that of the current numerical ephemerides of the Moon and the planets of series DE/LE calculated by NASAs JPL).

Hartmann and Wenzel (1994, 1995) used a different approach to obtain the expansion of the tide-generating potential. They used the JPL DE/LE 200 numerical ephemeris (Standish and Williams, 1981) to get the coordinates of attractive bodies. The authors prepared a table of the tide-generating potential for a 300-year period from 1850 to 2150, with a small interval, and performed Fourier analysis of the data. This approach is based on the more accurate initial ephemerides of attractive bodies as compared with the technique used by Roosbeek (1996). A classical shortcoming of any spectral method is insufficient separation of close-frequency components if the spectral analysis is carried out on the data calculated (or measured) over a rather short time period. Nevertheless, Hartmann and Wenzel's analytical expansion of the tide-generating potential was the most accurate at that time (1995).

A similar method was applied by Kudryavtsev (2004) but he used a new technique of the expansion of an arbitrary tabulated function of the coordinates of the Moon, the Sun and the planets into Poisson's series. The technique implies that the amplitudes and frequencies of the terms of the series are high-degree time polynomials (as opposed to classical Fourier's analysis, where the amplitudes and frequencies of the final expansion are constants). The coordinates of perturbing bodies were taken from the most accurate long-term DE/LE-406 numerical ephemeris (Standish, 1998). The new expansion (KSM03) includes 26 753 terms in the original format (or 28 806 terms in the standard HW95 format) with amplitudes no less than $10^{-8}\,m^2/s^2$.

The theories of the Moon's motion and the Earth's rotation employ the following fundamental arguments: the mean anomaly of the Moon $l = s - p$; the mean anomaly of the Sun $l' = h - p_S$; the mean elongation of the Moon from the ascending node $F = s - N$; the mean elongation of the Moon from the Sun $D = s - h$; and the mean longitude of the ascending node of the Moon N. The arguments depend on the mean longitudes s and h and perigees p and p_S of the Moon and the mean Sun, respectively. All the basic arguments are almost linear functions of time. Reduced to the fundamental epoch of 2000, January 1, 12 h TDB, they are described by the following expressions (Simon *et al.*, 1994):

$$\begin{aligned}
l &= 134^\circ.96340251 + 477198^\circ.8675T + 31''.8792T^2 \\
&\quad + 0''.051635T^3 - 0''.0002447T^4 \\
l' &= 357^\circ.52910918 + 35999^\circ.05028T - 0''.5532T^2 \\
&\quad - 0''.000136T^3 - 0''.00001149T^4 \\
F &= 93^\circ.27209062 + 483202^\circ.0172T - 12''.7512T^2 \\
&\quad - 0''.001037T^3 + 0.00000417T^4 \\
D &= 297^\circ.85019547 + 445267^\circ.1114T - 6''.3706T^2 \\
&\quad + 0''.006593T^3 - 0''.00003169T^4 \\
N &= 125^\circ.044555501 - 1934^\circ.1363T + 7''.4722T^2 \\
&\quad + 0''.007702T^3 - 0''.00005939T^4
\end{aligned} \tag{5.15}$$

where T is the time in Julian centuries (36 525 ephemeris days per century) starting from the fundamental epoch of 2000, January 1, 12 h TDB, that is from the Julian Date JD 2451545.0.

Doodson used linear combinations of the following six independent variables as arguments of some harmonics of his expansion of the tide-generating potential:

$\tau = 180^\circ + H$ is the mean Grinvich lunar time, H is the hour angle of the Moon;

s is the mean longitude of the Moon;

h is the mean longitude of the Sun;

p is the mean longitude of the lunar perigee;

$N' = -N$, where N is the mean longitude of the ascending node of the Moon's orbit;

p_S is the mean longitude of the solar perigee.

Doodson's variables are expressed through the fundamental arguments (5.15) as follows:

$$\begin{aligned}
\tau &= ST + \pi - s \\
s &= F + N \\
h &= s - D \\
p &= s - l \\
N' &= -N \\
p_s &= s - D - l'
\end{aligned} \tag{5.16}$$

The last five variables (5.16) are the mean longitudes measured along the ecliptic from the average point of the vernal equinox. Each of Doodson's variables is an almost linear time function. So, the mean longitudes reduced to the fundamental Epoch of 2000, January 1, 12 h TDB, are described by the following expressions:

$$\begin{aligned}
s &= 218^\circ.3166456 + 481267^\circ.8809T - 5''.279T^2 \\
&\quad + 0''.0067T^3 - 0''.0000552T^4 \\
h &= 280^\circ.46607 + 36000^\circ.7698T + 1''.089T^2 \\
&\quad + 0''.000057T^3 - 0''.0000235T^4 \\
p &= 83^\circ.3532431 + 4069^\circ.0134T - 37''.1582T^2 \\
&\quad - 0''.0449T^3 - 0''.000189T^4 \\
N &= 125^\circ.044555501 - 1934^\circ.1363T + 7''.4722T^2 \\
&\quad + 0''.007702T^3 - 0''.0000594T^4 = -N' \\
p_s &= 282^\circ.9369609 + 1^\circ.71952T + 1''.6422T^2 \\
&\quad - 0''.000079T^3 - 0.00001T^4
\end{aligned} \tag{5.17}$$

Here, as in (5.15), T is the time in Julian centuries. In (5.17), the mean longitudes are taken equal to the right ascensions.

The mean solar time t can be determined if we equate sidereal times ST calculated with the Moon ($ST = \tau + s$) and with the Sun ($ST = t + h$):

$$t = \tau + s - h \tag{5.18}$$

Doodson chose the variables for the expansion of the tide-generating potential so that the corresponding angular velocities were positive (therefore the changed sign of the longitude of the ascending angle N) and arranged in the decreasing order. So, upon performing differentiation on the variables (5.17) with respect to time and keeping only the constant term, we have the following angular velocities expressed in °/mean solar hour:

$$\begin{aligned}
\dot{s} &= 0.5490165 \\
\dot{h} &= 0.0410686 \\
\dot{p} &= 0.0046418 \\
\dot{N}' &= 0.0022064 \\
\dot{p}_s &= 0.00000196 \\
\dot{\tau} &= 14.492052109 \\
\dot{ST} &= 15.0410686
\end{aligned} \tag{5.19}$$

The fundamental periods given in Table 5.1 correspond to these angular velocities of the variables commonly encountered in the tidal theories:

The tropical month and the tropical year are the time intervals between two successive passages of the Moon and the Sun, respectively, through the vernal equinox. The nodical (draconic) month determines the time interval between two successive transits of the Moon through its ascending node. The anomalistic month and anomalistic year are the time intervals between two successive transits of the Moon and the Sun, respectively, through the perigee and perihelion. The synodic

Table 5.1 The fundamental frequencies in the Earth–Moon–Sun motion.

Variable	Frequency, °/hour	Period, mean solar day	Definition
τ	14.492052109	1.035050	The mean lunar day (24 h 50.47 min)
s	0.549016530	27.321582	The mean tropical month of the Moon
H	0.041068639	365.242199	The mean tropical year of the Sun
P	0.004641837	8.847 yr	The mean period of revolution of the lunar perigee
N'	0.002206413	18.613 yr	The mean period of regression of the lunar nodes
p_s	0.000 001 961	20 940 yr	The revolution period of the Earth orbital perihelion
$s-N$	0.551222943	27.21222	The lunar draconic month
$s-p$	0.544 374 693	27.55 455	The lunar anomalistic month
$h-p_S$	0.041066678	365.25964	The solar anomalistic year
$s-h$	0.507947891	29.53059	The lunar sinodic month
$s-2h+p$	0.471521089	31.81194	The evection period
$2(s-h)$	1.015895782	14.765330	The variation period

month determines the period of recurrence of lunar phases. The periods of evection and variation correspond to cycles of the change of the lunar orbit's eccentricity under the influence of the gravitational attraction of the Sun.

The declination δ, the hour angle H and the distance c/d of the Moon and the Sun are complicated functions of time t. Therefore, the factors with these parameters in the expression for the tide-generating potential U are expanded into the trigonometrical series (theoretically infinite). In the general case, it is usually represented in the form (Lambeck, 1988; Moritz and Müller, 1987):

$$U=\sum_{n=2}^{\infty}\sum_{m=2}^{n}P_{nm}(\sin\varphi)\sum_{j}C_{nmj}\cos\left[\omega_{nmj}t+\beta_{nmj}+m\lambda+(n-m)\frac{\pi}{2}\right] \quad (5.20)$$

where C_{nmj} and β_{nmj} are constants; ω_{nmj} is the angular velocity; $P_{nm}(\sin\varphi)$ are Legendre's polynomials. This expansion differs from Doodson's expansion in the signs of coefficients C_{2mj} and the phase angle $(n-m)\frac{\pi}{2}$ (specifically, the signs of C_{20j} and C_{21j} and the angle $(2-m)\frac{\pi}{2}$).

5.2.1 Semidiurnal Waves

Let us deduce the basic harmonics of the tide-generating potential U_2 for $R=a$ by using Melchior's approach (Melchior, 1983). Let us first consider the semidiurnal waves that are implied by the first term (5.14):

$$G\left(\frac{c}{d}\right)^3\cos^2\varphi\cos^2\delta\cos 2H=G_\varphi\left(\frac{c}{d}\right)^3\cos^2\delta\cos 2H \quad (5.21)$$

For the Moon, the hour angle $H = \tau$. The declination changes from the maximal value $\approx 23°26' \pm 5°$ to the minimal value $\approx -(23°26' \pm 5°)$ over the tropical month (27.321 days). Therefore, $\cos^2\delta$ can be approximated as:

$$\cos^2\delta \approx 1 + n\cos 2s \tag{5.22}$$

Because of the ellipticity of the lunar orbit, the distance between the Earth and the Moon varies over a month from 356 400 km to 406 700 km. This taken into account, we can write:

$$\left(\frac{c}{d}\right)^3 \approx 1 + m\cos(s-p) \tag{5.23}$$

Substituting (5.22) and (5.23) in (5.21), we have:

$$\begin{aligned} &G_\varphi[1 + m\cos(s-p)](1 + n\cos 2s)\cos 2\tau \\ &= G_\varphi\cos 2\tau + G_\varphi m\cos(s-p)\cos 2\tau \\ &+ G_\varphi n\cos 2s\cos 2\tau + G_\varphi m\, n\cos(s-p)\cos 2s\cos 2\tau \end{aligned} \tag{5.24}$$

Here, the basic wave (carrier) $G_\varphi\cos 2\tau$ is separated. As the product of cosines can be written as:

$$\cos\alpha\cos\beta = \frac{1}{2}\cos(\alpha+\beta) + \frac{1}{2}\cos(\alpha-\beta) \tag{5.25}$$

the last three terms (5.24) give the following pairs of the side harmonics that are equidistant, in terms of the arguments, from the basic wave:

$$\cos(s-p)\cos 2\tau = \frac{1}{2}\cos[2\tau-(s-p)] + \frac{1}{2}\cos[2\tau+(s-p)] \tag{5.26}$$

$$\cos 2s\cos 2\tau = \frac{1}{2}\cos 2(\tau-s) + \frac{1}{2}\cos 2(\tau+s) \tag{5.27}$$

$$\begin{aligned} \cos 2s\cos(s-p)\cos 2\tau &= \frac{1}{4}\cos[2\tau+(3s-p)] + \frac{1}{4}\cos[2\tau-(3s-p)] \\ &+ \frac{1}{4}\cos[2\tau+(s+p)] + \frac{1}{4}\cos[2\tau-(s+p)] \end{aligned} \tag{5.28}$$

All the above presented formulae are valid for the solar tides as well, if the lunar constants and arguments are replaced by the solar ones.

Let us specify the symbolic designations introduced by Darwin and explain the sense of the above waves.

$M_2 \equiv G_\varphi^M\cos 2\tau$ is the principal lunar semidiurnal wave (carrier). It corresponds to the semidiurnal tide that is created by the Moon moving uniformly along the circular orbit in the equatorial plane, at the velocity equal to the true moon's velocity. The period of wave M_2 is equal to half a lunar day (12 hour 25 min), which corresponds to an hourly velocity of 28°.984.

$S_2 \equiv G_\varphi^S\cos 2t$ is the principal solar semidiurnal wave (carrier) similar to the lunar wave M_2. The period of this wave is equal to half a mean solar day, that is 12 hour, and the angular frequency is 30° per hour.

$N_2 \equiv G_\varphi^M m\cos[2\tau-(s-p)]$ and $L_2 \equiv G_\varphi^M m\cos[2\tau+(s-p)]$ are two lunar elliptic semidiurnal waves arising due to the variation in the distance c/d. As follows from Table 5.1, the angular velocity $s-p=0°.544\,h^{-1}$. Consequently, the angular velocities ω and the periods T of waves N_2 and L_2 are as follows:

$$\text{For } N_2\text{: } \omega = 28°.984-0°.544 = 28°.440, \quad T = 12 \text{ hour } 39 \text{ min},$$

$$\text{For } L_2\text{: } \omega = 28°.984+0°.544 = 29°.528, \quad T = 12 \text{ hour } 11 \text{ min}.$$

The amplitude of tides reaches its maximum when the Moon is at perigee ($s-p=0°$), and its minimum when the Moon is at apogee ($s-p=180°$). The velocity of the Moon moving along the orbit is greater at perigee and smaller at apogee. Therefore, the amplitude and period of wave N_2 is greater than the amplitude and period of wave L_2. Because of this, wave N_2 is referred to as the principal elliptic wave and wave L_2, as the small elliptic wave.

Similar to the lunar waves N_2 and L_2, there are the following two waves T_2 and R_2 generated by the Sun:

$T_2 \equiv G_\varphi^S m^S\cos[2t-(h-p_S)]$ is the large elliptic semidiurnal wave whose angular velocity is $30° - 0°.041 + 0° = 29°.9589\,h^{-1}$, and period, 12 hour 1 min;

$R_2 \equiv G_\varphi^S m^S\cos[2t+(h-p_S)]$ is the small elliptic semidiurnal wave whose angular velocity is $30°.0411\,h^{-1}$ and period, 11 hour 59 min;

${}^M K_2 \equiv \frac{n_M}{2} G_\varphi^M\cos 2(\tau+s)$ is the lunar declinational semidiurnal wave (that is arising from the change of declination). The angular velocity of wave ${}^M K_2$: $\omega = 28°.984 + 1°.098 = 30°.082\,h^{-1}$ is easily determined from the data in Table 5.1. The period is 11 hour 58 min, which is equal to half a sidereal day;

${}^S K_2 \equiv \frac{n_S}{2} G_\varphi^S\cos 2(t+h)$ is the solar declinational semidiurnal wave whose angular velocity and period are the same as the velocity and period of the lunar wave ${}^M K_2$. These waves cannot be separated, therefore their combination $({}^M K_2 + {}^S K_2) = K_2$ is considered, referred to as the lunisolar semidiurnal wave.

Other semidiurnal waves are related to perturbations of the lunar orbit caused by the Sun. The ellipticity of the lunar orbit increases (decreases) when the Sun crosses the major (minor) axis of the lunar orbit. If the lunar perigee had not moved, the Sun would have passed through each axis of the lunar orbit precisely every half a year and the frequency of the eccentricity variation would have been $(s-2h)$. However, the lunar perigee moves from the west to the east with frequency p. Therefore, the frequency of actual changes of the lunar orbit's eccentricity (≈20%) is $(s-p)-2(h-p)$. This effect is referred to as evection. Its variable part can be approximated by the following expression: $[1-\cos(s-2h+p)]$.

Under the influence of the gravitational attraction of the Sun, the eccentricity of the lunar orbit increases at the full and new moon and decreases at the first quarter and last quarter moon. In this case, the frequency of the eccentricity variation is $2(s-h)$, that is half a synodic month. This effect is referred to as variation. It is approximated by the following expression: $[1-\cos 2(s-h)]$.

If multipliers $[1 - \cos(s - 2h + p)]$ and $[1 - \cos 2(s - h)]$ are substituted in (5.24) and transformations (5.25) are performed we have a new series of pairs of the harmonics corresponding to waves M_2, N_2, L_2, $^M K_2$, and so forth. Here, we only mention two waves of evection for M_2, named as:

λ_2 with argument $2\tau + (s - 2h + p)$

ν_2 with argument $2\tau - (s - 2h + p)$ and wave of variation for M_2, named as:

μ_2 with argument $2\tau - 2(s - h)$

The amplitudes of the evectional and variational harmonics of waves N_2, L_2 and $^M K_2$ are small and are not addressed here.

5.2.2 Diurnal Waves

Diurnal waves are described by the second term of expression (5.14):

$$G\left(\frac{c}{d}\right)^3 \sin 2\varphi \sin 2\delta \cos H = G_{2\varphi}\left(\frac{c}{d}\right)^3 \sin 2\delta \cos H \tag{5.29}$$

Here, the basic source of waves is the variations of declination δ and distance c/d of the Moon and the Sun. Let us consider the Moon case, following Melchior (Melchior, 1983, 1971). Using the sine formula of spherical trigonometry, it can be written: $\sin\delta \sin 90° = \sin s \sin \varepsilon$, where ε is the inclination of the Moon's orbit. Consequently $\sin\delta = \sin\varepsilon \sin s$. The average value of multiplier $\sin 2\delta$ is equal to zero and $\cos\delta \approx 1$. Therefore, multiplier $\sin 2\delta$ in (5.29) can be approximated by expression:

$$\sin 2\delta \approx 0 + 2\sin\varepsilon \sin s \tag{5.30}$$

Substituting (5.23) and (5.30) in (5.29) and taking into account that $H = \tau$, we have

$$\begin{aligned}
&2G_{2\varphi}[1 + m\cos(s - p)]\sin\varepsilon \sin s \cos\tau \\
&= 2G_{2\varphi}\sin\varepsilon \sin s \cos\tau + 2G_{2\varphi} m \sin\varepsilon \cos(s - p)\sin s \cos\tau \\
&= G_{2\varphi}\sin\varepsilon[\sin(\tau + s) - \sin(\tau - s)] \\
&+ \frac{1}{2}G_{2\varphi}\sin\varepsilon\{\sin[(\tau + s) + (s - p)] + \sin[(\tau + s) - [s - p]]\} \\
&- \frac{1}{2}G_{2\varphi}\sin\varepsilon\{\sin[(\tau - s) - (s - p)] + \sin[(\tau - s) + (s - p)]\}
\end{aligned} \tag{5.31}$$

As the average value of $\sin 2\delta = 0$, there is no basic lunar diurnal wave (carrier) M_1 with argument τ (angular velocity 14°.492) in the expansion (5.31). For the same reason, there also is no basic solar wave (carrier) S_1 with argument t and a period of 1 day. There are two declinational waves and four elliptic waves:

$^M K_1$ is the lunar declinational wave with argument $(\tau + s) = ST =$ 14°.492 + 0°.549 = 15°.041;

O_1 is the principal lunar declinational wave with argument $(\tau - s) =$ 14°.492 – 0°.549 = 13°.943.

The amplitude of wave O_1 is a little greater than that of wave $^M K_1$, so O_1 is referred to as the principal wave. The period of wave $^M K_1$ is exactly equal to one sidereal day.

Each of wave O_1 and ${}^{M}K_1$ has two side elliptic waves:

Q_1, with argument $(\tau - s) - (s - p) = 13^\circ.943 - 0^\circ.544 = 13^\circ.399$ and a period of 26 hour 52 min;
$\varepsilon\ (O_1)$, with argument $(\tau - s) + (s - p) = 13^\circ.943 + 0^\circ.544 = 14^\circ.487$ and a period of 24 hour 51 min;
J_1, with argument $(\tau + s) + (s - p) = 15^\circ.041 + 0^\circ.544 = 15^\circ.585$ and a period 23 hour 6 min;
$\varepsilon({}^{M}K_1)$ with argument $(\tau + s) - (s - p) = 15^\circ.041 - 0^\circ.544 = 14^\circ.497$ and a period of 24 hour 50 min.

The angular velocities of waves $\varepsilon\ (O_1)$ and $\varepsilon({}^{M}K_1)$ are very close, for the difference between their arguments is only $2p$. The periods of waves $\varepsilon\ (O_1)$ and $\varepsilon({}^{M}K_1)$ are practically the same as the period of the absent wave M_1, therefore their combination is usually designated by symbol M_1.

In the solar tides case, the expansion of the type (5.31) gives the following waves that are similar to the above-described lunar diurnal waves:

${}^{S}K_1$ the solar declinational wave with argument $(t + h) = ST = 15^\circ.000 + 0^\circ.041 = 15^\circ.041$ and the period exactly equal to sidereal day 23 h 56 min;
P_1 the principal solar declinational wave P_1, with argument $(t - h) = 15^\circ.000 - 0^\circ.041 = 14^\circ.959$ and a period of 24 hour 4 min.

Thus, the solar wave ${}^{S}K_1$ cannot be distinguished from the lunar wave ${}^{M}K_1$. This is why their combination K_1, named the lunisolar diurnal wave, is addressed.

Each of the waves, ${}^{S}K_1$ and P_1, has a pair of side harmonics arising from the ellipticity of the Earth's orbit:

π_1, with argument $(t - h) - (h - p_S) = 14^\circ.959 - 0^\circ.041 = 14^\circ.918$ and a period of 24 hour 9 min;
$\varepsilon\ (P_1)$, with argument $(t - h) + (h - p_S) = 14^\circ.959 + 0^\circ.041 = 15^\circ.000$ and a period of 24 hour;
ψ_1, with argument $(t + h) + (h - p_S) = 15.041 + 0^\circ.041 = 15^\circ.082$ and a period of 23 hour 52 min;
$\varepsilon({}^{S}K_1)$, with argument $(t + h) - (h - p_S) = 15^\circ.041 - 0^\circ.041 = 15^\circ.000$ and a period of 24 hour.

Variable p_S is very small, for its period is 20 940 years. Consequently, the angular velocities and periods of waves $\varepsilon\ (P_1)$ and $\varepsilon({}^{S}K_1)$ are practically the same as the missing main solar wave S_1. So, the waves $\varepsilon\ (P_1)$ and $\varepsilon({}^{S}K_1)$ are usually considered one wave that is designated by S_1.

Longitudes s and h are counted along the ecliptic, and right ascensions α, along the equator, therefore $\alpha \approx s - 0.043 \sin 2s$. These relationships taken into account, we get new waves:

the lunar declinational second order wave OO_1 with argument $\tau + 3s = 14^\circ.492 + 1^\circ.647 = 16^\circ.139$ and a period of 22 hour 18 min;
the solar declinational diurnal wave φ_1 with argument $t + 3h = 15^\circ.000 + 0^\circ.123 = 15^\circ.123$ and a period of 23 hour 48 min.

The influence of the evection and the variations give rise to the diurnal waves ρ_1, ϑ_1, χ_1, σ_1, SO_2, and τ_1 that have small amplitudes.

5.2.3 Long-Period Waves

Long-period waves are described by the third term of expression (5.14):

$$3G\left(\frac{c}{d}\right)^3\left(\sin^2\varphi-\frac{1}{3}\right)\left(\sin^2\delta-\frac{1}{3}\right)=\frac{3}{2}G_{3\varphi}\left(\frac{1}{3}-\cos 2\delta\right) \tag{5.32}$$

In this case, the waves result from the variation of the declination and distances c/d of the Moon and the Sun.

The Moon's declination δ changes with a period of 27.322 day, which corresponds to an angular velocity of $0^\circ.549\,\mathrm{h}^{-1}$. As the expression (5.32) contains $\cos 2\delta$, the period of the lunar declinational waves is equal to half a period of δ, that is 13.66 days. This wave is denoted with M_f, which means the lunar fortnightly wave. The wave has argument $2s$ and an angular velocity of $1^\circ.098\,\mathrm{h}^{-1}$.

The Sun's declination varies with a period of one year, so the period of the solar declinational wave is 0.5 year. This wave is designated by symbol S_{sa} and has argument $2h$ and an angular velocity of $0^\circ.082\,\mathrm{h}^{-1}$.

Elliptic waves are created due to the variation of the distance c/d. Among these waves, the most appreciable is the lunar monthly wave M_m. Its argument is equal to $s-p=0^\circ.5490-0^\circ.0046=0^\circ.5444$, and the period, 27.55 days.

The greatest among the solar elliptic waves is the solar annual wave S_A with argument

$$h-p_S=0^\circ.0411\,\mathrm{h}^{-1}$$

The nodes of the lunar orbit slowly move along the ecliptic from East to West, making a complete revolution in 18.6 years. The perigee of the lunar orbit moves from West to East with a period of 8.85 years. Therefore, all the Moon-related multipliers of the three tide types slowly vary with time. Also, the parameters of the observable tides continuously vary.

5.2.4 General Classification of Tidal Waves

The arguments of all the above-described tidal waves represent linear combinations of six independent Doodson's variables with the constant coefficients assuming the values 0, ±1, ±2, ±3, ... Based on this, Doodson introduced argument numbers that allowed all tidal waves to be classified. For this purpose, in his expression for the argument of waves, he put the Doodson variables in order of decreasing angular velocities and calculated the respective coefficients by using the six code numbers d_r:

$$d_1\tau + (d_2-5)s + (d_3-5)h + (d_4-5)p + (d_5-5)N' + (d_6-5)p_S \qquad (5.33)$$

All d_r are positive integers. As the arrangement of the variables in (5.33) is unchangeable, a tidal wave can be characterized by a six-digit argument number consisting of six code numbers $d_1d_2d_3d_4d_5d_6$.

The first code number d_1 is always equal to the order m of the analyzed spherical harmonic and consequently can assume values 0, 1 and 2 in the above-mentioned expansion. Other code numbers may vary from 8 to 2. For example, according to (5.33), the argument number of the solar elliptic wave π_1 with argument $t-2h+p_S=\tau+s-3h+p_S$ is 162 556.

As Doodson's variables in (5.33) are arranged in order of decreasing angular velocities, the angular velocity increases with the argument number of the wave. A systematic classification of tidal waves can be based on this rule. The above-described tidal waves are listed in Table 5.2 in order of increasing argument numbers (the second column), and therefore in order of ascending angular velocities (the fourth column), or in order of descending periods (the fifth column). Note that the first code number is equal to 0 for the long-period waves; 1, for the diurnal waves; and 2, for the semidiurnal tidal waves. The sixth column presents the amplitudes of the waves calculated by Hartmann and Wenzel (1994, 1995).

5.3 Theory of Tidal Variations in the Earth's Rotation Rate

Tidal deformations of the Earth's figure result in changes in the components of the Earth's inertia tensor. This results in changes in the rate of the Earth's rotation. Jeffreys was the first to notice it (Jeffreys, 1928). Woolard calculated the basic harmonics of the tidal variations in the time UT1 based on his fundamental research of the tide-generating potential (Woolard, 1959). There are more accurate calculations of the tidal effects on the Earth's rotation rate (Wahr, Sasao and Smith, 1981; Yoder, Williams and Parke, 1981). Let us consider the principles of the theory of the tide-generated instability of the Earth's rotation.

Let us consider the fluctuations in the Earth's rotation rate with periods of 1 to 365 days. For these periods, the angular momentum of the Earth and its atmosphere relative to the axis of rotation is constant:

$$I_{33}\omega_3 = \text{const} \qquad (5.34)$$

So, differentiating (5.34) and using increments, we have:

$$\nu \equiv \frac{\delta\omega_3}{\omega_3} = -\frac{\delta I_{33}}{I_{33}} \qquad (5.35)$$

Table 5.2 The main tidal waves.

Tide symbol	Doodson' Number	Argument a_j	Frequency ω_j (degree/hour)	Period (days)	Amplitude C_j, ($cm^2 s^{-2}$)	Tide name
Long-period tides						
M_0	055 555	0	0.000000		−5944	Constant flattening from Moon
S_0	055 555	0	0.000000		−2751	Constant flattening from Sun
S_a	056 554	$h - p_s$	0.0410687	365.25	−138	Solar annual
Ss_a	057 555	$2h$	0.082137	182.62	−855	Solar semiannual
Ms_m	063 655	$s - 2h + p$	0.471521	31.81	−186	
M_m	065 455	$s - p$	0.544375	27.55	−973	Lunar monthly
Ms_f	073 555	$2\ (s - h)$	1.015896	14.77	−161	Lunisolar fortnightly
M_f	075 555	$2s$	1.098033	13.66	−1841	Lunar fortnightly
	075 565	$2s + N'$	1.100239	13.63	−763	
Mt_m	085 455	$3s - p$	1.642408	9.13	−353	
Diurnal tides						
$2Q_1$	125 755	$\tau - 3s + 2p$	12.854286	1.1669	130	Lunar elliptic second order
σ_1	127 555	$\tau - 3s + 2h$	12.92714	1.1604	157	Lunar variational
Q_1	135 655	$(\tau - s) - (s - p)$	13.398661	1.1195	981	Lunar large elliptic: with O_1
ρ_1	137 455	$(\tau - p) + 2(h - s)$	13.471515	1.1135	186	Lunar large evectional
O_1	145 555	$\tau - s$	13.943036	1.0758	5125	Principal lunar declinational
M_1	155 655	$(\tau + s) - (s - p)$	14.496694	1.0347	−403	Lunar small elliptic: with ${}^MK_1$
π_1	162 556	$(t - h) - (h - p_S)$	14.917865	1.0055	139	Solar elliptic: with P_1
P_1	163 555	$t - h$	14.958931	1.0027	2380	Principal solar declinational
S_1	164 556	$(t + h) - (h - p_S)$	15.000002	1.0	−57	Solar elliptic: with ${}^SK_1$
${}^MK_1$	165 555	$\tau + s = ST + 12^h$	15.041069	0.9973	−4925	Lunar declinational
${}^SK_1$	165 555	$t + h = ST + 12^h$	15.041069	0.9973	−2280	Solar declinational

	165 565	$\tau + s + N'$	15.043275	0.9971	−978	Lunar
ψ_1	166 554	$(t + h) + (h - p_S)$	15.082135	0.9946	−57	Solar elliptic: with $^{S}K_1$
φ_1	167 555	$t + 3h$	15.123206	0.9918	−102	Solar declinational
J_1	175 455	$(\tau + s) + (s - p)$	15.585443	0.9624	−403	Lunar elliptic: with $^{M}K_1$
OO_1	185 555	$\tau + 3s$	16.139102	0.9294	−220	Lunar declinational second order
Semidiurnal tides						
$2N_2$	235 755	$2\tau - 2\ (s - p)$	27.895355	0.5377	+313	Lunar elliptic second order: with M_2
μ_2	237 555	$2\tau - 2\ (s - h)$	27.968208	0.5363	+378	Lunar variational
N_2	245 655	$2\tau - (s - p)$	28.439730	0.5274	+2366	Principal lunar elliptic: with M_2
ν_2	247 455	$2\tau - (s - 2h + p)$	28.512583	0.5261	+449	Lunar large evectional
M_2	255 555	2τ	28.984104	0.5175	+12356	Principal lunar
λ_2	263 655	$2\tau + (s - 2h + p)$	29.455625	0.5092	−91	Lunar small evectional
L_2	265 455	$2\tau + (s - p)$	29.528479	0.5080	−349	Lunar small elliptic: with M_2
T_2	272 556	$2t - (h - p_S)$	29.958933	0.5007	+336	Solar large elliptic: with S_2
S_2	273 555	$2t$	30.000000	0.5	+5738	Principal solar
R_2	274 554	$2t + h - p_S$	30.041067	0.4993	−48	Solar small elliptic: with S_2
$^{M}K_2$	275 555	$2\ (\tau + s) = 2ST)$	30.082137	0.4986	+1067	Lunar declinational
$^{S}K_2$	275 555	$2(t + h) = 2ST$	30.082137	0.4986	+495	Solar declinational
M_3	355 555	3τ	43.476156	0.3450	−150	Principal lunar

The Earth's inertia tensor is as follows:

$$I_{ij} = \iiint\limits_V \rho(x_l^2\delta_{ik} - x_i x_k)\,\mathrm{d}V \tag{5.36}$$

The trace of this tensor is as follows:

$$I_{11} + I_{22} + I_{33} = \iiint\limits_V (x_2^2 + x_3^2 + x_1^2 + x_3^2 + x_1^2 + x_2^2)\,\rho\,\mathrm{d}V = 2\iiint\limits_V r^2\rho\,\mathrm{d}V \tag{5.37}$$

The shape of the Earth changes due to the lunisolar tide-generating potential. In this context, however, the trace of the inertia tensor of the incompressible Earth does not change. We will prove it by following Moritz and Müller (Moritz and Müller, 1987). Differentiating (5.37) and using increments, we have:

$$\delta(I_{11} + I_{22} + I_{33}) = 2\iiint\limits_V \delta(r^2\rho\,\mathrm{d}V) = 2\iiint\limits_V f(r)Y_n(\theta,\lambda)\,\mathrm{d}V \tag{5.38}$$

This expression takes into account that the reaction of an incompressible body to the nth-degree harmonic of the tide-generating potential is the nth-degree spherical harmonic $Y_n(\theta, \lambda)$. Here, $f(r)$ is some function of radius r; θ and λ are the complement of latitude and longitude, respectively; $\mathrm{d}V = r^2 \sin\theta\, \mathrm{d}r\, \mathrm{d}\theta\, \mathrm{d}\lambda$ is a volume element.

Owing to the orthogonality of the spherical harmonics, the integral taken through the spherical surface of $Y_n(\theta, \lambda)$ is equal to zero, so we have:

$$\delta(I_{11} + I_{22} + I_{33}) = 2\int\limits_0^R f(r)r^2\,\mathrm{d}r \int\limits_{\lambda=0}^{2\pi}\int\limits_{\theta=0}^{\pi} Y_n(\theta,\lambda)\sin\theta\,\mathrm{d}\theta\,\mathrm{d}\lambda = 0 \tag{5.39}$$

The figure of the Earth is close to an ellipsoid of revolution, that is $I_{11} \approx I_{22}$ and $\delta I_{11} \approx \delta I_{22}$. Consequently, (5.39) gives:

$$\delta I_{11} = \delta I_{22} = -\frac{1}{2}\delta I_{33} \tag{5.40}$$

The tidal deformation of the Earth results in an additional potential δW, which is proportional to the perturbing tide-generating potential U_{nm}. Considering only the zonal terms of the second degree, we have

$$\delta W = kU_{20} \tag{5.41}$$

where k is the Love number.

Let us express the additional potential through the terrestrial parameters. For this purpose, we use the expression for the Earth's gravitational potential W:

$$W = \frac{GM}{r}\left[1 - \left(\frac{a}{r}\right)^2 J_2 P_{20}(\cos\theta) - \ldots\right] \tag{5.42}$$

Here, G is the gravitational constant; M is the Earth's mass; $J_2 = (I_{33} - I_{11})/Ma^2$ is the dynamical form factor of the Earth; a is the equatorial radius. Let us differentiate (5.42) and use increments, addressing only the harmonic of the second degree.

$$\delta W = -GM\frac{a^2}{r^3}\delta J_2 P_{20}(\cos\theta) \tag{5.43}$$

Here,

$$\delta J_2 = \frac{1}{Ma^2}(\delta I_{33} - \delta I_{11}) = \frac{3}{2Ma^2}\delta I_{33} \tag{5.44}$$

Thus, the increment of the potential δW on the Earth's surface ($r = R$) can be expressed as:

$$\delta W = -\frac{3}{2}\frac{G}{R^3}P_{20}(\cos\theta)\,\delta\, I_{33} \tag{5.45}$$

Equating (5.41) and (5.45), we obtain:

$$\delta I_{33} = -\frac{2}{3}k\frac{R^3}{G}\frac{U_{20}}{P_{20}(\cos\theta)} \tag{5.46}$$

It is known that the tide-generating potential is described by series (5.20). Changes in the angular velocity of the Earth's rotation are caused only by the zonal components of the potential U_{no}. Among the zonal components, the component U_{20} of the second degree is dominating. It exceeds by 62 times the magnitude of the nearest harmonic U_{30}. Hereafter, we only consider the effect of this harmonic:

$$U_{20} = P_{20}(\cos\theta)\sum_j c_j \cos a_j \tag{5.47}$$

Substituting (5.47) into (5.46) and taking into account (5.35) gives the following expression for the changes in the Earth's rotation rate:

$$\nu = \frac{2}{3}k\frac{R^3}{GI_{33}}\sum_j c_j \cos a_j \tag{5.48}$$

The phases a_j of the tide-generating harmonics are expressed through independent fundamental arguments (5.15).

At the present time, the time services use the series of the tidal harmonics including 62 periodic components (Yoder, Williams and Parke, 1981) for the reducing time scale UT1.

Table 5.3 lists the arguments, periods and amplitudes (amplitude coefficients of cosines) of the harmonics to be used in calculating the tidal fluctuations in the length of day and the angular velocity of the Earth's rotation. The periods are in days, and the deviations of the day duration $\delta\Pi$ and the angular velocity $\delta\omega = \nu\cdot\omega$, in 10^{-5} s and 10^{-14} rad/s, respectively.

Derived from Table 5.3, the tidal oscillations in the velocity of the Earth's rotation (2) during period October 1, 2006 – December 31, 2007 are depicted in Figure 3.11. One can see that the tidal variations contribute most to the variations in the Earth's rotation with periods of less than one month. The deviation of the observed curve (1)

Table 5.3 Zonal tides term in the length of day $\delta\prod$, and the angular velocity of the Earth $\delta\omega$ (IERS Conventions, 2003).

No.	Argument					Period (days)	$\delta\Pi$ (10^{-5} s)	$\delta\omega$ (10^{-14} rad/s)
	l	*l'*	*F*	*D*	*N*			
1	1	0	2	2	2	5.64	0.26	−0.22
2	2	0	2	0	1	6.85	0.38	−0.32
3	2	0	2	0	2	6.86	0.91	−0.76
4	0	0	2	2	1	7.09	0.45	−0.38
5	0	0	2	2	2	7.10	1.09	−0.92
6	1	0	2	0	0	9.11	0.27	−0.22
7	1	0	2	0	1	9.12	2.84	−2.40
8	1	0	2	0	2	9.13	6.85	−5.78
9	3	0	0	0	0	9.18	0.12	−0.11
10	−1	0	2	2	1	9.54	0.54	−0.46
11	−1	0	2	2	2	9.56	1.30	−1.10
12	1	0	0	2	0	9.61	0.50	−0.42
13	2	0	2	−2	2	12.81	−0.11	0.09
14	0	1	2	0	2	13.17	−0.12	0.10
15	0	0	2	0	0	13.61	1.39	−1.17
16	0	0	2	0	1	13.63	14.86	−12.54
17	0	0	2	0	2	13.66	35.84	−30.25
18	2	0	0	0	−1	13.75	−0.10	0.08
19	2	0	0	0	0	13.78	1.55	−1.31
20	2	0	0	0	1	13.81	−0.08	0.07
21	0	−1	2	0	2	14.19	0.11	−0.09
22	0	0	0	2	−1	14.73	−0.20	0.17
23	0	0	0	2	0	14.77	3.14	−2.65
24	0	0	0	2	1	14.80	0.22	−0.19
25	0	−1	0	2	0	15.39	0.21	−0.17
26	1	0	2	−2	1	23.86	−0.13	0.11
27	1	0	2	−2	2	23.94	−0.26	0.22
28	1	1	0	0	0	25.62	−0.10	0.08
29	−1	0	2	0	0	26.88	−0.11	0.09
30	−1	0	2	0	1	26.98	−0.41	0.35
31	−1	0	2	0	2	27.09	−1.02	0.86
32	1	0	0	0	−1	27.44	−1.23	1.04
33	1	0	0	0	0	27.56	18.99	−16.03
34	1	0	0	0	1	27.67	−1.25	1.05
35	0	0	0	1	0	29.53	−0.11	0.09
36	1	−1	0	0	0	29.80	0.12	−0.10
37	−1	0	0	2	−1	31.66	−0.24	0.20
38	−1	0	0	2	0	31.81	3.63	−3.07
39	−1	0	0	2	1	31.96	−0.26	0.22
40	1	0	−2	2	−1	32.61	−0.04	0.03
41	−1	−1	0	2	0	34.85	0.16	−0.13
42	0	2	2	−2	2	91.31	0.04	−0.03
43	0	1	2	−2	1	119.61	−0.02	0.01
44	0	1	2	−2	2	121.75	0.98	−0.83
45	0	0	2	−2	0	173.31	−0.09	0.08

Table 5.3 (*Continued*)

No.	Argument					Period (days)	$\delta\Pi$ (10^{-5} s)	$\delta\omega$ (10^{-14} rad/s)
	l	*l'*	*F*	*D*	*N*			
46	0	0	2	−2	1	177.84	−0.42	0.35
47	0	0	2	−2	2	182.62	16.88	−14.25
48	0	2	0	0	0	182.63	0.07	−0.06
49	2	0	0	−2	−1	199.84	−0.02	0.01
50	2	0	0	−2	0	205.89	0.17	−0.14
51	2	0	0	−2	1	212.32	−0.01	0.01
52	0	−1	2	−2	1	346.60	0.01	−0.01
53	0	1	0	0	−1	346.64	−0.02	0.01
54	0	−1	2	−2	2	365.22	−0.14	0.12
55	0	1	0	0	0	365.26	2.69	−2.27
56	0	1	0	0	1	386.00	0.02	−0.02
57	1	0	0	−1	0	411.78	0.00	0.00
58	2	0	−2	0	0	1095.17	−0.01	0.01
59	−2	0	2	0	1	1305.47	−0.02	0.02
60	−1	1	0	1	0	3232.85	0.00	0.00
61	0	0	0	0	2	−3399.18	0.15	−0.13
62	0	0	0	0	1	−6798.38	−15.62	13.18

from the calculated curve (2) are associated with hydrometeorological processes (see Sections 6.4 and 7.6).

5.4 Precession and Nutations of the Earth's Axis

The theory of tides is inseparably linked with the theory of precession and nutation of the Earth's rotation axis. But whereas the theory of tides employs the reference system Ox_i rotating along with the Earth, the precession and nutation theory is based on the inertial reference system $O\xi_i$ oriented on fixed celestial objects.

5.4.1 Lunisolar Moment of Forces

The absolute value of the moment of forces **L** affecting the Earth due to the gravitational attraction of the Moon is equal to the moment of forces affecting the Moon due to the gravitational attraction of the Earth (Melchior, 1971; Melchior, 1983), that is

$$\mathbf{L} = -\mu\left[\mathbf{d}\cdot\frac{\partial W}{d\partial\delta}\right] \tag{5.49}$$

where μ and δ are the mass and declination of the Moon, respectively; **d** is the geocentric radius-vector of the Moon; W is the Earth's gravitational potential described by the following expression:

$$W = \frac{GM}{r}\left[1 - J_2\left(\frac{a}{r}\right)^2\left(\frac{3}{2}\sin^2\varphi - \frac{1}{2}\right) + \ldots\right] \tag{5.50}$$

Here G is the gravitational constant; M is the mass of the Earth; r is the geocentric radius; $J_2 = \frac{C-A}{Ma^2}$ is the dynamical form factor of the Earth; a is the equatorial radius; C and A are the polar and equatorial moments of inertia of the Earth, respectively; φ is latitude (here it is identified with δ). Bearing in mind that $\varphi = \delta$, and $r = d$, we differentiate W. The result is substituted into (5.49):

$$L = \frac{3}{2}\frac{G\mu}{d^3}(C-A)\sin 2\delta \tag{5.51}$$

The vector of the moment of forces **L** is perpendicular to the plane of the meridian of the Moon, therefore its longitude is 90° less than the longitude of the Moon. The vector is directly related to the tesseral part of the tide-generating potential. It can be described as a complex combination of the projections L_1 and L_2 on the equatorial coordinate axes (Moritz and Muller, 1992)

$$\underline{L} = L_1 + iL_2 = (C-A)\Omega^2 \sum_j B_j e^{-i(\omega_j t + \beta_j)} \tag{5.52}$$

where Ω is the mean angular velocity of the Earth's rotation; ω_j is the angular velocity of the jth tidal wave; β_j is the initial phase; B_j is the dimensionless factor.

The lunisolar moment of forces **L** determines the motion of the poles and the precession and nutation of the Earth's rotation axis. Let us address these phenomena.

5.4.2
Motion of the Earth's Poles

The polar motion is due to the movement of the instantaneous axis of rotation of the Earth. For a perfectly rigid Earth model, the motion is described by Equation 4.17.

Let us substitute expression (5.52) into Equations 4.17 and neglect atmospheric and oceanic effects ($\underline{\tilde{\chi}} = 0$). So we have:

$$\frac{i}{\sigma_r}\frac{d\underline{m}}{dt} + \underline{m} = \frac{i\underline{L}}{\Omega^2(C-A)} = i\sum_j B_j e^{-i(\omega_j t + \beta_j)} \tag{5.53}$$

It is more convenient to transform this equation as follows:

$$\frac{d\underline{m}}{dt} - i\sigma_r \underline{m} = \sigma_r \sum_j B_j e^{-i(\omega_j t + \beta_j)} \tag{5.54}$$

and to assume that $\underline{m} = \underline{m}_0$ for $t = t_0$. First, we find the solution of the homogeneous equation: $\underline{m} = \underline{C}e^{i\sigma_r t}$

Then, for the assumed initial condition and the arbitrary constant $\underline{C} = \underline{C}(t)$ being varied, we find the general solution of the nonhomogeneous Equation 5.54:

$$\underline{m} = \mathrm{i}\sum_j \frac{\sigma_r}{\omega_j + \sigma_r} B_j e^{-\mathrm{i}(\omega_j t + \beta_j)} + \underline{m}_0 e^{\mathrm{i}\sigma_r t} \tag{5.55}$$

The solution (5.55) shows that the polar motion of the perfectly rigid Earth is composed of a slow motion with an amplitude of $\underline{m}_0$ and a period of $2\pi/\sigma_r \approx 304$ days and a number of rapid small-amplitude movements (about $1/10\underline{m}_0$) with periods of the diurnal lunisolar tides.

For elastic Earth, the polar motion is described by the Equation 4.29. Substituting the expression for the moment of forces (5.52) into (4.29) and neglecting functions of the atmospheric and oceanic angular moments gives: ($\underline{\chi} = \dot{\underline{\chi}} = 0$),

$$\frac{\mathrm{i}}{\sigma_0}\frac{d\underline{m}}{dt} + \underline{m} = \frac{\mathrm{i}\kappa \underline{L}}{\Omega^2(C-A)} = \mathrm{i}\kappa \sum_j B_j e^{-(\omega_j t + \beta_j)} \tag{5.56}$$

where σ_0 is the Chandler frequency; $k = 1.43$ is the parameter of the Earth's elasticity. Let us transform (5.56) to the following form:

$$\frac{d\underline{m}}{dt} - \mathrm{i}\sigma_0 \underline{m} = \sigma_0 \kappa \sum_j B_j e^{-\mathrm{i}(\omega_j t + \beta_j)} \tag{5.57}$$

Let $\underline{m} = \underline{m}_0$ at the initial moment $t = 0$. Then, solving the nonhomogeneous differential Equation 5.57 in the same manner as (5.54) gives:

$$\underline{m} = \mathrm{i}\sum_j \frac{\kappa\sigma_0}{\omega_j + \sigma_0} B_j e^{-\mathrm{i}(\omega_j t + \beta_j)} + \underline{m}_0 e^{\mathrm{i}\sigma_0 t} \tag{5.58}$$

As is shown in Section 3.3, the Chandler frequency σ_0, and parameter κ are equal:

$$\sigma_0 = \Omega(C - A - \mathfrak{R})/(A + \mathfrak{R})$$
$$\kappa = (C - A)/(C - A - \mathfrak{R})$$

Bearing in mind that $\omega_j \approx \Omega$ and $\sigma_r = \Omega(C - A)/A$, we have:

$$\frac{k\sigma_0}{\omega_j + \sigma_0} = \frac{C - A}{C} = \frac{\sigma_r}{\omega_j + \sigma_r}$$

that is, the multipliers of the amplitudes B_j in expressions (5.58) and (5.55) coincide. Consequently, the effect of the lunisolar diurnal tides on the elastic Earth's polar motion is the same as the effect of the lunisolar diurnal tides on the polar motion of the perfectly rigid Earth. In both cases, the lunisolar moment of forces causes many low-amplitude (≈ 0.6 m) polar movements with periods of $2\pi/\omega_j$ as the diurnal tides'. The right parts of expressions (5.58) and (5.55) only differ in the second terms describing the slow polar motion with the Chandler and Eulerian frequencies, respectively.

Along with the above-discussed movement of the instantaneous axis of the Earth's rotation, the lunisolar moment of forces causes the movement of the angular momentum vector and the axis of the Earth's figure. These movements are described in detail in the book (Moritz and Muller, 1992).

5.4.3
Precession and Nutations

It can be shown (Moritz and Müller, 1987) that the right part of Poisson's Equation 4.71 is expressed through the moment of forces $\underline{L}$ as follows:

$$\frac{d\theta}{dt}+i\sin\theta\frac{d\psi}{dt}=\frac{1}{C\Omega}\left(\frac{1}{\sin\theta}\frac{\partial U}{\partial\psi}-i\frac{\partial U}{\partial\theta}\right)=-\frac{i}{C\Omega}\underline{L}e^{i\Omega t} \tag{5.59}$$

Substituting the value of $\underline{L}$ from (5.52) into (5.59) gives:

$$\frac{d\theta}{dt}+i\sin\theta\frac{d\psi}{dt}=-i\frac{C-A}{C}\Omega\sum_j B_j e^{-i(\Delta\omega_j t+\beta_j)} \tag{5.60}$$

where $\Delta\omega_j=\omega_j-\Omega$; θ is the nutation angle, that is the obliquity of the ecliptic to the equatorial plane, and ψ is the precession angle corresponding to the ecliptic longitude.

The frequency spectrum of the tidal waves in the expansion of the tide-generating potential is symmetrical with respect to the central frequency Ω. There are as many waves to the left of the Earth's rotation frequency Ω as to the right. Let us number the frequencies so as the frequency ω_j equal Ω has index $j=0$. Then, $\Delta\omega_0=0$ and two components may be isolated in the right part of (5.60) corresponding to the cases $j=0$ and $j\neq 0$:

$$\frac{d\theta}{dt}+i\sin\theta\frac{d\psi}{dt}=-i\frac{C-A}{C}B_0\Omega-i\frac{C-A}{C}\sum_{j\neq 0}\Omega B_j e^{-i(\Delta\omega_j t+\beta_j)} \tag{5.61}$$

The multiplier of $\sin\theta$ can be considered time-constant.

Let the nutation and precession angles be equal to $\theta(t_0)$ and $\psi(t_0)$, respectively, at the initial moment of time $t=t_0$. Then integration of Equation 5.61 gives:

$$\Delta\theta+i\Delta\psi\sin\theta=-i\frac{C-A}{C}B_0\Omega(t-t_0)+\frac{C-A}{C}\sum_{j\neq 0}\frac{\Omega}{\Delta\omega_j}B_j e^{-i[\Delta\omega_j(t-t_0)+\beta_j]} \tag{5.62}$$

where $\Delta\theta=\theta(t)-\theta(t_0)$, $\Delta\psi=\psi(t)-\psi(t_0)$. Let us separate the real and imaginary parts in (5.62):

$$\Delta\theta=\frac{C-A}{C}\sum_{j\neq 0}\frac{\Omega}{\Delta\omega_j}B_j\cos\left[\Delta\omega_j(t-t_0)+\beta_j\right] \tag{5.63}$$

$$\Delta\psi\sin\theta=-\frac{C-A}{C}B_0\Omega(t-t_0)+\frac{C-A}{C}\sum_{j\neq 0}\frac{\Omega}{\Delta\omega_j}B_j\sin\left[\Delta\omega_j(t-t_0)+\beta_j\right] \tag{5.64}$$

Expression (5.63) describes the nutation angle $\Delta\theta$ changing with time. One can see that it has almost periodic changes that represent superposition of theoretically infinite series of separate harmonics with frequencies of $\Delta\omega_j$. There are no secular changes $\Delta\theta$.

Expression (5.64) describes the precession angle $\Delta\psi$ changing with time. The first term of the right part of (5.64) increases linearly with time. It is the precession of the axis of the Earth's rotation. The second term describes almost periodic fluctuations in the precession angle – similar to the fluctuations in the nutation angle $\Delta\theta$. This is why they are combined into $\Delta\theta + i\Delta\psi \sin\theta$, which is referred to as nutation. The term $\Delta\theta$ is nutation in the inclination, whereas $\Delta\psi \sin\theta$ is nutation in the longitude.

Thus, the nutation frequency is the tidal frequency minus the frequency of the Earth's rotation Ω. This can be easily understood if one bears in mind that whereas tides are observed with respect to the reference points located on the Earth and rotating with it at the angular velocity Ω, the Earth's axis precession and nutation are measured with respect to the fixed celestial objects, that is with respect to an inertial reference system. Since the tidal frequencies are close to frequency Ω, the frequencies of the nutational components are very low.

Two tidal waves, whose frequencies are equidistant from the frequency Ω, generate one nutational wave with the frequency $\Delta\omega = \omega - \Omega$. Besides the above representation of the nutational wave in the form (5.62) (nutation in longitude and nutation in inclination), it can be approximated by the sum of two circular movements in opposite directions, with identical periods and different amplitudes:

$$\Delta\theta + i\Delta\psi\sin\theta = A^p e^{i\Delta\omega t} + A^r e^{-i\Delta\omega t} \tag{5.65}$$

The counterclockwise motion of the axis (as viewed from the Celestial North Pole) which coincides with the direction of the Earth's rotation, is referred to as the direct one and the clockwise motion – as the reverse one. Positive frequency $\Delta\omega_j$ corresponds to the direct motion, whereas negative frequency $-\Delta\omega_j$ of the nutational harmonics corresponds to the opposite motion. As a result of the composition of the two movements, the axis of the Earth's rotation moves along an elliptic trajectory.

The denominator in expressions (5.62)–(4.64) for nutation includes the difference $\Delta\omega$; therefore the nearer the tidal frequency ω is to the angular velocity Ω, the greater the amplitude of nutation. It is clearly demonstrated by Table 5.4 that contains the basic nutational harmonics and their amplitudes (Melchior, 1968, 1975).

The lunisolar wave K_1 is related to the nutation of zero frequency. This is the lunisolar precession. As a result of the precession, the angle ψ decreases by $50''$, 34 per year; the Moon is responsible for 2/3 of this rate, the Sun, for 1/3.

The principal harmonic of nutation is generated by the tidal waves whose argument numbers are 165 545 and 165 565. They are so insignificant that they are not even included in Table 5.2. The harmonic's amplitude is $-17''.2327 \times \sin 23^\circ.44 = -6''.85$ (in longitude) and $+9''.21$ (in inclination). Next in amplitude is the semiannual nutation caused by the tidal waves P_1 and φ_1. Also, significant is the two-week nutation caused by the lunar tidal waves O_1 and OO_1.

At the present time, 106 harmonics are used in calculating the nutation (IERS Standards, 1989). The first 44 harmonics of this expansion are listed in Table 5.5.

The amplitudes of the harmonics listed in Table 5.5 differ from those in Table 5.4 because Table 5.5 is calculated in view of features of the structure of the Earth (Wahr, Sasao and Smith, 1981).

Table 5.4 Correspondence between tides and nutations.

Argument number	Tide symbol	Nutation argument, frequency	Period (days)	Amplitude, arcs		Term and origin
				$\Delta\psi$	$\Delta\theta$	
135 655	Q_1	$3s-p=1.642408$	9.1	−0.″030	+0.″013	9.1 d, Moon
195 455	—					
145 555	O_1	$2s=1.09803$	13.7	−0.2276	+0.098	Fortnightly, Moon
185 555	OO_1					
155 655	M_1	$s-p=0.544375$	27.6	+0.″071	−0.0007	Lunar monthly
175 455	J_1					
162 556	π_1	$3h-p_S=0.123204$	122	−0.0517	+0.022	Solar terannual
168 554	—					
163 555	P_1	$2h=0.082137$	183	−1.3171	+0.573	Solar semiannual
167 555	φ_1					
164 556	S_1	$h-p_S=0.041067$	365	+0.148	0.0074	Solar annual
166 554	ψ_1					
165 575	—	$2N=0.004413$	3399	+0.207	−0.0897	9.3 yr, Moon
165 545	—	$N=0.002206$	6798	−17.206	+9.205	Principal term, Moon
165 565	—					
165 555	K_1		∞	—	—	Precession, Moon and Sun

5.5 Introduction to the Theory of Atmospheric Tides

Atmospheric tides are forced oscillations caused by the lunisolar gravitational perturbations and the thermal air movement due to the temperature contrasts between the daylight and night hemispheres. Certainly, the atmosphere can also freely oscillate, but then there should be a mechanism to support such free oscillations.

Newton and Laplace contributed most to the development of the theory of tides. The latter formulated a tidal equation (Laplace's tidal equation) that is true for both the atmosphere and the ocean. The main contribution to the theory of the equation was made by the English astronomer Hough (Hough,1897, 1898). The theory of solutions to Laplace's equation is discussed in many works. Hereafter, we will follow Lamb (1932), Dynamic meteorology (1937), Dikiy (1969), and Chapman and Lindzen (1970).

The equations of tides in the atmosphere can be represented in the form of the equations of tides in an ocean of constant depth. So, we first write the traditional equations of the tides in the ocean of constant depth h on the rotating Earth of constant radius a (Izekov and Kochin, 1937; Lamb, 1932):

$$\frac{\partial u_\lambda}{\partial t}+2\omega\cos\theta v_\theta=-\frac{1}{a\sin\theta}\left(\frac{1}{\rho}\frac{\partial P}{\partial\lambda}+\frac{\partial U}{\partial\lambda}\right) \tag{5.66}$$

Table 5.5 The main components for nutation in longitude $\Delta\psi$ and obliquity $\Delta\varepsilon$, referred to the mean equator and equinox of date, with T measured in Julian centuries from epoch J2000.0.

Argument					Period (days)	Longitude (0.0001″)		Obliquity (0.0001″)	
l	*l′*	*F*	*D*	Ω					
0	0	0	0	1	−6798.38	−172 064	−174.7T	92 052	9.1T
0	0	2	−2	2	182.62	−13 171	−1.7 T	5730	−3.0 T
0	0	2	0	2	13.66	−2276	−0.2 T	978	−0.5 T
0	0	0	0	2	−3399.19	2075	0.2 T	−897	0.5 T
0	1	0	0	0	365.26	1476	−3.6 T	74	−0.2 T
0	1	2	−2	2	121.75	−517	1.2 T	224	−0.7 T
1	0	0	0	0	27.55	711	0.1 T	−7	0.0 T
0	0	2	0	1	13.63	−387	−0.4 T	201	0.0 T
1	0	2	0	2	9.13	−301	0.0 T	129	−0.1 T
0	−1	2	−2	2	365.22	216	−0.5 T	−96	0.3 T
0	0	2	−2	1	177.84	128	0.1 T	−69	0.0 T
−1	0	2	0	2	27.09	124	0.0 T	−53	0.0 T
1	0	0	−2	0	31.81	−157	0.0 T	−1	0.0 T
1	0	0	0	1	27.67	63	0.1 T	−33	0.0 T
−1	0	0	0	1	−27.4	−58	−0.1 T	31	0.0 T
−1	0	2	2	2	9.56	−60	−0.0 T	26	−0.0 T
1	0	2	0	1	9.12	−52	−0.0 T	26	0.0 T
−2	0	2	0	1	1305.48	46	0.0 T	−24	−0.0 T
0	0	0	2	0	14.77	63	0.0 T	−1	0.0 T
0	0	2	2	2	7.10	−39	0.0 T	16	0.0 T
2	0	0	−2	0	205.89	−48	0.0 T	0	0.0 T
2	0	2	0	2	6.86	−31	0.0 T	13	0.0 T
1	0	2	−2	2	23.94	29	0.0 T	−12	0.0 T
−1	0	2	0	1	26.98	20	0.0 T	−11	0.0 T
2	0	0	0	0	13.78	29	0.0 T	−1	0.0 T
0	0	2	0	0	13.61	26	0.0 T	−1	0.0 T
0	0	2	−2	0	−173.31	22	0.0 T	−0	0.0 T
0	−2	0	0	0	−182.63	−17	0.1 T	0	0.0 T
0	2	2	−2	2	91.31	−16	0.1 T	7	−0.0 T
−1	0	0	2	1	31.96	15	0.0 T	−8	0.0 T
0	1	0	0	1	386.00	−14	−0.0 T	9	0.0 T
1	0	0	−2	1	31.66	−13	−0.0 T	7	0.0 T
0	−1	0	0	1	346.64	−13	0.0 T	6	0.0 T
2	0	−2	0	0	1095.18	−11	0.0 T	0	0.0 T
−1	0	2	2	1	9.54	−10	−0.0 T	5	0.0 T
1	0	2	2	2	5.64	−8	0.0 T	3	0.0 T
0	−1	2	0	2	14.19	−7	0.0 T	3	0.0 T
0	0	2	2	1	7.09	−7	−0.0 T	3	−0.0 T
1	1	0	−2	0	34.85	−7	0.0 T	0	0.0 T
0	1	2	0	2	13.17	8	−0.0 T	−3	−0.0 T
−2	0	0	2	1	199.84	−6	−0.0 T	3	0.0 T
0	0	0	2	1	14.80	−6	−0.0 T	3	0.0 T
2	0	2	−2	2	12.81	6	0.0 T	−3	0.0 T
1	0	0	2	0	9.61	6	0.0 T	0	0.0 T

$$\frac{\partial v_\theta}{\partial t} - 2\omega\cos\theta\, u_\lambda = -\frac{1}{a}\left(\frac{1}{\rho}\frac{\partial P}{\partial\theta} + \frac{\partial U}{\partial\theta}\right) \qquad (5.67)$$

$$0 = \frac{\partial P}{\partial z} - g\rho \qquad (5.68)$$

Here, u_λ and v_θ are the velocity components directed from west to east and from north to south, respectively; $\theta = 90^\circ - \varphi$ is the colatitude of latitude φ, λ is the east longitude; z is the depth; P and ρ are the pressure and the density, respectively; ω is the angular velocity of the Earth's rotation; U is the potential of the tide-generating forces, the vertical component of which can be ignored, as compared with the gravity g.

Let ζ (θ, λ, t) be the elevation of a point on the ocean's surface above the average ocean level. Let us integrate Equation 5.68 between the ocean's perturbed surface ζ, where $P = P_0$, and the constant level z:

$$P = \rho g\zeta - \rho gz + P_0 \qquad (5.69)$$

If we denote

$$-\frac{U}{g} = \bar{\zeta}(\theta, \lambda, t) \qquad (5.70)$$

and take (5.69) into account, we have:

$$\frac{\partial u_\lambda}{\partial t} + 2\omega\cos\theta\, v_\theta = -\frac{g}{a\sin\theta}\frac{\partial(\zeta - \bar{\zeta})}{\partial\lambda} \qquad (5.71)$$

$$\frac{\partial v_\theta}{dt} - 2\omega\cos\theta\, u_\lambda = -\frac{g}{a}\frac{\partial(\zeta - \bar{\zeta})}{\partial\theta} \qquad (5.72)$$

Note that the velocities u_λ and v_θ in the above-described tide-generating movements do not depend on z, that is, they are horizontal movements.

The next equation to describe the tides is the equation of continuity:

$$\frac{\partial\zeta}{\partial t} + \frac{h}{a\sin\theta}\left[\frac{\partial(v_\theta\sin\theta)}{\partial\theta} + \frac{\partial u_\lambda}{\partial\lambda}\right] = 0 \qquad (5.73)$$

Equations 5.71–5.73 are the equations of the theory of the oceanic tides. Here, $\bar{\zeta}(\theta, \lambda, t)$ is considered a prescribed function, and u_λ, v_θ and ζ, unknowns.

Let us now derive equations for the atmospheric tides. Let the pressure, density and temperature corresponding to equilibrium of the atmosphere be denoted by $\bar{P}$, $\bar{\rho}$ and $\bar{T}$, respectively. These variables depend only on the height z and are interrelated as follows: $\bar{P} = \bar{\rho}R\bar{T}$ and $\frac{\partial\bar{P}}{\partial z} = -g\bar{\rho}$, where R is the gas constant. The former formula is an equation of state, and the latter is an equation of hydrostatic equilibrium. Thus, there is one free variable, for example $\bar{T}(z)$.

We consider small fluctuations in the atmosphere around its equilibrium, so:

$$P = \bar{P} + P'; \quad \rho = \bar{\rho} + \rho'; \quad T = \bar{T} + T' \qquad (5.74)$$

where P', ρ', T' are the small deviations of the pressure, density and temperature from their average values $\bar{P}$, $\bar{\rho}$, and $\bar{T}$, respectively. Linearizing the traditional equations of atmosphere dynamics (that is, rejecting products of small deviations) gives a system of equations for the perturbations of the first order:

$$\frac{\partial u_\lambda}{\partial t} + 2\omega \cos\theta\, v_\theta = -\frac{1}{a \sin\theta}\left(\frac{1}{\bar{\rho}}\frac{\partial P'}{\partial \lambda} + \frac{\partial U}{\partial \lambda}\right) \tag{5.75}$$

$$\frac{\partial v_\theta}{\partial t} - 2\omega \cos\theta\, u_\lambda = -\frac{1}{a}\left(\frac{1}{\bar{\rho}}\frac{\partial P'}{\partial \theta} + \frac{\partial U}{\partial \theta}\right) \tag{5.76}$$

$$0 = -\frac{\partial P'}{\partial z} - g\rho' \tag{5.77}$$

$$\frac{\partial \rho'}{\partial t} + w\frac{d\bar{\rho}}{dz} + \bar{\rho}\chi = 0 \tag{5.78}$$

where w is the vertical component of the velocity, and χ is the three-dimensional divergence:

$$\chi = \frac{1}{a \sin\theta}\frac{\partial u_\lambda}{\partial \lambda} + \frac{1}{a \sin\theta}\frac{\partial}{\partial \theta}(v_\theta \sin\theta) + \frac{\partial w}{\partial z} \tag{5.79}$$

Here, (5.75) and (5.76) are linearized equations of horizontal motion; (5.77) is the equation of hydrostatic equilibrium; (5.78), the continuity equation.

The heat inflow equation closes the system of equations (Chapman and Lindzen, 1970):

$$\frac{\partial P'}{\partial t} + w\frac{d\bar{P}}{dz} = \kappa g H \frac{d\rho}{dt} + (\kappa - 1)\,\bar{\rho} J \tag{5.80}$$

where $\kappa = c_p/c_v = 1.4$; $c_p = 0.2405$ cal/(g degree), and $c_v = 0.1719$ cal/(g degree) are the heat capacities of air at constant pressure and constant volume, respectively; $H = R\bar{T}/g$ is the equivalent height of the homogeneous atmosphere; J is the heat inflow caused by an external source.

The atmosphere is a thin skin of air that covers the Earth. Its thickness is negligibly small in comparison with the Earth's radius. Therefore, in the first approximation, the atmosphere can be considered two-dimensional, with its characteristics averaged over its thickness. Formally, we can derive two-dimensional equations if we assume that the vertical velocity is equal to zero and other characteristics do not depend on the height:

$$\frac{\partial \bar{u}_\lambda}{\partial t} + 2\omega \cos\theta\, \bar{v}_\theta = -\frac{1}{\bar{\rho}\, a \sin\theta}\frac{\partial P'}{\partial \lambda} - \frac{1}{a \sin\theta}\frac{\partial U}{\partial \lambda} \tag{5.81}$$

$$\frac{\partial \bar{v}_\theta}{\partial t} - 2\omega \cos\theta\, \bar{u}_\lambda = -\frac{1}{\bar{\rho}\, a}\frac{\partial P'}{\partial \theta} - \frac{1}{a}\frac{\partial U}{\partial \theta} \tag{5.82}$$

$$\frac{1}{g}\frac{\partial}{\partial t}(P' - gT') + \frac{H}{a \sin\theta}\left[\frac{\partial}{\partial \theta}(\bar{v}_\theta \sin\theta) + \frac{\partial \bar{u}_\lambda}{\partial \lambda}\right] = 0 \tag{5.83}$$

Comparison of the derived equations with Equations 5.71–5.73 for the oceanic tides shows that there is a perfect analogy between them. The following variables reciprocally correspond to each other:

$$u \leftrightarrow \bar{u}_\lambda \quad v \leftrightarrow \bar{v}_\theta \quad \zeta - \bar{\zeta} \leftrightarrow \frac{1}{g}(P' + U);$$

$$\zeta \leftrightarrow \frac{1}{g}(P' - gT'); \quad \bar{\zeta} \leftrightarrow -T' - \frac{U}{g}; \quad h \leftrightarrow H$$

Consequently, from the point of view of the tidal theory, the atmosphere can be regarded as an ocean of constant depth H, which is affected by the perturbing potential $-g\bar{\zeta} = U + gT'$ that partly consists of the lunisolar tide-generating potential U and partly of the potential gT' related to the nonhomogeneous heating of the atmosphere. For an isothermal atmosphere, whose air particles' state changes adiabatically, the depth H is equal to:

$$H = \frac{\kappa RT}{g} = \kappa H_0 = 1.4 \cdot 8.0 = 11.2\,\text{km}$$

where H_0 is the height of the homogeneous atmosphere. For isothermal changes, $\kappa = 1$ and $H = H_0 = 8\,\text{km}$. Calculation for the real atmosphere gives $H \approx 10\,\text{km}$.

The tide-generating forces are periodic functions of time and longitude. Therefore, we assume that changes in wind u_λ, v_θ, w, pressure P', density ρ', and gravitational potential U are periodic functions of time and longitude:

$$x = \underline{x}(\theta, z)e^{i(\sigma t + s\lambda)} \tag{5.84}$$

where $\underline{x}(\theta, z)$ is the complex function of latitude and height, σ is the excitation frequency, s is the wave number ($s = 0 \pm 1, \pm 2, \ldots$.). Observed diurnal variations of wind and pressure confirm such an approximation (Hsu and Hoskins, 1989). Using (5.84), we derive the expressions for velocities $\underline{u}_\lambda$ and $\underline{v}_\theta$:

$$\underline{u}_\lambda = \frac{-\sigma}{4a\omega^2(f^2 - \cos^2\theta)}\left(\frac{\cos\theta}{f}\frac{\partial}{\partial\theta} + \frac{s}{\sin\theta}\right)\left(\frac{P'}{\bar{\rho}} + U\right) \tag{5.85}$$

$$\underline{v}_\theta = \frac{i\sigma}{4a\omega^2(f^2 - \cos^2\theta)}\left(\frac{\partial}{\partial\theta} + \frac{s\,\text{ctg}\theta}{f}\right)\left(\frac{P'}{\bar{\rho}} + U\right) \tag{5.86}$$

where $f = \sigma/2\omega$ is the dimensionless frequency.

Substituting (5.85) and (5.86) into (5.79) gives:

$$\chi - \frac{\partial w}{\partial z} = \frac{i\sigma}{4a^2\omega^2}F\left(\frac{P'}{\bar{\rho}} + U\right) \tag{5.87}$$

where F is the Laplace operator in the following form:

$$F \equiv \frac{1}{\sin\theta}\frac{\partial}{\partial\theta}\left(\frac{\sin\theta}{f^2 - \cos^2\theta}\frac{\partial}{\partial\theta}\right) - \frac{1}{f^2 - \cos^2\theta}\left(\frac{s}{f}\frac{f^2 + \cos^2\theta}{f^2 - \cos^2\theta} + \frac{s^2}{\sin^2\theta}\right) \tag{5.88}$$

The main unknown function in the tidal theory is:

$$G = -\frac{1}{\kappa \bar{P}}\frac{dP}{dt} \tag{5.89}$$

Using Equations 5.87, 5.89, 5.80, 5.78 and 5.77, we can formulate one equation to find G:

$$H\frac{\partial^2 G}{\partial z^2} + \left(\frac{\partial H}{\partial z} - 1\right)\frac{\partial G}{\partial z} = \frac{g}{4a^2\omega_0^2}F\left[G\left(\frac{\partial H}{\partial z} + \gamma\right) - \frac{\gamma J}{\kappa g H}\right] \tag{5.90}$$

Here, the term $\sigma/g\,(\partial^2 U/\partial z^2)$ is omitted because of its vanishing smallness. Factor $\gamma \equiv (\kappa - 1)/\kappa = 2/7$.

The method of separation of variables is used to solve Equation 5.90. Let us assume that

$$G(\theta, \lambda, z, t) = \sum_n K_n(z)\Psi_n(\theta)e^{i(\sigma t + s\lambda)} \tag{5.91}$$

and the system of functions $\Psi_n(\theta)$ is complete within the interval $0 \leq \theta \leq \pi$ if n runs through all the values. Then, the function J can be expanded into a series in functions $\Psi_n(\theta)$:

$$J(\theta, \lambda, z, t) = \sum_n J_n(z)\Psi_n(\theta)e^{i(\sigma t + s\lambda)} \tag{5.92}$$

Substituting (5.91) and (5.92) into (5.90) gives the following system of equations to determine functions $K_n(z)$ and $\Psi_n(\theta)$:

$$F(\Psi_n(\theta)) + \frac{4a^2\omega_0^2}{gh_n}\Psi_n(\theta) = 0 \tag{5.93}$$

$$H\frac{d^2 K_n}{dz^2} + \left(\frac{dH}{dz} - 1\right)\frac{dK_n}{dz} + \frac{1}{h_n}\left(\frac{dH}{dz} + \gamma\right)K_n = \frac{\gamma J_n}{\kappa g H h_n} \tag{5.94}$$

where h_n is the constant of separation of variables, F is the Laplace operator.

Equation 5.93 is Laplace's tidal equation. The boundary conditions for the equation are as follows: the functions $\Psi_n(\theta)$ are bounded at the poles, that is, for $\theta = 0$ and π. Given the boundary conditions, Laplace's equation determines a system of the eigenfunctions and eigenvalues h_n.

The boundary conditions for Equation 5.94 are as follows: w is zero on the Earth surface, and the solutions are bounded at infinity.

The general solution of the equations is a sum of the solutions:

$$G(z, \theta, \lambda, t) = \sum_{l,s,n} K_n^{l,s}(z)\Psi_n^{l,s}(\theta)\exp[i(\sigma_l t + s\lambda)] \tag{5.95}$$

Analogous expressions can be derived for other unknowns. The fluctuations under study are characterized by numbers l and s. A set of waves with wave numbers $s = 0$, $s = \pm 1$, $s = \pm 2 \ldots$ corresponds to every excitation frequency σ_l. For given l and s, the equation has an infinite number of solutions $\Psi_n^{l,s}(\theta)$, each of which is related to the respective depth h_n. Variables h_n are the eigenvalues of operator F. The eigenfunctions $\Psi_n^{l,s}(\theta)$ are the Hough functions.

Similar to σ, the frequency $f = \sigma/2\omega_0$ can be positive or negative. Since the operator F (5.88) includes the ratio s/f, so Equation (5.93) has a solution for $-s$ and $-f$ provided that it has a solution for s and f. In other words, the solutions always form pairs exp $(\pm\, \mathrm{i}(\sigma t + s\lambda))\Psi_n\, (\theta)$. Let s be positive and σ be either positive or negative. The real solutions can be represented as cos $(\sigma t + s\lambda)\Psi_n$ (cosθ) and sin $(\sigma t + s\lambda)\Psi_n$ (cosθ). In that case, the waves propagate from east to west, if $s/f > 0$, and from west to east, if $s/f < 0$.

The wave corresponding to the azimuthal number $s = 0$ is a standing wave, that is, it is a function of universal time. The wave corresponding to $s = 1$ travels westward, along with the Sun, and is proportional to a function of local time.

According to experimental data (Chapman and Lindzen, 1970), the amplitude of the wave with $s = l$ is maximal in spite of the fact that, for every frequency l, there is a set of waves with respective s. Thus, for example, the main wave of the lunar semidiurnal tide is the wave with $s = 2$. The amplitudes of the waves with numbers $s = 0$ and $s = -2$ are $\sim 1/10$ of the amplitude of the main wave.

The main wave of the solar diurnal tide is the wave with $s = 1$. The global mode of that type is caused only by the planetary distribution of water vapor. Thus, if we consider only the basic tidal modes (that is, the diurnal tides with $s = 1$ and the semidiurnal tides with $s = 2$), details of the water-vapor distribution may be ignored, and we can use the zone-averaged climatic data to calculate the function of heating.

The semidiurnal tide is excited, in the first place, by ozone heating in the stratosphere. For the diurnal motion of the Sun, the thermal excitation is not described by a unique diurnal harmonic; it contains higher harmonics as well. The largest is the semidiurnal harmonic.

Laplace's tidal Equation 5.93 is studied in detail in a number of works (Siebert, 1961; Dikiy, 1969; Longuet-Higgins, 1968). Solutions of the equation are given by the Hough functions that can be represented in the form of a series in terms of the associated Legendre polynomials.

There are several methods to solve Equation 5.93. Here, we put Equation 5.93 in the form of a system of two second-order equations. The method was developed by Eckart (Eckart, 1960).

Let a new variable, $\mu = \cos\theta$ be introduced; now Equation 5.93 can be formulated as follows:

$$\left\{-\frac{\mathrm{d}}{\mathrm{d}\mu}\left[\frac{1}{f^2-\mu^2}\left(-(1-\mu^2)\frac{\mathrm{d}}{\mathrm{d}\mu}+\frac{s\mu}{f}\right)\right]+\frac{1}{f^2-\mu^2}\left(\frac{s\mu}{f}\frac{\mathrm{d}}{\mathrm{d}\mu}-\frac{s^2}{1-\mu^2}\right)\right\}$$

$$\Psi_n(\mu)+\frac{4a^2\omega_0^2}{gh_n}\Psi_n(\mu)=0 \qquad (5.96)$$

Such an equation is obtained for each pair of wave numbers (s, l). Eckart (1960) suggested that a new function $\xi_n(\mu)$, should be introduced. The definition of the function is the expression in brackets. Then, Equation (5.96) becomes a system of linear equations:

$$\begin{aligned}
&\frac{1}{f^2-\mu^2}\left(-(1-\mu^2)\frac{\mathrm{d}}{\mathrm{d}\mu}+\frac{s\mu}{f}\right)\Psi_n(\mu)=\xi_n(\mu)\\
&\left(-(1-\mu^2)\frac{\mathrm{d}}{\mathrm{d}\mu}+\frac{s\mu}{f}\right)\xi_n(\mu)=\left(\frac{s^2}{f^2}-(1-\mu^2)\Gamma\right)\Psi_n(\mu)
\end{aligned}\tag{5.97}$$

where $\Gamma = 4\alpha^2\omega_0^2/gh_n$.

As Hough showed, the eigenfunctions of the operator F asymptotically is either the Legendre function $P_n^s(\mu)$ (for large values of f) or the linear combination of two Legendre functions (for $\Gamma^{-1} \to \infty$). Naturally, solutions for other values of parameters, f and Γ, should be obtained by Galerkin's method, that is, in the form of a linear combination of the Legendre functions (Dikiy, 1969).

Let the operator on the left part of the Legendre equation be designated by L (the operator defines the Legendre polynomials). Then:

$$L=(1-\mu^2)\frac{\mathrm{d}^2}{\mathrm{d}\mu^2}-2\mu\frac{\mathrm{d}}{\mathrm{d}\mu}-\frac{s^2}{1-\mu^2}$$

Equation 5.97 are represented in the following form (Dikiy, 1969):

$$\begin{aligned}
&\left(L+\frac{s}{f}\right)\xi=\Gamma\left[\left(\frac{s}{f}+2\right)\mu-(1-\mu^2)\frac{\mathrm{d}}{\mathrm{d}\mu}\right]\Psi_n(\mu)\\
&\left[L-\frac{s}{f}+\frac{s^2}{f^2}+(f^2-1)\Gamma\right]\Psi_n(\mu)=\left[(1-\mu^2)\frac{\mathrm{d}}{\mathrm{d}\mu}+\mu\left(\frac{s}{f}-2\right)\right]\xi_n(\mu)
\end{aligned}\tag{5.98}$$

The solutions of system (5.98) are determined in the form of a series in terms of the normalized Legendre polynomials:

$$\Psi_n(\mu)=\sum_{n=s}^{\infty}a_n^{(m)}\sqrt{\frac{2n+1}{2}\frac{(n-s)!}{(n+s)!}}P_n^s(\mu)$$

$$\xi_n(\mu)=\sum_{n=s}^{\infty}a_n^{(m)}\Gamma\sqrt{\frac{2n+1}{2}\frac{(n-s)!}{(n+s)!}}P_n^s(\mu)$$

Substituting the series into system (5.98) and taking the following recurrent relationships

$$\mu P_n^s(\mu)=\frac{n-s+1}{2n+1}P_{n+1}^s(\mu)+\frac{n+s}{2n+1}P_{n-1}^s(\mu)$$

$$(1-\mu^2)\frac{dP_n^s(\mu)}{d\mu}=-\frac{n(n-s+1)}{2n+1}P_{n+1}^s(\mu)+\frac{(n+1)(n+s)}{2n+1}P_{n-1}^s(\mu)$$

$$LP_n^s(\mu)=-n(n+1)P_n^s(\mu)$$

into account gives the difference equation for factors α_n:

$$\Gamma^{-1}a_n=L_{n-2}a_{n-2}+M_na_n+L_na_{n+2},\quad n\geq s\tag{5.99}$$

where

$$L_n = \frac{1}{\frac{s}{f}-(n+1)(n+2)}\sqrt{\frac{(n-s+1)(n+s+1)(n-s+2)(n+s+2)}{(2n+1)(2n+3)(2n+3)(2n+5)}}$$

$$M_n = \frac{f^2-1}{\left(-\frac{s}{f}+n\right)\left(\frac{s}{f}-n-1\right)} + \frac{\left(\frac{s}{f}-n+1\right)(n-s)(n+s)}{\left(\frac{s}{f}+n\right)\left(\frac{s}{f}-n(n-1)\right)(2n-1)(2n+1)}$$
$$+\frac{\frac{s}{f}+n+2}{\left(\frac{s}{f}-n-1\right)\left(\frac{s}{f}-(n+1)(n+2)\right)}\frac{(n-s+1)(n+s+1)}{(2n+1)(2n+3)}$$

It is apparent that system (5.99) falls into two parts: for one of them, $n=s, s+2, \ldots$, for another, $n=s+1, s+3, \ldots$ Nontrivial solutions of the system exist provided that the determinant is zero. Consequently, we have the following condition for the first of the systems ($n=s, s+2, \ldots$):

$$\begin{vmatrix} M_s-\Gamma^{-1} & L_s & 0 & 0 & \cdot\;\cdot\;\cdot & \cdot \\ L_s & M_{s+2}-\Gamma^{-1} & L_{s+2} & 0 & \cdot\;\cdot\;\cdot & \cdot \\ 0 & L_{s+2} & M_{s+4}-\Gamma^{-1} & L_{s+4} & \cdot\;\cdot\;\cdot & \cdot \\ \cdot & \cdot & \cdot & \cdot & \cdot\;\cdot\;\cdot & \cdot \\ \cdot & \cdot & \cdot & \cdot & \cdot\;\cdot\;L_{N-2} & M_N-\Gamma^{-1} \end{vmatrix} = 0 \tag{5.100}$$

whereas we have the following condition for the second one ($n=s+1, s+3, \ldots$):

$$\begin{vmatrix} M_{s+1}-\Gamma^{-1} & L_{s+1} & 0 & 0 & \cdot\;\cdot\;\cdot & \cdot \\ L_{s+1} & M_{s+3}-\Gamma^{-1} & L_{s+3} & 0 & \cdot\;\cdot\;\cdot & \cdot \\ 0 & L_{s+3} & M_{s+5}-\Gamma^{-1} & L_{s+5} & \cdot\;\cdot\;\cdot & \cdot \\ \cdot & \cdot & \cdot & \cdot & \cdot\;\cdot\;\cdot & \cdot \\ \cdot & \cdot & \cdot & \cdot & \cdot\;\cdot\;L_{N-1} & M_{N+1}-\Gamma^{-1} \end{vmatrix} = 0 \tag{5.101}$$

Expressions (5.100)–(5.101) represent the characteristic equation $Ax=\lambda x$ in the matrix form. In that case, the determination of the eigenvalues and eigenvectors is facilitated by the symmetry and three-diagonal form of the matrix (Wilkinson and Reinsch, 1976). To solve the Laplace equation, Zharov (1996b) defined the values of parameter $\frac{s}{f}$ by using the frequencies of the gravitational and thermal excitation. He determined s and f, then the eigenvalues Γ^{-1}, h_n, the coefficients $a_n^{(m)}$ and, eventually, the Hough function Ψ_n (μ).

If the eigenvalue and the coefficients $a_n^{(m)}$ are obtained from system (5.100), the expansion contains functions P_s^s, P_{s+2}^s, P_{s+4}^s, ... If the coefficients are obtained from (5.101), the expansion includes functions P_{s+1}^s, P_{s+3}^s, P_{s+5}^s, ... In the former case, the functions Ψ_n (μ) are even, in the latter case, odd.

The wave corresponding to some wave number n, is referred to as a mode. Thus, the solution of the tidal equation falls into the symmetric and antisymmetric modes with respect to the equator of mode.

Let us now consider the vertical structure Equation 5.94. Usually, this equation is reduced to the canonical form, with variable z replaced by x using the formula:

$$x = \int_0^z \frac{dz}{H(z)}$$

and K_n replaced with y_n using the formula $K_n y_n \exp(x/2)$. Then, (5.94) takes the form (Chapman and Lindzen, 1970):

$$\frac{d^2 y_n}{dx^2} - \left[\frac{1}{4} - \frac{1}{h_n}\left(\gamma H + \frac{dH}{dx}\right)\right] y_n = \frac{\gamma J_n(x)}{\kappa g h_n} e^{-x/2} \tag{5.102}$$

Equation 5.102 is solved for every value of h_n.

Unknown parameters u, v, w, p are defined by (5.95), where $K_n(x)$ is equal to one of the functions $u_n(x)$, $v_n(x)$, $w_n(x)$, $p_n(x)$. Then:

$$u_n(x) = \frac{i\kappa g h_n e^{x/2}}{4a\omega_0^2}\left(\frac{dy_n}{dx} - \frac{1}{2}y_n\right) \cdot \frac{1}{f^2 - \cos^2\theta}\left(\frac{\cos\theta}{f}\frac{\partial}{\partial\theta} + \frac{s}{\sin\theta}\right) \tag{5.103}$$

$$v_n(x) = \frac{\kappa g h_n e^{x/2}}{4a\omega_0^2}\left(\frac{dy_n}{dx} - \frac{1}{2}y_n\right) \cdot \frac{1}{f^2 - \cos^2\theta}\left(\frac{\partial}{\partial\theta} + \frac{s\,\mathrm{ctg}\theta}{f}\right) \tag{5.104}$$

$$w_n(x) = -\frac{i\sigma U_n}{g} + \kappa h_n e^{x/2}\left(\frac{dy_n}{dx} + \left(\frac{H}{h_n} - \frac{1}{2}\right)y_n\right) \tag{5.105}$$

$$P_n(x) = \frac{P_0(0)}{H(x)}\left[-\frac{U_n}{g}e^{-x} + \frac{\kappa h_n}{i\sigma}e^{-x/2}\left(\frac{dy_n}{dx} - \frac{1}{2}y_n\right)\right] \tag{5.106}$$

Here, the expansion of the potential in the Hough functions is used:

$$U = \sum_n U_n(x)\,\Psi_n(\theta)$$

$P_0(0)$ is the pressure on the Earth surface.

Equation 5.102 is solved under the boundary conditions. The lower boundary ($x = z = 0$) condition is as follows: the vertical velocity is zero, that is $w = 0$. Then, we obtain from (5.102) at $x = 0$:

$$\frac{dy_n}{dx} + \left(\frac{H}{h_n} - \frac{1}{2}\right)y_n = \frac{i\sigma}{\kappa g h_n}U_n \tag{5.107}$$

Thus, for the gravitational tides, Equation 5.102 turns out to be homogeneous ($J_n = 0$), the excitation being taken into account only through the lower boundary condition (5.107). For the gravitational tides, the excitation energy is concentrated on the Earth surface.

The upper boundary condition usually is either the boundedness condition of $y_n(x)$ at $x \to \infty$ or the radiation condition (Chapman and Lindzen, 1970). If the height H is constant above some level x_{max} and $J_n = 0$, then (5.102) is of the form:

$$\frac{d^2 y_n}{dx^2} - \left(\frac{1}{4} - \frac{\gamma H}{h_n}\right) y_n = 0 \qquad (5.108)$$

If $h_n < 4\gamma H$, the solution of this equation is:

$$y_n = Ae^{i\alpha x} + Be^{-i\alpha x}$$

where $\alpha = \sqrt{\gamma H/h_n - 0,25}$. The wave $Ae^{i\alpha x}$ propagates downward and the wave $Be^{i\alpha x}$ – upward. If there are no sources of excitation above the level x_{max}, then $B = 0$.

For $h_n > 4\gamma H$, one of the solutions of Equation 5.108 represents a decaying exponential and another – a growing exponential, that is $y_n = C\exp(\pm kx)$, where $k = \sqrt{0.25 - \gamma H/h_n}$. We try the solutions bounded at infinity, therefore the desired solution is as follows: $y_n = C\exp(-kx)$.

To sum up, the solutions of the vertical structure equation can be obtained separately for the gravitational and thermal excitations.

6
The Air-Mass Seasonal Redistribution and the Earth's Rotation

6.1
Air-Mass Seasonal Redistribution

Sidorenkov and Stehnovskii have performed investigations of the air-mass seasonal redistribution in the atmosphere (Sidorenkov and Stekhnovsky, 1971, 1972; Stekhnovsky, 1962; Stekhnovsky and Sidorenkov, 1977). The initial data for them were the climatologicaly charts and tables of the sea-level atmospheric pressure (Stekhnovsky, 1960) and the mean monthly temperatures for January and July (Batyayeva, 1960) and the mean height of the ground above see level, z. The surface of the Earth was divided by the parallels separated by 4° into 46 latitudinal zones (44 four-degree zones in the belt from 88°N to 88°S and two two-degree polar caps). Each of these zones was in turn divided into trapezoids by the meridians at intervals of 4° in the belt from 40°N to 40°S, 6° in the belt from ±40° to ±72°, and 60° in the belt from ±72° to ±88°. Each of the two-degree polar caps was taken as a single trapezoid.

In each ith trapezoid of the jth latitude zone we recorded the long-term mean monthly pressure, reduced to the sea level and the normal gravity, $P_{ij}(0, g_{45°,0})$, for January and July. For those trapezoids that included dry land, we also recorded the mean monthly temperatures $T_{ij}(0)$ for January and July and the mean height of the ground above sea level, z_{ij}.

For the pressure measured at the basis of an air column is necessary to divide by the acceleration of gravity (referred to the height of the center of gravity of this column) in order to calculate the mass of a particular atmospheric column.

Since in practice we can use only the pressure reduced to the sea level and the normal gravity $P(\varphi,\lambda,0,g_{45°,0})$, we must perform the inverse operation of reducing the pressure to the Earth's surface and the actual gravity $P(\varphi,\lambda,z,g_{\varphi,z})$. Assuming that the atmosphere is polytropic up to the land maximum height, the mass m of an atmospheric unit column was calculated by the formula:

$$m(\varphi,\lambda) = \frac{P(\varphi,\lambda,z,g_{\varphi,z})}{g_{\varphi,h}} = \frac{P(\varphi,\lambda,0,g_{45°,0})}{g_{45^0,h}}\left[\frac{T(\varphi,\lambda,z)}{T(\varphi,\lambda,0)}\right]^{\bar{g}/R_B\gamma_1} \tag{6.1}$$

Here, φ and λ are the latitude and the longitude; z is the height of the Earth's surface above the see level; $h = 7$ km is the height of the center of gravity of the atmosphere; T

The Interaction Between Earth's Rotation and Geophysical Processes. Nikolay S. Sidorenkov

ISBN: 978-3-527-40875-7

is the absolute temperature; $R = 0.287 \times 10^7\,\text{erg}\,\text{g}^{-1}\,\text{deg}^{-1}$ is the gas constant; $\gamma_1 = 6/\text{km}$ is the mean temperature gradient; $g_{\varphi,z} = g_{45°,0}(1 - \alpha \cos 2\varphi)(1 - \beta z)$ is the acceleration of gravity at the given latitude φ and height z; $\alpha = 1/298.25$ is the flattening of the Earth; $\beta = 0.315 \times 10^{-3}\,\text{km}^{-1}$; $\bar{g} = 9.80618\,\text{m/s}^2$; $g_{45°,7\ \text{km}} = 9.78455\ \text{m/s}^2$. In expression (6.1) it has been kept in mind that as $z \ll h$, therefore

$$\frac{g_{\varphi,z}}{g_{\varphi,h} \cdot g_{45°,0}} = \frac{g_{45°,0}(1-\alpha\cos 2\varphi)(1-\beta z)}{g_{45°,0}(1-\alpha\cos 2\varphi)(1-\beta h)g_{45°,0}} \approx \frac{1}{g_{45°,h}}$$

Based on the respective calculations, a map of the long-term mean seasonal redistribution of the atmospheric air over the globe has been constructed (Stekhnovsky and Sidorenkov, 1977) (Figure 6.1). The map represents the isolines of air-mass increments (g/cm^2) from July to January.

This map validates the known regularities of the air-mass seasonal redistribution: a tendency for air-mass accumulation over continents in winter and over oceans in summer. In addition, the map demonstrates some special features of this redistribution. In particular, it shows that the air-mass seasonal redistribution occurs with the opposite sign over the largest mountains and the surrounding plains. Within vast expanses of Eurasia and northern Africa over which the air mass increases in winter, there are mountain areas over which the air-mass outflows (or inflows but little). Over the Plateau of Tibet, for example, the air-mass negative increments from July to

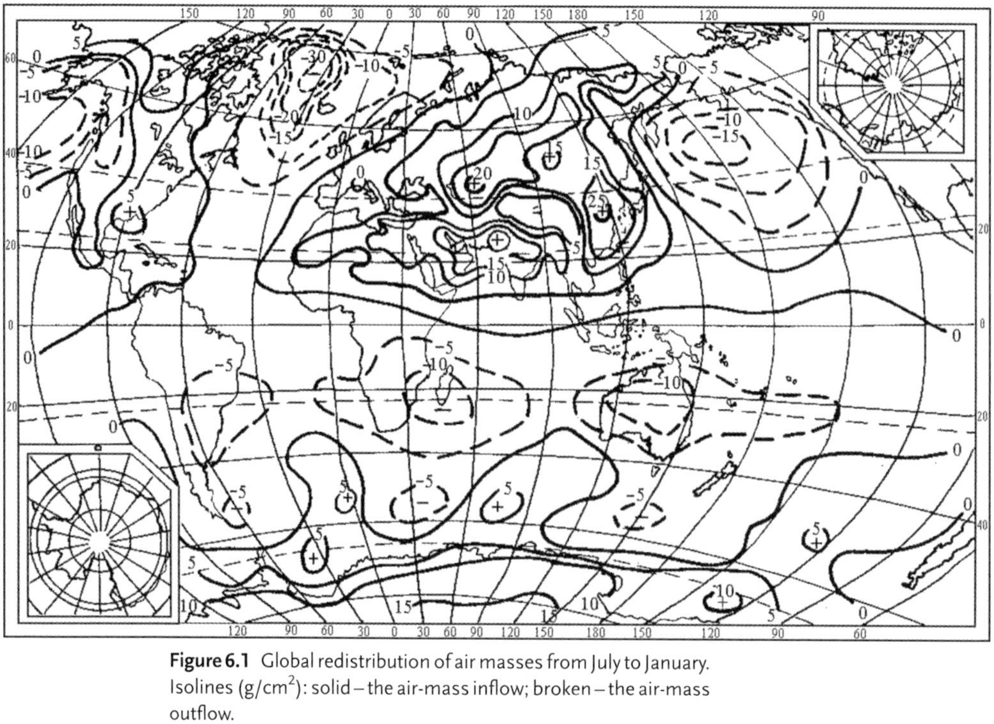

Figure 6.1 Global redistribution of air masses from July to January. Isolines (g/cm^2): solid – the air-mass inflow; broken – the air-mass outflow.

January are about $-16.3\,\mathrm{g/cm^2}$; over the Cordilleras (of the Northern Hemisphere), these increments are up to $-7\,\mathrm{g/cm^2}$, while over central areas of North America the air-mass increments are positive.

In the Southern Hemisphere (summer), the air-mass outflows over plains and inflows over the Andes narrow ridge. The air-mass inflow is also characteristic of central Antarctica, where it exceeds $15\,\mathrm{g/cm^2}$. This peculiar regime of the air-mass seasonal redistribution over mountain areas may be due to the changes in atmospheric pressure with height under the effect of air cooling/warming in the plain and mountain areas. With cooling, the air pressure grows over plains and drops over mountains, and vice versa: with warming, the atmospheric pressure decreases over plains and increases over mountains.

Over the oceans of the Northern Hemisphere, the center of the seasonal largest changes in air pressure ($>15\,\mathrm{g/cm^2}$) is located in the region of the Aleutian Low. The center of air-mass negative increments in the region of the Icelandic Low, which is shown in the earlier published maps, is absent in our map. Its absence may be due to the dominant effect of the air mass changes over Greenland.

In the region subjected to the influence of the Siberian High, the respective air-mass increments are not very significant (about $15\,\mathrm{g/cm^2}$) and are comparable in absolute value with the respective increments over the North Pacific Ocean. There are also centers of an intense air-mass inflow over the northeastern part of China (up to $25\,\mathrm{g/cm^2}$) and the Hindustan and Arabian peninsulas (about $20\,\mathrm{g/cm^2}$), both centers being due to the monsoon circulation.

In the subtropical zone of the Southern Hemisphere, the air-mass negative increments prevail, the centers of these changes being shifted to the continents' eastern parts. In the moderate zone of this hemisphere, the respective isolines present a complicated but well-defined pattern showing the alteration of the air-mass inflow and outflow. They reflect the seasonal shifts of frontal zones, along which the cyclones of moderate latitudes move.

Under close examination of the map one can see that in the Southern Hemisphere, the outflow of air mass prevails over its inflow in summer. Since the air mass cannot disappear, one may suppose that the atmospheric air outflows into the Northern Hemisphere.

The seasonal flow of air from one hemisphere to another was first postulated by Shaw (1936). Studying the deviations of the mean monthly fields of atmospheric pressure from its mean yearly field over the Northern Hemisphere, Shaw concluded that the air-mass redistribution occurs not only within one hemisphere, but also between the hemispheres. Rough estimates of the seasonal air exchange between the hemispheres are given in (Bonchkovskaya, 1956; Stekhnovsky, 1960, 1962). These estimates vary between 3.87×10^{15} and $5.95 \times 10^{15}\,\mathrm{kg}$.

According to our calculations (Sidorenkov and Stekhnovsky, 1971), in the Northern Hemisphere the air masses in January and July are 2.5778×10^{18} and 2.5740×10^{18} kg, respectively; and in the Southern Hemisphere they are 2.5794×10^{18} and $2.5839 \times 10^{18}\,\mathrm{kg}$, respectively. In the first case the difference between January and July is $+3.8 \times 10^{15}\,\mathrm{kg}$ ($1.5\,\mathrm{g/cm^2}$) and in the second case is $-4.5 \times 10^{15}\,\mathrm{kg}$ ($-1.8\,\mathrm{g/cm^2}$). Thus, the air mass in the Northern Hemisphere decreases by about $4 \times 10^{15}\,\mathrm{kg}$

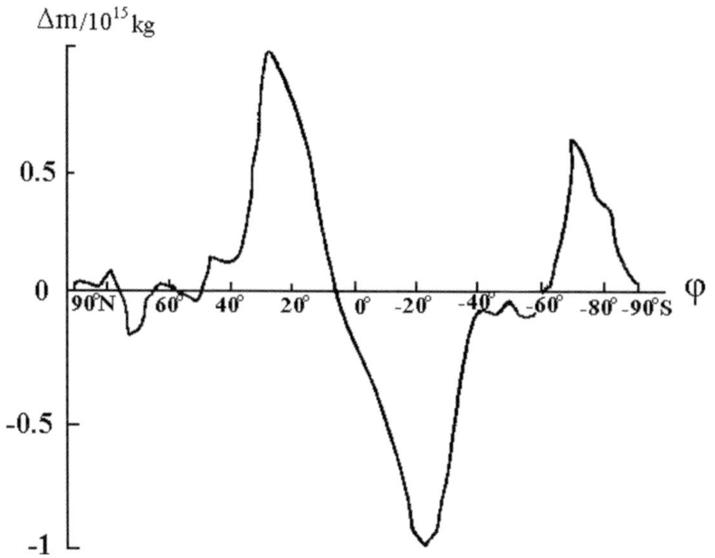

Figure 6.2 Increment of air mass from July to January as function of latitude.

between January and July, while the mass in the Southern Hemisphere increases by the same amount. These figures give the long-term mean seasonal exchange of air mass between the hemispheres.

The contributions of individual latitudinal zones to this process are illustrated by Figure 6.2, in which the abscissa is the latitudinal zones and the ordinate is the air-mass increments between January and July in each latitudinal zone. We see that in the subtropical zones, the air mass decreases markedly from winter to summer. This decrease is due to a significant air inflow into the subtropical zone of the opposite hemisphere, together with a less-marked air inflow to the high-latitude zones within the same hemisphere. It is interesting that the change of sign of the air-mass seasonal redistribution does not occur at the equator, but somewhat to the north (about 6°N). The total decrease in the air mass from January to July in the belt from 6 to 48°N is about 4.7×10^{15} kg, of which about 4.0×10^{15} kg (85%) flows from the zones between 32 and 12°, with a maximum at about 25°N. This decrease in the air mass is due to a slight flow of air to the high latitudes of the Northern Hemisphere (0.4×10^{15} kg, or 9%), and a very large southward flow (4.3×10^{15} kg, or 91%). In the subtropical zones of the Southern Hemisphere (8–36°S), the air-mass increment between January and July is about 5.2×10^{15} kg.

The belt of the air-mass negative increments extends from 6°N to 60°S, where the air-mass accumulation from January to July makes about 6.5×10^{15} kg. Apart from a considerable air-mass inflow from the subtropics of the Northern Hemisphere (4.3×10^{15} kg, or 67%), a relatively large quantity of air inflows from the Antarctic regions (2.1×10^{15} kg or 33%) into this belt.

Thus, the most intense air-mass exchange between hemispheres occurs across the 6°N parallel through which about 4.3×10^{15} kg of air flows southward from January to July and northward from July to January. About 1.1×10^{8} kg of air flows back and forth each year through each meter of the 6°N parallel in the active layer of the atmosphere. We recall that as a result of the monsoon activity, about 1–1.5 × 10^8 kg of air is transferred (in the active layer of the atmosphere) through each meter of the shore line from oceans to continents in winter and from continents to oceans in summer (Bonchkovskaya, 1956).

The results of calculations of the air-mass redistribution between continents and oceans are published in (Sidorenkov and Stekhnovsky, 1972). It is shown that over the continents of the Northern Hemisphere, as the land cools and the anticyclonic circulation intensifies from July to January, the air mass increases in a wide latitudinal zone (68°–8°N) and makes 5.8 × 10^{15} kg. At the same time, the air mass over oceans significantly decreases, most intensely in the zone of 52°–40°N, where the air-mass increment from July to January makes 1.28 × 10^{15} kg. In the zone of 0°–8°N, the pattern is reverse: the air-mass outflows from land to oceans, which is characteristic of the Southern Hemisphere during this season. Since the land area in the Southern Hemisphere makes up only 25% of the total area (that is, much less than in the Northern Hemisphere, where it makes up 50.7%) and since land areas are situated in the low and high latitudes, the air-mass seasonal redistribution between continents and oceans is insignificant there. It is approximately 6 times less than that in the Northern Hemisphere.

The air mass over the entire land of the globe increases from July to January by 4.57 × 10^{15} kg and decreases over oceans by 5.29 × 10^{15} kg. The difference between these values is likely to be due to the air-mass decrease from July to January (at the expense of the decreased air humidity, for example). The atmospheric mass in January is 5.157 × 10^{18} kg and that in July is 5.158 × 10^{18} kg. The mean planetary pressure of the atmosphere on the Earth's surface is 987.1 hPa in January and 987.3 hPa in July (Sidorenkov and Stekhnovsky, 1971).

Burlutskii (Sonechkin and Burlutskiy, 2005) has shown that changes in the hemisphere-average surface pressure for two hemispheres are specularly symmetric. If air pressure increases in the Northern Hemisphere, it decreases in the Southern hemisphere by the same value (the coefficient of correlation is −0.94). He has shown that changes in the global surface pressure are in close correlation with changes in the amount of water vapor in the global atmosphere. The mass of the entire dry atmosphere is constant. Changes in the atmospheric mass are mainly due to changes in the amount of water vapor in it. Hence, the variations in the mean global value of the surface pressure are the reliable indicators of variations in the global value of water vapor in the atmosphere, the latter value significantly affects the temperature of the surface layer of the atmosphere.

6.2 Components of the Inertia Tensor of the Atmosphere

The components of the inertia tensor of the atmosphere are the fundamental geophysical parameters to be known when developing a high-precision theory of the nutation of the Earth's rotation axis. Some of these parameters are needed for better understanding and interpreting the specific features of atmospheric circulation and the Earth's gravitational field. In this connection, let us calculate the components of the inertia tensor of the atmosphere.

We choose a coordinate system that is rigidly bound to the Earth and participates in all its motions. We place the origin of this system at the Earth's center of inertia. The x_3 axis points to the mean position of the North Pole, and the x_1 and x_2 axes, which lie in the plane of the equator, are directed at the Meridian of Greenwich and 90° to the east of it, respectively. As we know, the components of the inertia tensor of the atmosphere n_{ij} in the Cartesian system of coordinates may be described in the same way as N_{ij} in matrix (4.10) in Section 4.1. Since the Earth's atmosphere is spheroidal, it will be more convenient to use the spherical coordinate system for calculations. In this system, the coordinates are: $x_1 = r\sin\theta\cos\lambda$; $x_2 = r\sin\theta\sin\lambda$; $x_3 = r\cos\theta$; a volume element is $dV = r^2\sin\theta dr d\theta d\lambda$. Substituting these expressions in (4.10), we may find the inertia-tensor component in the spherical coordinates:

$$\begin{aligned}
n_{11} &= \iiint \rho r^4(1-\sin^2\theta\cos^2\lambda)\sin\theta\, d\lambda\, d\theta dr \\
n_{22} &= \iiint \rho r^4(1-\sin^2\theta\sin^2\lambda)\sin\theta\, d\lambda\, d\theta\, dr \\
n_{33} &= \iiint \rho r^4\sin^3\theta\, d\lambda\, d\theta\, dr \\
n_{12} &= -\iiint \rho r^4\sin^3\theta\cos\lambda\sin\lambda\, d\lambda\, d\theta\, dr \\
n_{13} &= -\iiint \rho r^4\sin^2\theta\cos\theta\cos\lambda\, d\lambda\, d\theta\, dr \\
n_{23} &= -\iiint \rho r^4\sin^2\theta\cos\theta\sin\lambda\, d\lambda\, d\theta\, dr
\end{aligned} \tag{6.2}$$

Here, ρ is the air density, r is the geocentric distance, θ is the colatitude, and λ is the eastern longitude; $n_{12} = n_{21}$; $n_{13} = n_{31}$; $n_{23} = n_{32}$. The integration over the height yields:

$$\int_{R+z}^{\infty} \rho r^4 dr = \frac{(R+h)^4}{g_{\theta,h}} \int_0^{P(\theta,\lambda,z,g_{\theta,z})} dp = \frac{(R+h)^4}{g_{\theta,h}} P(\theta,\lambda,z,g_{\theta,z}) \approx \frac{(R+h)^4}{g_{45^\circ,h}} P(\theta,\lambda,z,g_{45^\circ,0}) \tag{6.3}$$

where R is the radius of the sea-level surface, h is the height of the center of gravity of the given unit volume of the atmosphere, $g_{\theta,h}$ is the acceleration of gravity at the given θ and h, and $P(\theta,\lambda,z,g_{\theta,z})$ is the atmospheric pressure at the point with coordinates (θ,λ), at height z above sea level, for the acceleration of gravity $g_{\theta,z}$.

We approximate the radius of the center-of-gravity surface of the atmosphere using the expression:

$$R + h = (a+h)(1-\alpha\cos^2\theta) \tag{6.4}$$

and the acceleration of gravity, with the expression:

$$g_{\theta,h} = g_{90^\circ,0}(1 + 0.005288\cos^2\theta - 0.59\times 10^{-5}\sin^2 2\theta - 0.315\times 10^{-3}h) \tag{6.5}$$

Taken into account (6.3)–(6.5), expression (6.2) can be written

$$
\begin{aligned}
n_{11} &= A\int_0^{\pi}\int_0^{2\pi}(\sin\theta-\sin^3\theta\cos^2\lambda)(1-\alpha\cos^2\theta)^4 P d\lambda d\theta \\
n_{22} &= A\int_0^{\pi}\int_0^{2\pi}(\sin\theta-\sin^3\theta\sin^2\lambda)(1-\alpha\cos^2\theta)^4 P d\lambda d\theta \\
n_{33} &= A\int_0^{\pi}\int_0^{2\pi}\sin^3\theta\,(1-\alpha\cos^2\theta)^4 P d\lambda\, d\theta \\
n_{12} &= -\frac{1}{2}A\int_0^{\pi}\sin^3\theta(1-\alpha\cos^2\theta)^4\int_0^{2\pi}P\sin 2\lambda\, d\lambda\, d\theta \\
n_{13} &= -A\int_0^{\pi}\sin^2\theta\cos\theta\,(1-\alpha\cos^2\theta)^4\int_0^{2\pi}P\cos\lambda\, d\lambda\, d\theta \\
n_{23} &= -A\int_0^{\pi}\sin^2\theta\cos\theta(1-\alpha\cos^2\theta)^4\int_0^{2\pi}P\sin\lambda d\lambda d\theta
\end{aligned}
\tag{6.6}
$$

Here, $A = 10^3(a+h)^4/g_{45^\circ,h}$; P is the pressure in hectoPascals (hPa), and a and α are the equatorial radius and the flattening of the Earth.

The initial data for calculations were the data on atmospheric pressure described in Section 6.1.

The components of the inertia tensor were calculated by the numerical integration over the following expressions:

$$
\begin{aligned}
n_{11} &= A\left(\sum_{j=1}^{46}\Phi_j\Delta\lambda_j\sum_{i=1}^{N}P_{ij}-\sum_{j=1}^{46}X_j\sum_{i=1}^{N}\Lambda_i P_{ij}\right) \\
n_{22} &= A\left(\sum_{j=1}^{46}\Phi_j\Delta\lambda_i\sum_{i=1}^{N}P_{ij}-\sum_{j=1}^{46}X_j\sum_{i=1}^{N}K_i P_{ij}\right) \\
n_{33} &= A\sum_{j=1}^{46}X_j\Delta\lambda_j\sum_{i=1}^{N}P_{ij} \\
n_{12} &= -A\sum_{j=1}^{46}X_j\sum_{i=1}^{N}I_i P_{ij} \\
n_{13} &= -A\sum_{j=1}^{46}\Omega_j\sum_{i=1}^{N}\Pi_i P_{ij} \\
n_{23} &= -A\sum_{j=1}^{46}\Omega_j\sum_{i=1}^{N}O_i P_{ij}
\end{aligned}
\tag{6.7}
$$

where N is the number of trapezoids in the jth zone, $\Delta\lambda_j = \bar{\lambda} - \underline{\lambda} = 2\pi/N$, $A = 1.698795 \times 10^{28}\,\mathrm{kg\,m^2/hPa}$ (it was assumed that $a = 6378.14\,\mathrm{km}$, $h = 7\,\mathrm{km}$, and $g_{45^\circ,7\,\mathrm{km}} = 9.78455\,\mathrm{m/s^2}$), $\alpha = 1/298.25$, and P_{ij} is the pressure of the jth trapezoid of the ith zone, expressed in hPa,

$$\Phi_j = \int_{\underline{\theta}}^{\bar{\theta}} \sin\theta(1-\alpha\cos^2\theta)^4 d\theta \approx \left(-\cos\theta + \frac{4\alpha}{3}\cos^3\theta\right)\Bigg|_{\underline{\theta}}^{\bar{\theta}}$$

$$X_j = \int_{\underline{\theta}}^{\bar{\theta}} \sin^3\theta(1-\alpha\cos^2\theta)^4 d\theta \approx \left(-\cos\theta + \frac{1+4\alpha}{3}\cos^3\theta - \frac{4\alpha}{5}\cos^5\theta\right)\Bigg|_{\underline{\theta}}^{\bar{\theta}}$$

$$\Omega_j = \int_{\underline{\theta}}^{\bar{\theta}} \sin^2\theta\cos\theta(1-\alpha\cos^2\theta)^4 d\theta \approx \left(\frac{1-4\alpha}{3}\sin^3\theta + \frac{4\alpha}{5}\sin^5\theta\right)\Bigg|_{\underline{\theta}}^{\bar{\theta}}$$

$$\Lambda_i = \left(\frac{1}{2}\lambda + \frac{1}{4}\sin 2\lambda\right)\Bigg|_{\underline{\lambda}}^{\bar{\lambda}};\quad K_i = \left(\frac{1}{2}\lambda - \frac{1}{4}\sin 2\lambda\right)\Bigg|_{\underline{\lambda}}^{\bar{\lambda}};$$

$$I_i = \frac{1}{2}\sin^2\lambda\Bigg|_{\underline{\lambda}}^{\bar{\lambda}};\quad \Pi_i = \sin\lambda\Bigg|_{\underline{\lambda}}^{\bar{\lambda}};\quad O_i = -\cos\lambda\Bigg|_{\underline{\lambda}}^{\bar{\lambda}} \qquad (6.8)$$

The inertia tensor components are calculated for two "climatic" months – mean January and mean July. Since the rheological properties of land and ocean must be taken into account, the summations were made separately over the trapezoids occupied by ocean and the trapezoids occupied by land or closed bodies of water. In the latter case, calculations were carried out using the atmospheric-pressure data specified in two variants. In the first variant, we used the sea-level pressure $P_{ij} = P_{ij}(0, g_{45^\circ,0})$, and in the second was the pressure reduced to the average height of the Earth's surface in a particular trapezoid, $P_{ij} = P_{ij}(z, g_{45^\circ,0})$. The respective expressions and the description of details can be found in (Sidorenkov, 1973).

As we know, the atmospheric pressure varies with a one-year period. These variations are for the most part due to the variations of temperature over the year. The hemisphere-averaged pressure increases from summer to winter, and it decreases from winter to summer. The atmospheric-pressure annual variations are most conspicuous in the middle latitudes. In winter, when the ocean surface is considerably warmer than the land surface, the atmospheric pressure over oceans is lower than that over continents. The thermal situation is the reverse in summer, so that a lower pressure is registered over continents and a higher one over oceans. Since at most localities on the Earth's surface the pressure, like the temperature, varies harmonically over the year, and then, as we see from (6.6), the inertia tensor components should vary in a similar fashion:

$$n_{ij}(\oplus) = \bar{n}_{ij} + E_{ij}\cos(\oplus - \beta_{ij}) \qquad (6.9)$$

where $\bar{n}_{ij}$ is the annual mean value, E_{ij} and β_{ij} are the amplitude and phase of the annual variations of the component under consideration, $\oplus = 2\pi t/365.25$ day is the longitude of the mean sun reckoned from the beginning of the year, and t is the time in days. In most cases, the extreme values of the annual atmospheric-pressure variation coincide approximately in time with the temperature extremes. It may be assumed that for the Earth as a whole, they are observed in January and July. Consequently, $\bar{n}_{ij}$ and E_{ij}, are in first approximation determined as the half-sum and half-difference of their values in July and January, respectively:

$$\begin{aligned} \bar{n}_{ij} &\approx \frac{1}{2}\left[n_{ij}(I) + n_{ij}(VII)\right] \\ E_{ij} &\approx \frac{1}{2}\left[n_{ij}(I) - n_{ij}(VII)\right] \end{aligned} \qquad (6.10)$$

and the phase β equals zero.

Table 6.1 gives the annual-average values and the ranges of the annual variations of the atmosphere's inertia-tensor components, as calculated in this way. It also includes the absolute values of air mass and average atmospheric pressure, together with the ranges of their annual variations. All these characteristics are given separately for the regions of integration, which extend over land, ocean, and the entire planet. For land and the planet as a whole, the above calculations are made with and without account for land elevation above see level. The corrections for height are given in a separate column. The areas of integration are given in the last line.

Table 6.1 Mean values $\bar{n}_{ij}$ and amplitudes E_{ij} of annual variations in the atmospheric inertia tensor components (in units of 10^{28} kg m^2), atmospheric masses M (in units of 10^{15} kg), and pressures $\langle P \rangle$ (in hPa).

Element	Region Ocean	Land		Height correction	Planet as a whole	
		Sea-level pressure	Land-level pressure		Sea-level pressure	Land-level pressure
$\bar{n}_{11}$	8294	5867	5588	279	14161	13882
E_{11}	−4.6	19.9	8.9	11.1	15.3	4.3
$\bar{n}_{22}$	8935	5228	5008	219	14162	13943
E_{22}	−9.5	13.4	5.8	7.6	3.9	−3.7
$\bar{n}_{33}$	9374	4995	4753	242	14370	14128
E_{33}	−7.4	9.0	3.7	5.4	1.6	−3.8
$\bar{n}_{12}$	−164.2	164.2	179.1	−14.9	0.04	14.92
E_{12}	−0.17	−1.80	−1.39	−0.41	−1.63	−1.22
$\bar{n}_{13}$	301.0	−303.8	−301.2	−2.52	−2.76	−0.23
E_{13}	−1.40	−0.89	−0.76	−0.14	−2.30	−2.16
$\bar{n}_{23}$	320.0	−326.1	−284.7	−41.3	−6.06	35.28
E_{23}	−1.86	−6.97	−4.54	−2.43	−8.83	−6.40
Mass $\bar{M}$	3271	1979	1888	91.0	5250	5159
E_{M_a}	−2.64	5.23	2.28	2.95	2.59	−0.36
Pressure $\langle P \rangle$	1011.4	994.4	948.7	45.7	1004.9	987.5
E_P	−0.82	2.64	1.14	1.49	0.50	−0.07
$S \times 10^{-14}$ m^2	3.164	1.948			5.112	

The principal moments of inertia of the atmosphere are needed for estimating the instabilities of the Earth's rotation. We should recall that the mean moment of inertia of the atmosphere relative to the polar axis (the axial moment of inertia) is equal to $n_{33} = 1.413 \times 10^{32}\,\mathrm{kg\,m^2}$. The mean values of the inertia moments of the atmosphere over the Northern and Southern Hemispheres are $n_{33} = 0.702 \times 10^{32}\,\mathrm{kg\,m^2}$, and $n_{33} = 0.711 \times 10^{32}\,\mathrm{kg\,m^2}$, respectively. The equatorial inertia moments are equal to $n_{11} = 1.388 \times 10^{32}\,\mathrm{kg\,m^2}$ and $n_{22} = 1.394 \times 10^{32}\,\mathrm{kg\,m^2}$, respectively.

It is practically impossible to estimate (with whatever reliability) the errors of the characteristics given in Table 6.1. It can only be assumed that the errors of their absolute values are within the tenths of a per cent, while those of the amplitudes make a few per cent.

6.3 Estimations of Instabilities in the Earth's Rotation

As is shown in Section 4.4, the effect of air-mass seasonal redistribution on the Earth's rotation is described by the equations:

$$\begin{aligned}
&\frac{1}{\sigma_0}\frac{d\nu_1}{dt} + \nu_2 = \psi_2 = \frac{\tilde{n}_{23}}{C-A} \\
&\frac{1}{\sigma_0}\frac{d\nu_2}{dt} - \nu_1 = -\psi_1 = -\frac{\tilde{n}_{13}}{C-A} \\
&\frac{d\nu_3}{dt} = \frac{d\psi_3}{dt} = -\frac{1+k'}{C}\frac{d\tilde{n}_{33}}{dt}
\end{aligned} \tag{6.11}$$

Substituting the appropriate values of $\tilde{n}_{ij}$ from the last column of Table 6.1 into (6.11) and assuming $C - A = 2.63 \times 10^{42}\,\mathrm{g\,cm^2}$, $C = 8.04 \times 10^{44}\,\mathrm{g\,cm^2}$, and $\Omega = 7.29 \times 10^{-5}\,\mathrm{s^{-1}}$, we obtain, in units of 10^{-8},

$$\psi'_1 = -8.2\cos\oplus, \quad \psi'_2 = -24.3\cos\oplus, \quad \psi'_3 = 0.033\cos\oplus \tag{6.12}$$

The excitation function would have these components if the entire Earth (together with oceans) behaved as an absolutely rigid body. But actually, the Earth is deformed due to variations in the load onto its surface that result from the air-mass redistribution in the atmosphere. Here, the global ocean apparently responds to the seasonal atmospheric-pressure variations like an inverted barometer. Where the pressure increases, the sea level drops, and vice versa. If we consider only the effect of the air-mass redistribution in the atmosphere, the annual pressure variations at a certain depth in oceans, should be coordinate independent. We see from Table 6.1 that $\langle P_0 \rangle$ (the pressure averaged over the surface of the World Ocean), varies over the year in accordance with:

$$\langle P_0 \rangle = (1011.4 - 0.82\cos\oplus)\,\mathrm{hPa} \tag{6.13}$$

Also, we calculated the components of the atmosphere inertia tensor for this case of the pressure distribution over the World Ocean (the case of the "inverted

Table 6.2 Mean values of $\bar{c}_{ij}$ and amplitudes of annual variations D_{ij} of the components of the atmosphere inertia tensor, 10^{28} kg m^2, with the World Ocean behaving as the "inverted barometer".

Region	Component											
	$\bar{n}_{11}$	E_{11}	$\bar{n}_{22}$	E_{22}	$\bar{n}_{33}$	E_{33}	$\bar{n}_{12}$	E_{12}	$\bar{n}_{13}$	E_{13}	$\bar{n}_{23}$	E_{23}
Ocean	8301	−6.7	8942	−7.3	9359	−7.6	−164.5	0.13	300.3	−0.24	317.5	−0.26
Land	5588	8.9	5008	5.8	4753	3.7	179.1	−1.39	−301.2	−0.76	−284.7	−4.54
Entire planet	13890	2.2	13950	−1.4	14113	−3.9	14.6	−1.26	−0.9	−1.00	32.8	−4.79

barometer"). The atmospheric pressure over the entire ocean was assumed to be 1010.58 and 1012.22 hPa in January and July, respectively. The initial data on the pressure over land were taken as unchanged, as well as the calculation formulas (6.7). The results are given in Table 6.2.

Substituting the values $\tilde{n}_{13}$, $\tilde{n}_{23}$ and $\tilde{n}_{33}$ from Table 6.2 into expressions (6.11), we find that when the ocean responds to the atmospheric-pressure variations as an inverted barometer, the components of the excitation function have the following values (multiplied by 10^{-8}):

$$\psi_1 = -3.8\cos\oplus, \quad \psi_2 = -18.2\cos\oplus, \quad \psi_3 = 0.034\cos\oplus \tag{6.14}$$

The first two components that describe the motion of the excitation pole can also be written in the form of circular motion. Then using the formulas of transition (4.39)–(4.48), we obtain:

$$\psi_1 + \mathrm{i}\psi_2 = 9.2\mathrm{e}^{\mathrm{i}(\oplus + 258^\circ)} + 9.3\mathrm{e}^{-\mathrm{i}(\oplus + 102^\circ)} \tag{6.15}$$

In the 1980s, the calculations of the atmospheric excitation functions were organized both in operational and in a retrospective way, known as reanalysis. They were based on the analyses of the surface pressure and wind fields, which were performed using the global atmospheric models developed in the world's major meteorological centers: the U.S. National Centers for Environmental Prediction (NCEP), the European Center for Medium-range Weather Forecasting (ECMWF), the UK Meteorological Office (UKMO), and the Japan Meteorological Agency (JMA). These models are widely used in the calculations and correlation analysis of the time series of the pole's coordinates (Barnes *et al.*, 1983; Salstein *et al.*, 1993; Salstein, 2000; Nastula, Ponte, and Salstein, 2000; Wilson, 2000). We should underline that the atmosphere excitation functions that were obtained in the above centers with the use of the detailed and accurate atmospheric pressure fields, nevertheless, do not account fully for the distortions due to land topography.

The excitation functions of the form of $\psi_i = a_i\cos\oplus + b_i\sin\oplus$ describe the forced motion of the poles and the variations of the Earth's rotation velocity with an annual period. As is shown in Section 4.5, they can be found by the following formulas:

$$\nu_i + i\nu_2 = \frac{1}{T^2-1}\{[-(a_1 + Tb_2) - i(a_2 - Tb_1)]\cos\oplus + [(Ta_2 - b_1) - i(Ta_1 + b_2)]\sin\oplus\}$$

$$\nu_3 = \psi_3 = a_3\cos\oplus + b_3\sin\oplus \tag{6.16}$$

Substituting $T = 1.2$ year, and the coefficients of the excitation function components from (6.14) ($a_1 = -3.8$, $b_1 = 0$, $a_2 = -18.2$, $b_2 = 0$, $a_3 = 0.034$ $b_3 = 0$) into expressions (6.16), we obtain the final expressions for the annual motion of the pole and the variations of the Earth's rotation velocity that are caused by the redistribution of air mass in the atmosphere (units of 10^{-8}):

$$\nu_1 + i\nu_2 = 8.6\cos\oplus - 49.7\sin\oplus + i(41.4\cos\oplus + 10.4\sin\oplus) \tag{6.17}$$

$$\nu_3 = 0.034\cos\oplus \tag{6.18}$$

If the trajectory of the excitation pole is given in the form of circular movements (6.15), then the forced movement of the pole of rotation is:

$$\begin{aligned}\nu_1 + i\nu_2 &= -\frac{1}{T-1}|\psi^+|e^{i(\oplus+\lambda^+)} + \frac{1}{T+1}|\psi^-|e^{-i(\oplus+\lambda^-)}\\ &= 46.5e^{i(\oplus+78^\circ)} + 4.2e^{-i(\oplus+102^\circ)}\end{aligned} \tag{6.19}$$

By comparison, if the effect of the "inverted barometer" in the ocean is not accounted for (that is, if the excitation functions (6.12) are used instead of (6.14)), then the parameters of the poles' motion and of the Earth's rotation irregularities would be:

$$\begin{aligned}\nu_1 + i\nu_2 &= 18.6\cos\oplus - 66.4\sin\oplus + i(55.3\cos\oplus + 22.4\sin\oplus)\\ &= 64.2e^{i(\oplus+71^\circ)} + 5.9e^{-i(\oplus+109^\circ)}\end{aligned} \tag{6.20}$$

$$\nu_3 = 0.033\cos\oplus \tag{6.21}$$

6.4 Discussion of Results

Let us compare the above results with astronomical observations, by computing the annual variations of the length of day. According to astronomical observations, the Earth's rotational velocity varies with a 1-year period in the following manner (Sidorenkov, 1975a) (units of 10^{-8}):

$$\nu_3 = -0.42\cos\oplus - 0.19\sin\oplus = 0.46\cos(\oplus - 205^\circ) \tag{6.22}$$

Comparing (6.18) and (6.22), we conclude that the redistribution of masses in the atmosphere cannot account for the observed annual variations of the Earth's rotational velocity. The calculated amplitude of variations makes only 7% of the observed value, and the calculated phase is 250° larger than the observed phase (that

Table 6.3 Annual-motion trajectories of the Earth's North Pole of rotation, according to various authors (in units of 10^{-8} radian).

Author	Period	ν_1^C	ν_1^S	ν_2^C	ν_2^S	F	φ_1, deg	N	φ_2, deg	λ, deg
1. Astronomical observations										
Walker and Young	1899–1954	−31.0	−34.4	33.9	−22.3	46.3	228	40.6	326	132
Jeffreys	1899–1967	−23.3	−37.2	33.7	−13.6	43.9	238	36.4	338	125
Rykhlova										
series 1	1891–1960	−14.5	−40.2	30.1	−12.1	42.7	250	32.4	338	116
series 2	1846–1915	−31.5	−20.8	29.6	−16.0	37.7	213	33.6	332	137
Gaposhkin	1891–1970	−20.5	−38.6	27.9	−16.5	43.7	242	32.4	330	126
Korsun *et al.*										
series 1	1846–1889	−48.2	−14.3	21.4	−25.0	50.2	197	32.2	310	156
series 2	1890–1956	−22.0	−38.2	28.1	−16.3	44.1	240	32.5	330	128
series 3	1957–1971	−15.3	−42.8	36.7	−18.4	45.5	250	41.1	333	113
2. Air-mass redistribution										
Sidorenkov		8.6	−49.7	41.4	10.4	50.4	280	42.6	14	78
Wilson and Haubrich		7.9	−28.6	23.9	9.1	29.7	285	25.5	21	72
Kikuchi		16.7	−51.6	41.1	19.3	54.2	288	45.4	25	68
Munk and Hassan		6.8	−35.6	29.9	7.2	36.3	281	33.3	14	77
Munk and McDonald		8.2	−42.3	34.4	8.2	43.1	281	35.4	13	77
Byzova		12.5	−55.7	45.2	15.0	57.1	283	47.6	18	74
Rosenhead		9.8	−67.8	56.5	12.4	68.5	278	57.8	12	80
Jeffreys		−4.8	−49.9	43.4	−2.9	50.1	264	43.5	356	96

$\underline{\nu} = \nu_1^c \cos \oplus + \nu_1^s \sin \oplus + i(\nu_2^c \cos \oplus + \nu_2^s \sin \theta) = F \cos(\oplus - \varphi_1) + iN \cos(\oplus - \varphi_2)$, λ is the Eastern Longitude of the Pole at $\oplus = 0°$.

is the resultant effect is of the opposite sign). This is in good agreement with the results of previous investigations (Munk and Mcdonald, 1960; Lambeck, 1980).

Let us now turn to the annual motion of the pole of the Earth's rotation. The trajectory of the annual motion of the North Pole has been determined from astronomical observations several times. The results of most of these studies are summarized in (Munk and McDonald, 1960; Lambeck, 1980). In part I of Table 6.3 we list the values obtained from analysis of the longest series of the pole coordinates (Walker and Young, 1957; Rykhlova, 1971; Jeffreys, 1968; Gaposchkin, 1972; Korsun, Mayor, and Yatskiv, 1974; Vondrak and Pejovic, 1988).

The parameters of the pole of rotation circular movements, which correspond to the trajectories presented in Table 6.3, are given in Table 6.4. They are calculated by the formulas of transition (4.39)–(4.48) with the use of the values of ν_1^c, ν_1^s, ν_2^c and ν_2^s from Table 6.3.

We should note that the initial data used in the works of A. Walker, A. Young, and G. Jeffreys are the pole coordinates in the system of the International Latitude Service (ILS), which are known to be referred to several different starting epochs. L. Rykhlova used the pole coordinates referred to the mean pole of the observation epoch (the system of A. Orlov). E. Gaposhkin used mainly Rykhlova's data and partly the ILS data. That is why the trajectory of the North Pole of the Earth's rotation calculated by

Table 6.4 Parameters of circular movements of the Earth's rotation pole with the annual period (in units of 10^{-8} rad).

Author	Period	v^+	λ_v^+	v^-	λ_v^-	$v^+ + v^-$	$v^+ - v^-$	e_v	λ_v
1. Astronomical observations									
Walker and Young	1899–1954	43.3	128	4.4	177	47.7	38.9	0.18	156
Jeffreys	1899–1967	40.0	118	5.2	160	45.2	34.8	0.23	159
Rykhlova									
series 1	1891–1960	37.6	111	5.2	103	42.8	32.4	0.24	184
series 2	1846–1915	34.6	134	8.9	210	43.5	25.7	0.41	142
Gaposhkin	1891–1970	38.1	119	5.7	111	43.8	32.4	0.26	184
Korsun *et al.*									
series 1	1846–1889	40.7	154	12.1	197	52.8	28.6	0.46	158
series 2	1890–1956	38.3	120	5.8	119	44.1	32.5	0.26	180
series 3	1957–1971	43.2	113	3.4	63	46.6	39.8	0.15	205
Vondrak	1976–1985	40.7	115	2.4	94	43.1	38.3	0.11	190
2. Air-mass redistribution									
Vondrak	1977–1985	39.8	78	3.4	99	43.1	36.4	0.16	170
Wilson and Haubrich		27.6	72	2.4	105	30.0	25.2	0.16	164
Sidorenkov		46.5	78	4.2	102	50.8	42.3	0.17	168
Kikuchi		49.7	69	5.4	104	55.1	44.3	0.20	162
Munk and Hassan		33.5	78	2.9	94	36.4	30.6	0.19	172
Munk and McDonald		39.2	78	3.9	90	43.1	35.3	0.18	174
Byzova		52.3	75	5.4	103	57.7	46.9	0.19	166
Rosenhead		63.1	80	5.8	103	68.9	57.3	0.17	168
Jeffreys		46.8	95	3.4	106	50.4	43.4	0.14	174

Walker and Young is in good agreement with those calculated by Jeffreys and the trajectory calculated by Rykhlova agrees with those calculated by Gaposhkin. However, there are some distinctions between those trajectories. Rykhlova's results, which are obtained for the period of 1846–1915, are not very reliable, because they are based on observations from a very sparse network of observatories that existed in the nineteenth century and the beginning of the twentieth century. Korsun, Mayor, and Yatskiv (1974) used the homogeneous series of the pole's coordinates for 1846–1971, which was obtained in the Main Astronomical Observatory of the Ukrainian Academy of Sciences (Fedorov *et al.*, 1972). The authors of this work averaged the parameters of the pole's annual motion over three time intervals, which approximately correspond to the periods when the accuracy of determination of the poles' coordinates sharply increased. Their estimates of parameters for the 1890–1956 period are close to previous estimates. The respective estimates for other observation periods are markedly different. This is likely to be indicative of a nonstationary annual motion of the pole. J. Vondrak used the data of the Bureau International de l'Heure (BIH) for 1976–1985 (Vondrak and Pejovic, 1988).

The comparison of the trajectories of the pole of rotation, which we calculated from the meteorological data (the meteorological trajectory) and from the astronomical observations (the astronomical trajectory) (Figure 6.3) indicates that the two are similar, though the axes of the meteorological trajectory are approximately 1–2 times

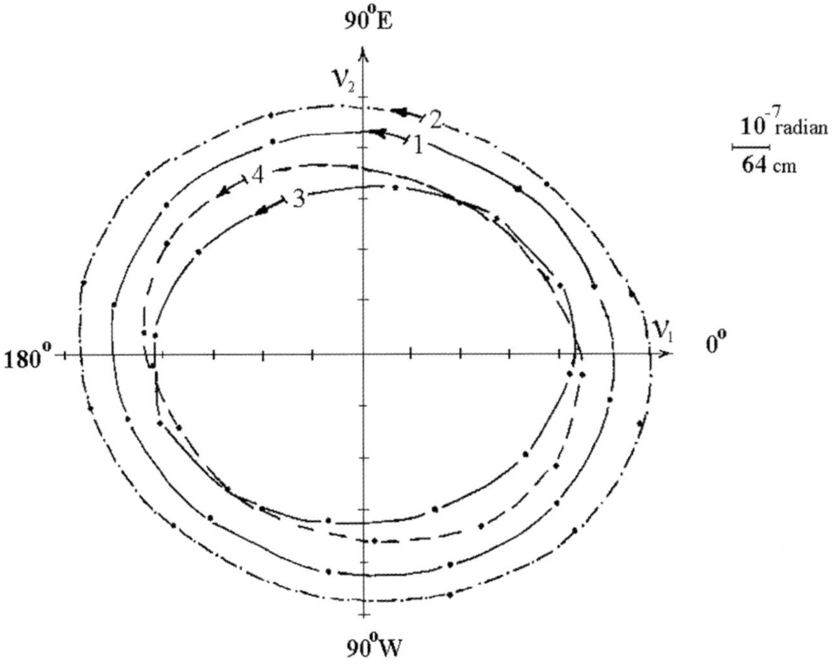

Figure 6.3 Trajectories of the motion of the Earth's pole of rotation as calculated from meteorological (1, 2) and astronomical (3, 4) data. (1) Sidorenkov (1973); (2) Byzova (1947); (3) Rykhlova (1971); (4) Jeffreys (1968).

longer than the axes of the trajectory observed by astronomers. In both cases, the pole moves along an ellipse with a small eccentricity in the positive direction (from west to east, or counterclockwise). The meteorological pole has a phase lag of about 45° behind the astronomical pole. Thus, at the beginning of the year (at $\oplus = 0°$), the meteorological pole is on the meridian of 78° E, and the astronomical pole at 120° E.

Let us compare all these values with the estimates obtained earlier by other authors from the data on the seasonal variations in the global atmospheric pressure.

The first attempt to estimate the influence of the air-mass distribution on the motion of the Earth's poles was made in (Spitaler, 1901). This estimate was very rough and is only of historical interest. The first detailed calculation of this effect was carried out by Jeffreys (1916), who used Bartholomew's meteorological atlas (1899). Rosenhead (1929) performed similar study using the data from Shaw's Handbook of Meteorology (1928). More detailed calculations were made in (Byzova, 1947) on the basis of the Ocean Atlas maps (Morskoy atlas, 1953) of seasonal air-mass transport.

Munk and Hassan (1961) used for their calculations the data on the atmospheric pressure (not reduced to the sea level) from a set of continental meteorological stations situated mainly in the Northern Hemisphere. Wilson and Haubrich (1976) reestimated the effect of air-mass redistribution on the poles' motion. They used the pressure series from 1325 meteorological stations for 1901–1970. These data for 1901–1940 nearly coincide with the data used in (Munk and Hassan, 1961).

Kikuchi (1971) investigated the effect of air-mass shifts on the Earth's rotation by expanding the atmospheric-pressure field in a spherical-function series. The basic data were the climatic values of the mean monthly atmospheric pressure at sea level that were published in 1968 by the Meteorological Research Institute of the Japanese Meteorological Agency.

In the papers cited above, the effect of air-mass distribution was estimated in different ways, and the results were expressed in arbitrary units. For comparison, we reduced the available results to a unique system, namely: the values of coefficients were converted to radians; the argument was the longitude of the mean sun reckoned from the beginning of the year; the ocean's response to the atmospheric pressure variations was assumed to be of the type of an inverted barometer. The coefficients for the expression of the trajectory of the excitation pole, which were determined in the above way by various authors, are given in Table 6.5.

Table 6.5 Trajectories of the annual motion of the North excitation pole, according to various authors (in units of 10^{-8} radian).

Author	Parameter								
	a_1	b_1	a_2	b_2	A	ξ_1, deg	B	ξ_2, deg	λ, deg
1. Air-mass redistribution									
Sidorenkov	−3.8	0	−18.2	0	3.8	180	18.2	180	258
Wilson and Haubrich	−3.1	0	−10.5	−0.3	3.1	180	10.5	181	254
Kikuchi	−6.5	−2.4	−20.9	−0.7	6.9	200	20.9	182	253
Munk and Hassan	−1.8	0.2	−12.9	−1.0	1.8	174	12.9	184	262
Munk and McDonald	−1.7	−0.9	−16.3	−1.6	1.9	209	16.4	186	264
Byzova	−5.5	−1.4	−21.6	0	5.7	194	21.6	180	256
Rosenhead	−5.0	−0.1	−25.0	0.6	5.0	181	25.0	178	259
Jeffreys	−1.4	2.2	−16.4	2.9	2.6	121	16.7	170	265
2. Correction for land elevation									
Sidorenkov	0.52	0	9.24	0	0.52	0	9.24	0	87
Rosenhead	0.49	0.06	13.91	2.96	0.49	7	14.20	12	88
Munk and McDonald	−0.10	−0.62	9.12	5.16	0.63	99	10.48	30	92
Jeffreys	0.67	−0.13	16.17	3.17	0.68	349	16.48	11	88
3. Astronomical observations									
Walker and Young	−4.3	6.3	−7.4	14.9	7.6	124	16.6	116	240
Jeffreys	−6.9	3.2	−11.0	14.3	7.6	155	18.0	128	238
Rykhlova series 1	0	−4.2	−18.2	5.3	4.2	270	19.0	164	270
Rykhlova series 2	−12.3	14.6	4.6	21.8	19.1	130	22.3	78	159
Gaposhkin	−0.7	−5.1	−18.4	8.1	5.2	262	20.1	156	268
Korsun *et al.*, 1846–1889	−18.1	−11.4	−4.2	32.8	18.6	148	33.1	83	167
Korsun *et al.*, 1890–1956	−3.4	−4.5	−17.8	10.0	17.9	242	20.4	150	262
Korsun *et al.*, 1957–1971	6.8	1.3	−14.6	0	16.1	11	14.6	180	295

$\psi = a_1 \cos \oplus + b_1 \sin \oplus + i(a_2 \cos \oplus + b_2 \sin \oplus) = A \cos(\oplus - \xi_1) + iB \cos(\oplus - \xi_2)$; λ is the Eastern Longitude of the Pole at $\oplus = 0°$.

Table 6.6 Parameters of the excitation pole circular movements with an annual period (in units of 10^{-8} rad).

Author	Parameter							
	ψ^+	λ_ψ^+	ψ^-	λ_ψ^-	$\psi^+ + \psi^-$	$\psi^+ - \psi^-$	e_ψ	λ_ψ
1. Air-mass redistribution								
Sidorenkov	9.2	258	9.3	102	18.5	−0.1	1.01	78
Wilson and Haubrich	5.4	252	5.4	105	10.8	0	1.00	74
Kikuchi	9.9	249	12.0	104	21.9	−2.1	1.10	72
Munk and Hassan	6.7	258	6.4	94	13.1	0.3	0.98	82
Munk and McDonald	7.8	258	8.6	90	16.4	−0.8	1.05	84
Byzova	10.5	255	11.8	103	22.3	−1.4	1.06	76
Rosenhead	12.6	260	12.8	103	25.4	−0.2	1.01	79
Jeffreys	9.4	275	7.4	106	16.8	2.0	0.88	84
2. Astronomical observations								
Walker and Young	8.7	308	9.6	177	18.3	−0.9	1.05	66
Jeffreys	8.0	297	11.3	161	19.3	−3.3	1.17	68
Rykhlova series 1	7.5	291	11.5	103	19.0	−4.0	1.21	94
Rykhlova series 2	6.9	313	19.6	210	26.5	−12.7	1.48	52
Gaposhkin	7.6	299	12.5	111	20.1	−4.9	1.24	94
Korsun *et al.*, 1846–1889	8.1	334	26.6	197	34.7	−18.5	1.53	68
Korsun *et al.*, 1890–1956	7.7	300	12.8	119	20.5	−5.1	1.25	90
Korsun *et al.*, 1957–1971	8.6	293	7.5	63	16.1	1.1	0.93	115

The parameters of the excitation pole circular movements, which correspond to the trajectories in Table 6.5, are given in Table 6.6. They are calculated by the formulas of transition (4.39)–(4.48) from Section 4.5, with the use of the values of a_1, a_2, b_1, and b_2 from Table 6.5.

Figure 6.4 illustrates the motion of the North pole of excitation, as estimated by various authors. We see that according to all estimations the (meteorological) excitation pole moves along the ellipses with the eccentricity close to unity. The minor axes of these ellipses do not exceed 4×10^{-8} (25 cm on the Earth's surface). The direction of the excitation pole motion is estimated as negative (from east to west) in (Kikuchi and Byzova) and positive in (Jeffreys, Munk and Hassan). This indicates that the minor axis of the respective ellipse is rather small to be determined from available meteorological data. The major axis of the ellipse is determined with greater confidence. It makes $\sim 38 \times 10^{-8}$ rad, or 240 cm on the Earth's surface. The excitation pole is the farthest from its mean position at $\oplus \approx 0°$ (the beginning of January) and $\oplus \approx 180°$ (the beginning of July), which correspond to $\sim$100°W and $\sim$80°E, respectively. This type of motion of the meteorological excitation pole indicates that, as we should expect, the seasonal variations of the air mass above the Eurasian continent is the basic factor in this motion.

To estimate the excitation pole movement by astronomical observations, we used (4.57) to calculate the trajectory of its motion from the parameters given in part 1 of

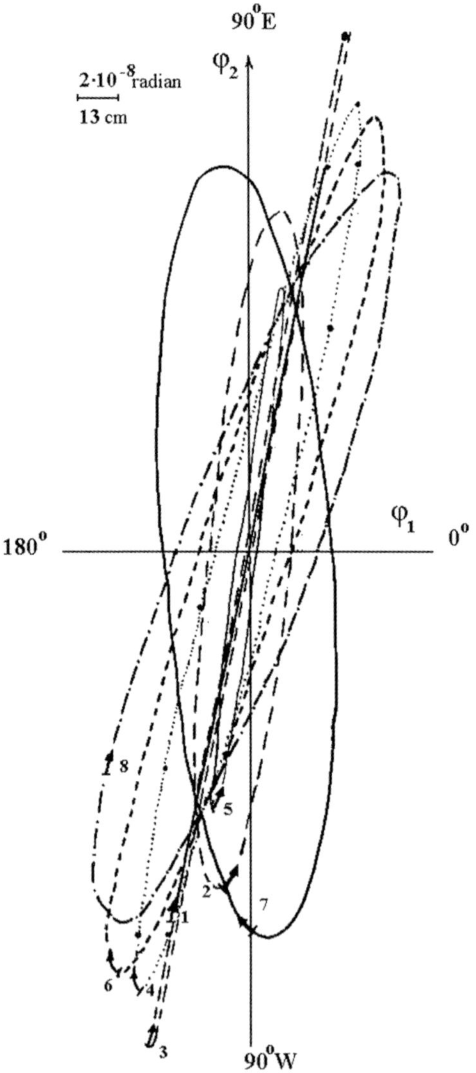

Figure 6.4 Trajectories of the Earth's excitation pole as obtained from meteorological (1–6) and astronomical (7, 8) data. (1) Sidorenkov (1973); (2) Jeffreys (1916); (3) Rosenhead (1929); (4) Byzova (1947); (5) Munk and Hassan (1961); (6) Kikuchi (1971); (7) Rykhlova (1971); (8) Jeffreys (1968).

Table 6.3. The results are given in part 3 of Table 6.5. The parameters of circular movements of the excitation pole, which are calculated from the data of astronomical observations, are given in Table 6.6. By way of illustration, Figure 6.4 shows the astronomical trajectories of the excitation pole from Rykhlova's (series I) and Jeffreys' (1968) data.

Thus, according to both the astronomical and the meteorological data, the excitation pole moves along an ellipse, the semiaxes of the ellipse being similar in both cases.

In order to understand why the conclusions of astronomers on the excitation pole motion are so conflicting, let us analyze the parameters of circular movements of the Earth's rotation pole (Table 6.4). It is seen that the observed radius of the positive movement is approximately 10 times larger than that of the negative movement. With this, the standard errors of determination of both radii from astronomic observations are similar. This is easily deduced with the help of formulas (4.44)–(4.45) (for example), which are written for radii $|\underline{\nu}^+|$ and $|\underline{\nu}^-|$. Consequently, the relative error of determination of the parameters of the positive circular movement is smaller by ten times than that for the negative circular movement. With this, from Equation 4.59 at $T = 1.2$ year we have:

$$\underline{\psi}^+ = -0.2\underline{\nu}^+; \quad \underline{\psi}^- = 2.2\underline{\nu}^- \tag{6.23}$$

Hence, the positive circular movement of the pole of rotation (more reliably determined) is transmitted to the excitation pole, being decreased by 5 times; and the respective negative movement (less reliably determined) is transmitted, being increased by 2.2 times. It is clear that the trajectory of the excitation pole is less precise than the trajectory of the pole of rotation.

In view of the above, it is more advantageous to compare the calculated and observed effects of the air-mass redistribution in the atmosphere with the help of the trajectories of the Earth's pole of rotation rather than the inertial pole. We used expressions (4.52), (4.53) and the geophysical estimates obtained by various authors (Table 6.5 and Table 6.6) to compute the parameters of the trajectory of the Earth's pole of rotation. They are given in part 2 of Tables 6.3–6.4 and, in part, in Figure 6.3. All these trajectories agree closely with regard to the orientation of the ellipse axes and the initial phase angles. The only difference is between the amplitudes of the motion, which is probably due to the errors in the initial values of atmospheric pressure.

Thus, the air-mass redistribution in the atmosphere is evidently not responsible for the observed variations in the length of day, but it does cause the annual variations of the Earth's excitation pole in the plane of the 80°E and 100°W meridians. The amplitude of oscillations is 19×10^{-8} radian (or 120 cm on the Earth's surface) and the initial phase is 180°. Seasonal variations in the air mass over Eurasia are the controlling factor in this motion. The excitation pole annual variations cause the motions of the Earth's pole of rotation with the same period. The trajectory of the pole of rotation, as calculated from meteorological data, agrees closely with the observed trajectory in the orientation of the ellipse axes and the direction of the motion. In both cases the ellipses major axes lie in the plane of the 10°W–170°E meridians, and the motion has the positive direction (from west to east); however, the initial phase angles and the lengths of the axes differ. According to the estimates of all the above authors, the "meteorological" pole of the Earth's rotation lags 30–60° behind the astronomical pole. At the beginning of the year, they are situated at 80°E and 115–130°E, respectively. The axes of the trajectory of the "meteorological" pole of rotation are slightly (by a factor of about 1.2) longer than those of the astronomical trajectory. The calculated major semiaxis is 50×10^{-8} radian, or 320 cm; the observed value is 43×10^{-8} radian, or 270 cm.

The totality of these data suggests that the observed annual motion of the Earth's poles of rotation is due not only to the redistribution of air masses in the atmosphere, but also to other phenomena. These probably include the seasonal redistribution of water masses in the World Ocean, seasonal variations in the amount of ground water on land, and so forth.

7
Angular Momentum of Atmospheric Winds

7.1
Functions of the Angular Momentum of the Atmosphere

Angular momentum is the integral of movement, which is associated with the isotropy of space. It has an important property of additivity. This means that the angular momentum of the entire system consisting of several parts (the interaction of which can be neglected) is equal to the sum of angular momentums of all these parts (Landau and Lifshitz, 1965). The angular momentum of a closed system is constant; it cannot arise or disappear but can only redistribute between individual parts of the system.

Angular momentum is one of the most important integrals of motion of the atmosphere. The investigation of this additive integral of motion and its time-dependent variations helps us to better understand the nature of the atmosphere general circulation and the astronomically observed nonuniformity of the Earth's rotation.

Let us take (as discussed in Section 4.1) the system of the movable Cartesian coordinates (Ox_i), which is invariably linked with the Earth and rotates with it, with the angular velocity ω, with respect to the inertial system of coordinates ($O\xi_i$). Both coordinate systems have the same origin – the mass center of the Earth O. Axes (Ox_i) coincide with the major axes of the ellipsoid of inertia of the unperturbed Earth. Let us choose the orthogonal basis $\{e_1, e_2, e_3\}$ of the rotating system so that the first two orts determine the plane of the equator and the third ort is directed along the middle axis of rotation Ox_3.

The atmosphere rotates as a solid body together with the Earth and, besides, moves with respect to the Earth's surface. Hence, its absolute angular momentum $\boldsymbol{H}$ is the sum of two terms:

$$\boldsymbol{H} = \int_A \boldsymbol{r} \times (\boldsymbol{\omega} \times \boldsymbol{r})\, \rho \mathrm{d}V + \int_A (\boldsymbol{r} \times \boldsymbol{u}) \rho \mathrm{d}V \tag{7.1}$$

Here $\boldsymbol{r} = \sum_{i=1}^{3} x_i \boldsymbol{e}_i \equiv x_i \boldsymbol{e}_i$ is the radius-vector of the atmosphere's unit volume $\mathrm{d}V$ under consideration; $\boldsymbol{\omega} = \omega_i \boldsymbol{e}_i$ is the vector of the angular velocity of the Earth's

The Interaction Between Earth's Rotation and Geophysical Processes. Nikolay S. Sidorenkov

ISBN: 978-3-527-40875-7

rotation; $\boldsymbol{u} = u_i\boldsymbol{e}_i$ is the vector of wind velocity; ρ is the air density; and A is the volume of the entire atmosphere. The first term is the translational angular momentum of the atmosphere $\boldsymbol{H}_\Omega = H_{\Omega i}\boldsymbol{e}_i$, which originates due to the rotation of the atmosphere (as a solid body) together with the Earth with the angular velocity ω. The second term, $\boldsymbol{h} = h_i\boldsymbol{e}_i$, is called the relative angular momentum of the atmosphere. It characterizes the movement of atmospheric air with respect to the Earth's surface, that is winds. Therefore, this term is often called the angular momentum of winds. As is shown in Section 4.1,

$$H_{\Omega i} = \omega_k \int_A (x_l^2\delta_{ik} - x_i x_k)\rho \mathrm{d}V = n_{ik}\omega_k \tag{7.2}$$

$$h_i = \int_A \varepsilon_{ijk} x_j u_k \rho \mathrm{d}V \tag{7.3}$$

where n_{ik} are the components of the tensor of inertia; δ_{ik} is Kronecker's symbol ($\delta_{ik} = 1$ at $i = k$ and $\delta_{ik} = 0$ at $i \neq k$); ε_{ijk} is Levi-Civita's alternator equal to: $+1$ if subscripts are in even order: 1, 2, 3, 1, 2, 3, ...; -1 if subscripts are in odd order: 1, 3, 2, 1, 3, ...; and 0 if any two subscripts are equal: $i = j$, $i = k$, $j = k$; x_i and u_i are the Cartesian coordinates and the components of the velocity of volume $\mathrm{d}V$.

The absolute angular momentum of the atmosphere changes because of the variations in the components of the tensor of inertia n_{ik} (the variations resulting from the redistribution of air masses) and the variations in the components of the relative angular momentum h_i (that is, of the wind fluctuations). As is shown in Sections 4.3 and 4.4, the effect of these variations on the instability of the Earth's rotation can be assessed with the help of the effective functions χ_i of the angular momentum of the atmosphere (Barnes *et al.*, 1983):

$$\begin{aligned}
\chi_1 &= \frac{1}{\Omega(C-A)}(\Omega n_{13} + \kappa h_1) \\
\chi_2 &= \frac{1}{\Omega(C-A)}(\Omega n_{23} + \kappa h_2) \\
\chi_3 &= \frac{1}{C\Omega}(\Omega(1+k')n_{33} + h_3)
\end{aligned} \tag{7.4}$$

Here, C and A are the solid Earth's polar and equatorial moments of inertia, respectively; Ω is the mean angular velocity of the Earth's diurnal rotation; $\kappa = 1.43$ and $k' = -0.3$ are the parameters accounting for the rotational and loading deformations of the Earth (see Section 4.3). The $\tilde{\chi}_i$ and χ_i values are proportional to the components of the absolute angular momentum of the atmosphere. As distinct from the functions of the atmosphere's angular momentum $\tilde{\chi}_i$ (which are used in the case of the absolutely solid Earth (see Section 4.2)), the χ_i values (in which the Earth's rotational and loading deformations are accounted for) are called the effective functions of the atmospheric angular momentum. They are composed of two terms, the first of which (that contains the n_{ik} values) describes the effect of the air-mass redistribution and the second one (that contains the h_i values) accounts for the effect of winds in the entire atmosphere. Therefore, functions χ_i are usually written as the sum of the term of pressure χ^{P} (because the n_{ik} values are calculated by the data on the surface atmospheric pressure) and the term of wind χ^{W} (because the angular momentum h_i is calculated by the data on winds in the entire atmosphere):

$$\chi = \chi^{\mathrm{P}} + \chi^{\mathrm{W}} \tag{7.5}$$

The components of the atmosphere's tensor of inertia n_{13}, n_{23}, and n_{33}, are written in the explicit form in expression (6.2). Below, we write the expressions for the components of the angular momentum of winds:

$$\begin{aligned} h_1 &= \int_A (x_2 u_3 - x_3 u_2)\, \rho\, \mathrm{d}V \\ h_2 &= \int_A (x_3 u_1 - x_1 u_3)\, \rho\, \mathrm{d}V \\ h_3 &= \int_A (x_1 u_2 - x_2 u_1)\, \rho\, \mathrm{d}V \end{aligned} \tag{7.6}$$

The functions of the angular momentum h_i are calculated in the spherical system of coordinates (R, λ, φ), in which the Cartesian geocentric coordinates x_i have the following form:

$$\begin{aligned} x_1 &= R \cos\varphi \cos\lambda \\ x_2 &= R \cos\varphi \sin\lambda \\ x_3 &= R \sin\varphi \end{aligned} \tag{7.7}$$

Differentiating (7.7) and introducing the components of velocity in the spherical system of coordinates $u_R = \mathrm{d}R/\mathrm{d}t$, $u_\varphi = R\mathrm{d}\varphi/\mathrm{d}t$ and $u_\lambda = R\cos\varphi\, \mathrm{d}\lambda/\mathrm{d}t$, we obtain the expressions for velocities u_i:

$$\begin{aligned} u_1 &= u_R \cos\varphi \cos\lambda - u_\varphi \sin\varphi \cos\lambda - u_\lambda \sin\lambda \\ u_2 &= u_R \cos\varphi \sin\lambda - u_\varphi \sin\varphi \sin\lambda + u_\lambda \cos\lambda \\ u_3 &= u_R \sin\varphi + u_\varphi \cos\varphi \end{aligned} \tag{7.8}$$

Here, φ is the geographical latitude; λ is the eastern longitude; R is the geocentric radius; u_φ, u_λ, and u_R are the components of velocity of the south, west, and vertical wind, respectively. Substituting the x_i values from (7.7) and u_i values from (7.8) into (7.6), we obtain:

$$h_1 = \int_A R(u_\varphi \sin\lambda - u_\lambda \sin\varphi \cos\lambda) \rho \mathrm{d}V \tag{7.9}$$

$$h_2 = -\int_A R(u_\varphi \cos\lambda + u_\lambda \sin\varphi \sin\lambda) \rho \mathrm{d}V \tag{7.10}$$

$$h_3 = \int_A R u_\lambda \cos\varphi\, \rho \mathrm{d}V \tag{7.11}$$

The spherical element of volume $dV = R^2 \cos\varphi \, d\lambda \, d\varphi \, dz$. Using the equation of hydrostatic equilibrium, we can write:

$$\rho dz = -dp/g \tag{7.12}$$

where dp is the increment of pressure, and g is the gravity acceleration.

With account for expressions (7.9–7.12), the terms of wind in the effective functions of the angular momentum can be written in the following form:

$$\chi_1^W = \frac{-1.43R^3}{\Omega(C-A)g}\int_0^P\int_{-\frac{\pi}{2}}^{\frac{\pi}{2}}\int_0^{2\pi}(u_\lambda \sin\varphi\cos\varphi\cos\lambda - u_\varphi\cos\varphi\sin\lambda)d\lambda\, d\varphi\, dp \tag{7.13}$$

$$\chi_2^W = \frac{-1.43R^3}{\Omega(C-A)g}\int_0^P\int_{-\frac{\pi}{2}}^{\frac{\pi}{2}}\int_0^{2\pi}(u_\lambda \sin\varphi\cos\varphi\sin\lambda + u_\varphi\cos\varphi\cos\lambda)d\lambda\, d\varphi\, dp \tag{7.14}$$

$$\chi_3^W = \frac{R^3}{\Omega C g}\int_0^P\int_{-\frac{\pi}{2}}^{\frac{\pi}{2}}\int_0^{2\pi} u_\lambda \cos^2\varphi\, d\lambda\, d\varphi\, dp \tag{7.15}$$

When calculating χ_i, we assume: $C = 7.04 \times 10^{37}\,\text{kg m}^2$ is the polar moment of inertia of the crust and mantle (it is assumed that the Earth's core does not experience the observed instabilities of rotation of the solid Earth at time scales less than two years); $C - A = 0.00333 \times C$, where A is the mantle's equatorial moment of inertia; $R = 6.37 \times 10^6$ m is the mean radius of the Earth; $g = 9.81\,\text{m s}^{-2}$ is the gravity acceleration; $\Omega = 7.29 \times 10^{-5}\,\text{s}^{-1}$ is the mean angular velocity of the Earth's rotation; P is the atmospheric pressure at the Earth's surface; u_λ and u_φ are the velocities of the west and south wind, respectively; φ is the latitude; λ is the eastern longitude. The data on winds are at best available up to a level of 10 hPa (≈31 km).

7.2 Climatic Data

In 1976, the author thoroughly investigated the angular momentum of zonal winds (Sidorenkov, 1976), using the most complete series of aerological data available at the World Meteorological Centre at that time and generalized by I.G. Guterman (Guterman, 1975, 1976, 1978). Our investigation is still of interest now and is presented below. We have analyzed 12 meridional sections for the long-term monthly means of zonal winds, which had been constructed by I.G. Guterman by way of averaging (over 1957–1965) the data from the World network of the aerological stations. These climatic sections provide the velocities of zonal wind at the isobaric surfaces of: 850, 700, 500, 300, 200, 100, 50, and 30 hPa for the latitudes of: 0°, ±10°, ±15°, ±20°, ±30°, ±35°, ±40°, ±50°, ±60°, ±70°, and ±80°.

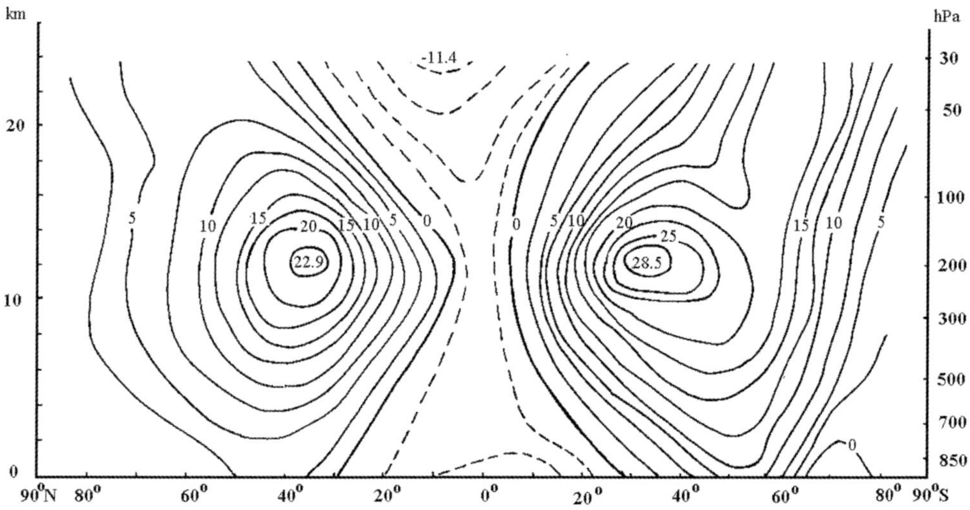

Figure 7.1 Meridional section of the mean annual zonal atmospheric circulation. Isotachs in m/s.

The zonal wind velocity at any point undergoes the seasonal variations that are usually approximated by the expression

$$u_\lambda = \bar{u}_\lambda + A\cos(\oplus - \varphi_1) + B\cos 2(\oplus - \varphi_2) \tag{7.16}$$

where $\bar{u}_\lambda$ is the annual average velocity; A, B and φ_1, φ_2 are the amplitudes and initial phases of the annual and semiannual harmonics, respectively; $\oplus = 2\pi t/365.25$ is the Sun's mean longitude, and t is the time in days. Using the method of least squares, we calculated the unknowns $\bar{u}_\lambda$, A, B, φ_1, and φ_2 for each nodal point of the section, using average values of u_λ for twelve calendar months. The results were used to construct sections of $\bar{u}_\lambda$, A, and B, which are shown in Figures 7.1–7.3.

It follows from Figure 7.1 that westerly winds predominate in the atmosphere on average over the year. Easterlies blow only near the equator. The boundaries between

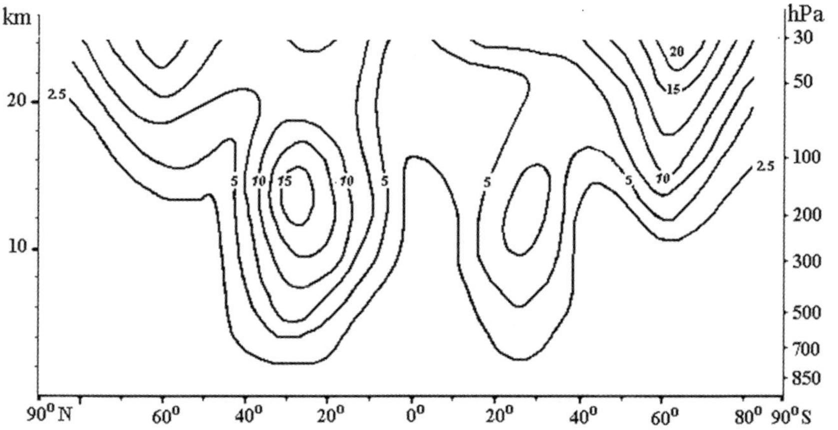

Figure 7.2 Meridional section of the annual amplitudes of the zonal wind velocity. Isolines in m/s.

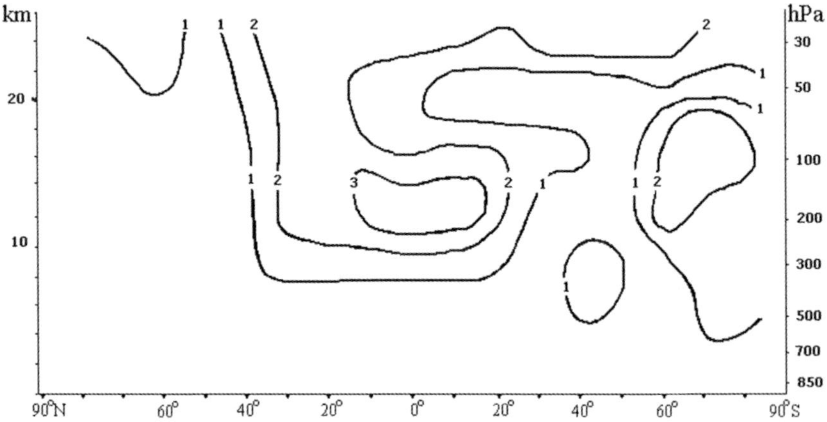

Figure 7.3 Meridional section of the semiannual amplitudes of the zonal wind velocity. Isolines in m/s.

west and easterly winds (the zero isotachs) at sea level lie at latitudes ±30°. With the ascent to the 200-hPa level, the zone of easterlies narrows toward the equator, down to the narrowest zone limited by the latitudes ±6°. Above 200-hPa level the zone of easterlies spreads out into middle latitudes and at the 30-hPa level, it extends from 40° N to 25° S. The highest westerly wind velocities are observed at the 200-hPa level in the subtropics, at around 35° of the northern and southern latitudes. These velocities attain 23 and 29 m/s in the Northern and the Southern Hemispheres, respectively. These high westerly velocities reflect the presence of the subtropical jet streams in the upper troposphere. The strongest easterlies are encountered in the equatorial stratosphere. The velocities range up to 12 m/s at 30 hPa and latitude 10° N. The velocities of the easterly winds do not exceed 4 m/s in the troposphere around the equator.

The amplitudes of the annual wind-velocity variations attain their maxima below the tropopause, near the zone of the subtropical jet streams (Figure 7.2). The values are 15 and 8 m/s in the Northern and Southern Hemispheres, respectively. Higher maxima are encountered in the stratosphere at latitudes ±60°, where easterlies blow in summer and westerlies in winter. The amplitude distributions of the semiannual variations are given in Figure 7.3. One can see that these amplitudes are much smaller than the annual amplitudes. Only in the latitudinal zone from 35° N to 40° S and above the 300 hPa level do they vary within the interval of 1–3 m/s.

The distribution of the long-term means of zonal winds u_λ in the atmosphere (up to a height of ~31 km) in January and July is shown in Figure 7.4. The averaging of data is performed with the help of the NCEP/NCAR reanalyses for 1948–1998. One can see that the subtropical jet stream in the troposphere of the Northern Hemisphere features the velocities of about 40 and 20 m/s in January and July, respectively. In the Southern Hemisphere, the difference between the wind velocities in these months is much less: they are about 30 and 40 m/s in January and July, respectively.

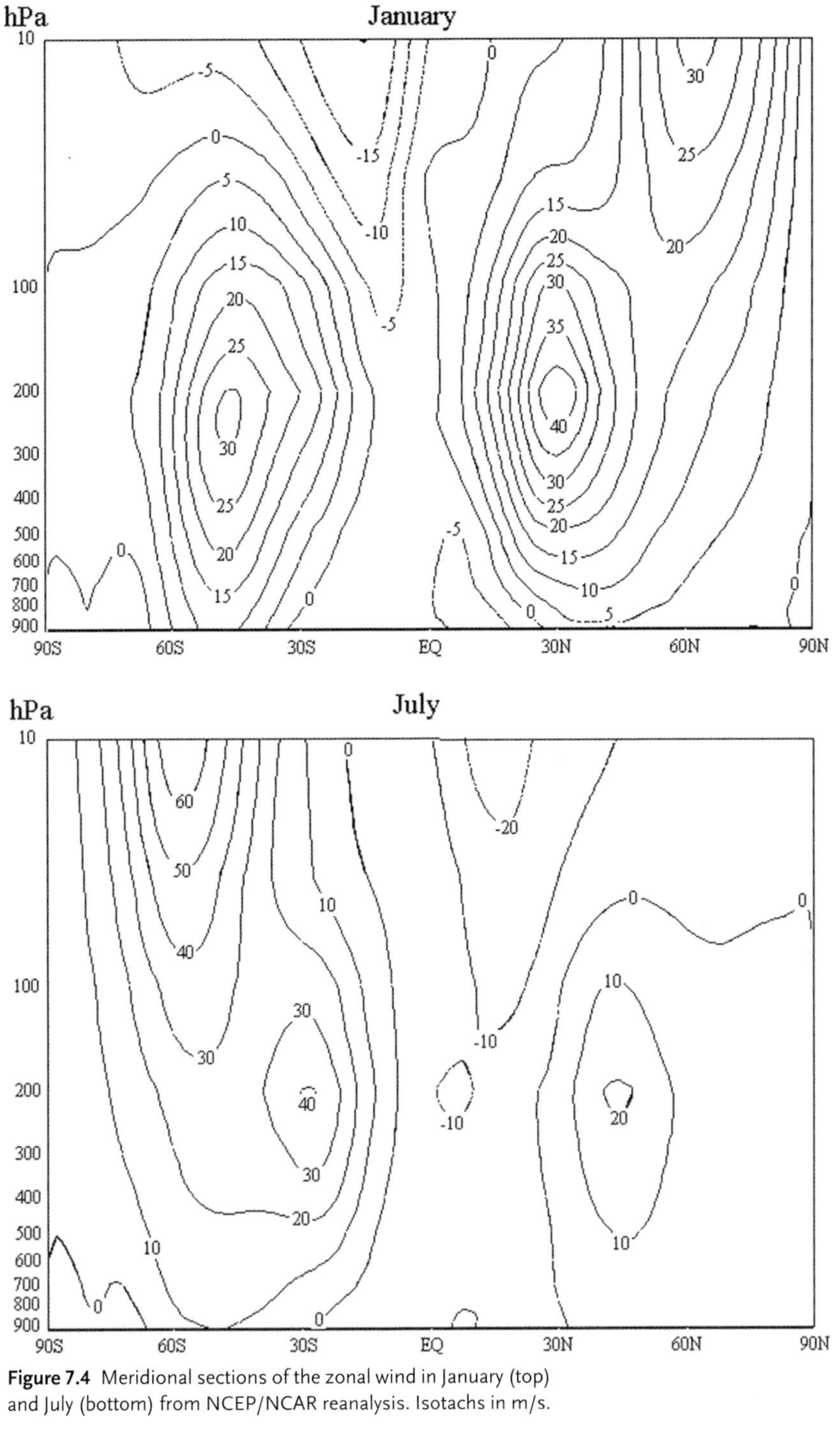

Figure 7.4 Meridional sections of the zonal wind in January (top) and July (bottom) from NCEP/NCAR reanalysis. Isotachs in m/s.

The data on the distribution of wind velocities in the atmospheric layers above 35 km are only episodic. They are mostly obtained with the help of rocket sounding and indirect methods (observations of the movement of artificial smoke clouds, meteoric traces, and so on). In winter, strong westerly winds are observed in the upper stratosphere. They are particularly strong in the middle latitudes at a height of the stratopause (65 km), where the wind velocities attain 100 m/s on average. In summer, easterly winds dominate above 15–20 km, their maximum velocity – about 70 km – being observed in the subtropics at a height of 55 km. Thus, the annual difference in wind velocities in the upper stratosphere can attain 170 m/s; in the zone of jet streams, which are observed near the tropopause (10–13 km) and contain the bulk of the angular momentum h_3, this difference is 30 and 16 m/s in the Northern and Southern Hemispheres, respectively.

Over the equator, in a layer from 18 to 30 km, the quasibiennial oscillation (QBO) in wind direction is traced (see Section 9.2): during one period (about 17 months) easterly winds blow there and during another period (about 11 months) westerly winds. The full cycle can last from 20 to 30 months, averaging about 28 months. Above 35 km over the equator, the semiannual oscillation is observed. In the transitional seasons (in spring and autumn), westerly winds prevail, whereas in winter and summer easterly winds do so.

Having the meridional sections of the zonal wind, we calculated the angular momentum h_3 of the zonal wind, converting from height to pressure as a variable, $\rho \mathrm{d}R = -\mathrm{d}P/g$. Here, P is the atmospheric pressure and g is the acceleration of gravity, which is approximated as

$$g_{\theta,z} = g_{90^\circ,0}(1+0.005288\cos^2\theta)(1-0.315\times10^{-6}z) \tag{7.17}$$

In (7.17) $g_{\theta,z}$, is the acceleration of gravity at latitude $\varphi = 90^\circ - \theta$ and height z, which is reckoned from sea level in meters. The level-surface radius was represented in the form $R = R_0 + z$, where $R = a\,(1-\alpha\cos^2\theta)$ is the radius of the geoid, a is the equatorial radius, and α is the flattening of the Earth. Thus,

$$h_3 = \frac{2\pi\times10^3}{g_{90^\circ,0}}\int_0^{\pi}\sin^2\theta'\frac{(1-a\cos^2\theta')^3}{(1+0.005288\cos^2\theta')}\int_0^{P}\frac{(a+z)^3 u_\lambda \mathrm{d}P}{(1-0.315\times10^{-6}z)}\mathrm{d}\theta \tag{7.18}$$

Going from integration to summation, we have

$$h_3 = D\sum_{j=1}^{N}\Phi_j\sum_{k=1}^{Q}\frac{(a+z_k)^3}{(1-0.315\times10^{-6}z_k)}\Delta P_k(u_\lambda)_{kj} \tag{7.19}$$

where j and k are the zone and layer numbers, respectively, $\Delta P_k = \underline{P} - \bar{P}$, and $\underline{P}$ and $\bar{P}$ are the atmospheric pressures at the lower and upper boundaries of the

kth layer. These boundaries were placed at the 1000, 775, 600, 400, 250, 150, 75, 40 and 20 hPa levels; N and Q are the total numbers of zones and layers, $D = 2\pi \times 10^3/g_{90°,0} = 0.6424321$, $g_{90°,0} = 9.78031\,\text{m/s}^2$, and $a = 6378.14$ km. The heights z_k were put equal to 1, 3, 5.5, 9, 12, 16, 20, and 24 km, respectively.

$$\Phi_j = \int_{\theta_-}^{\theta^-} \sin^2\theta' \frac{(1-\alpha\cos^2\theta')}{(1+0.005288\cos^2\theta')} d\theta$$

$$= (0.498082\Delta\theta - 0.25 \sin 2\theta + 0.00048 \sin 4\theta)\Big|_{\underline{\theta}}^{\bar{\theta}} \qquad (7.20)$$

The zone boundaries were the latitudes ±5, ±12.5, ±17.5, ±25, ±32.5 ±37.5, ±45, ±55, ±65, ±75, and ±85°.

The results of calculations are given in Table 7.1. The mean monthly values of h_3 are given separately for: (i) the entire Earth; (ii) the Northern and (iii) the Southern Hemispheres. Values of h_3 are given separately for (a) westerly and easterly winds; (b) westerly winds only; and (c) easterly winds only. Table 7.1 gives also the harmonic-analysis parameters for all of these quantities: the means $\bar{h}_3$ for the year, the amplitudes of variations with the annual A and semiannual B periods, and the initial phases φ_1 and $2\varphi_2$ of these variations.

It is seen that the entire-atmosphere h_3 is always positive, that is, the atmosphere rotates from west to east faster than the Earth. The angular momentum h_3 is the largest ($145 \times 10^{24}\,\text{kg m}^2\,\text{s}^{-1}$) in May and December and the smallest (91×10^{24} kg $\text{m}^2\,\text{s}^{-1}$) in August. These values of the atmosphere's h_3 would be obtained in the case of a rigid-body rotation if it made one revolution with respect to the Earth's surface within 70 days in May and December and 113 days in August. The variations of h_3 are small from December through May. The annual h_3 variation for the entire atmosphere can be approximated formally by the expression

$$\begin{aligned} h_3 &= (128 + 11.6 \sin \oplus + 18.6 \cos \oplus - 8.8 \sin 2\oplus - 8.2 \cos 2\oplus) \times 10^{24} \\ &= [128 + 22\cos(\oplus - 32°) + 12\cos(2\oplus - 227°)] \times 10^{24}\ \text{kg m}^2\,\text{s}^{-1} \end{aligned} \qquad (7.21)$$

The angular momentum variations taken separately for the Northern and Southern Hemispheres have a more distinct annual period. The amplitude A of the annual variations in the Northern Hemisphere is almost twice as large as that in the Southern Hemisphere (46 as against 25) (Table 7.1). The initial phases φ_1 differ by 170° in the Northern and Southern Hemispheres, that is, the variations are out of phase and compensate each other. The total compensation does not occur because of differences in amplitudes A. The semiannual variations have almost equal initial phases in both hemispheres. As a result, even though the semiannual amplitudes are smaller than the annual amplitudes (in each hemisphere taken alone), they add for the Earth as a whole and assume the same order of magnitude as the annual variations. The semiannual variations manifest themselves most distinctly in the stratosphere of lower latitudes.

Table 7.1 The relative angular momentum h_3 of the atmosphere, $10^{24}\,\mathrm{kg\,m^2\,s^{-1}}$. $h = \bar{h} + A\cos(\oplus - \varphi_1) + B\cos 2(\oplus - \varphi_2)$.

Indicator	I	II	III	IV	V	VI	VII	VIII	IX	X	XI	XII	$\bar{h}$	A	φ_1	B	$2\varphi_2$
Westerly and easterly winds																	
Entire Earth	138	143	141	144	147	122	102	91	108	129	132	145	128	22	32	12	227
Northern Hemisphere	83	87	78	63	47	16	−6	−8	8	36	53	73	44	46	20	6	209
Southern Hemisphere	55	56	64	81	100	106	109	99	100	93	79	73	84	25	190	6	244
Westerly winds																	
Entire Earth	164	165	165	167	168	154	146	138	148	155	161	171	158	12	30	6	237
Northern Hemisphere	96	97	90	76	60	38	27	27	38	53	72	88	63	36	17	2	203
Southern Hemisphere	68	68	75	91	108	115	119	111	110	102	89	83	95	24	191	5	249
Easterly winds																	
Entire Earth	−26	−22	−24	−23	−21	−32	−43	−47	−40	−26	−29	−26	−30	10	34	6	216
Northern Hemisphere	−14	−10	−13	−13	−13	−23	−33	−35	−29	−17	−18	−16	−20	10	30	4	211
Southern Hemisphere	−12	−12	−11	−10	−8	−9	−10	−12	−11	−9	−11	−10	−10	1	165	2	232

7.3 Axial Angular Momentum (Reanalysis Data)

In the 1990s, the National Center for Environmental Prediction (NCEP) and the National Center for Atmospheric Research (NCAR) performed an important project of the reanalyses of all past observational data bases and the global analyses of the six-hourly atmospheric fields, gradually passing from the present time into the past down to 1948 including (Kalnay *et al.*, 1996). The data sets of wind and pressure fields were created for 28 isobaric surfaces up to 10 hPa (about 31 km). On the basis of these data, the AAM h_i and EFAAM components χ_i^W were calculated separately for the Northern and Southern Hemispheres (Salstein *et al.*, 1993; Salstein, 2000; Zhou, Salstein and Chen 2006). The calculations were carried out with an interval of six hours (00:00, 06:00; 12:00, and 18:00 UTC) for the entire period of records.

The projection h_3 of the atmospheric relative angular momentum onto the Earth's rotation axis is called the angular momentum of atmospheric winds. It follows from (7.11) that h_3 is determined by the zonal wind velocity. For this reason, h_3 is referred to hereafter as the angular momentum of zonal winds. Since zonal motions prevail in the atmosphere, h_3 is hundreds of times higher than the equatorial components h_1 and h_2 of the wind angular momentum.

The angular momentum h_3 is the major integral of motion of the atmosphere. It characterizes the intensity of the atmospheric zonal circulation (Sidorenkov, 1991a; Sidorenkov and Svirenko, 1991; Hide *et al.*, 1997). The higher the value of h_3, the stronger the westerly winds and the weaker the easterly winds in the atmosphere. The lower the value of h_3, the weaker the westerly winds and the stronger the easterly winds. Since the atmosphere's moment of inertia n_{33} relative to the Earth's rotation axis varies little, in the first approximation h_3 determines the angular velocity α of the atmosphere rotating as a solid body relative to the Earth's surface:

$$\alpha = h_3/n_{33} \tag{7.22}$$

The value of $n_{33} = 1.413 \times 10^{32}$ kg m^2 (Table 6.1), which is 500 thousand times less than the moment of inertia of the mantle. Therefore, by the angular momentum conservation law, the variations in α must be 500 thousand times larger than the seasonal variations in the mantle's rotation velocity ν_3.

The series of h_3 over 1948–2006 provides a unique opportunity to study temporal variations of the atmospheric zonal circulation in the range from 6 hours to 50 years.

The values of h_3 measured at a fixed observation time were averaged over a long period to obtain a long-term average of h_3 or a normal at this observation time. Performing these computations for 00:00, 06:00, 12:00, and 18:00 UTC, we produced the normals characterizing the diurnal oscillations in h_3. Next, the daily means $\bar{h}_3$ were subtracted from the observed h_3 and were additionally averaged over seasons to obtain the angular momentum deviations h'_3. Figure 7.5 shows the diurnal oscillation of h'_3 for winter, spring, summer, and fall. It can be seen that h_3 has two maxima (at 00:00 and 12:00) and two minima (at 06:00 and 18:00). The minimum at 18:00 is weak and disappears in fall. The diurnal range of h_3 is 1.1×10^{24}, that is, about 2% of the seasonal range.

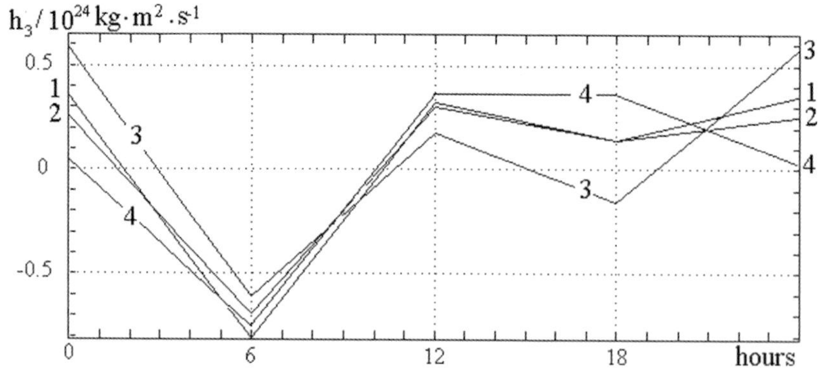

Figure 7.5 Diurnal oscillations in zonal-wind angular momentum h_3' in winter (1), spring (2), summer (3), and fall (4).

The spectrum of observed h_3 exhibits the harmonics of diurnal tides, which will be discussed below (Section 7.5). The causes of diurnal oscillations will be considered in Section 8.6. Here, we only note that at 00:00 and 12:00 UTC the Sun is over the Pacific and Atlantic oceans, respectively, and at 06:00 and 18:00 UTC – over Asia and America, respectively.

The observed values of h_3 were averaged to obtain daily, monthly, and annual means. The monthly and annual means of h_3 over 1948–2000 are presented in Sidorenkov (2002a, 2002b). Figure 7.6 shows the temporal variations in the monthly means of h_3.

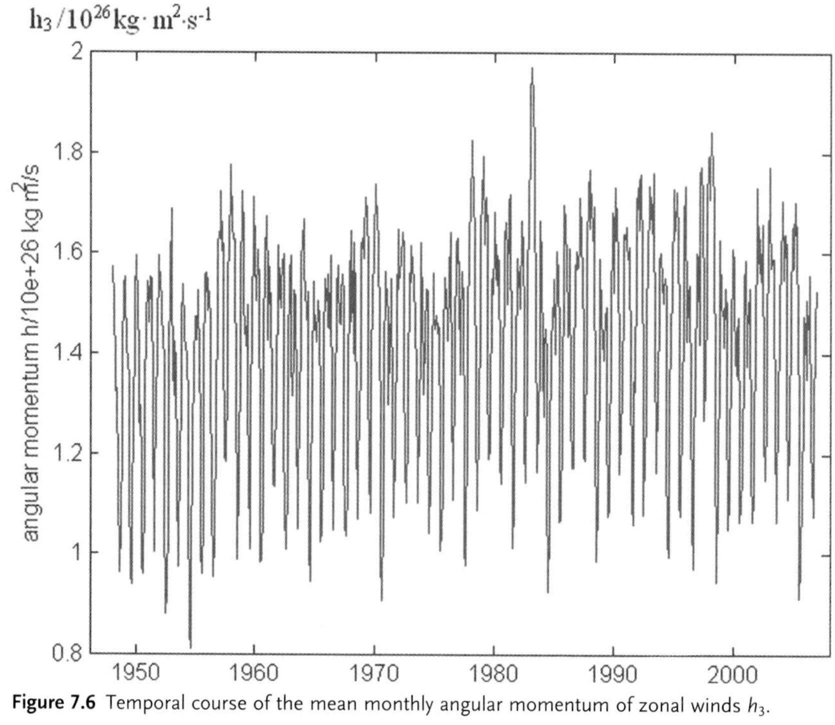

Figure 7.6 Temporal course of the mean monthly angular momentum of zonal winds h_3.

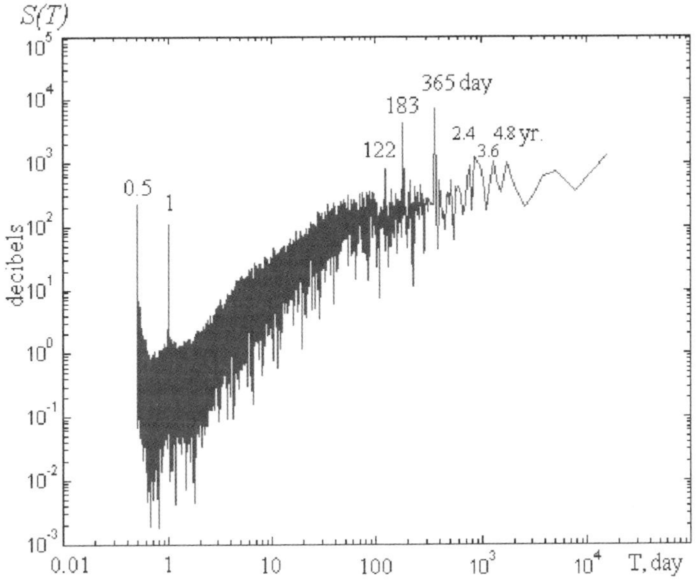

Figure 7.7 Power spectrum of the angular momentum of zonal winds h_3.

The power spectrum of this series (Figure 7.7) reveals the intense annual and semiannual harmonics, whose sum yields the seasonal variations. There are also small peaks at periods of 2.4, 3.6, and 4.8 years. It will be shown in Chapter 9 that they are associated with the interannual oscillations in the Earth–ocean–atmosphere system.

To extract the seasonal variations from the series of monthly means of h_3, their moving annual averages were subtracted from the series of h_3 and the resulting deviations were averaged over every month of the year for 1961–2006. The data for preceding years are less reliable. For this reason, they were not used in the averaging procedure. As a result, we obtained the long-term means (or the monthly normals) of h_3 (Table 7.2).

Figure 7.8 illustrates the behavior of the h_3 normals for the entire atmosphere and for the Northern and Southern Hemispheres separately. It can be seen that the angular momentum h_3 of the entire atmosphere has two maxima of 161×10^{24} (in April and December) and two minima: in July (108×10^{24}) and February (154×10^{24}). The minimum in July is much deeper than that in February. In the case of the solid-body rotation, the atmosphere would have the above values of h_3 if it made one revolution relative to the Earth's surface in time $T = 2\pi n_{33}/h_3 = 64$ days in April and December and in $T = 95$ days in July. The value of h_3 varies little from December to May. The difference between the maximum of h_3 in December (or April) and its minimum in July is $53 \times 10^{24}\,\mathrm{kg\,m^2\,s^{-1}}$.

Based on the reanalysis data over 1958–2000, the seasonal variation of h_3 in the entire atmosphere can be formally approximated by the expression

$$\begin{aligned} h_3 &= (144 + 7.8 \sin \oplus + 21.6 \cos \oplus - 9.2 \sin 2\oplus - 10.3 \cos 2\oplus) \times 10^{24} \\ &= [144 + 23 \cos(\oplus - 20^\circ) + 14 \cos(2\oplus - 222^\circ)] \times 10^{24}\,\mathrm{kg\,m^2\,s^{-1}} \end{aligned} \tag{7.23}$$

Table 7.2 Normals of the polar component h_3 and equatorial components h_1 and h_2 of the angular momentum of atmospheric winds for 1961–2000; $h = \bar{h} + A\cos(\oplus - \varphi_1) + B\cos 2(\oplus - \varphi_2)$.

Indicator	I	II	III	IV	V	VI	VII	VIII	IX	X	XI	XII	$\bar{h}$	A	φ_1	B	$2\varphi_2$
Polar component h_3, $10^{24}\,\mathrm{kg\,m^2\,s^{-1}}$																	
Entire Earth	157	154	157	161	156	132	109	109	125	146	160	161	144	23	20	14	222
Northern Hemisphere	105	105	96	78	50	16	−9	−6	15	46	74	94	56	57	16	7	206
Southern Hemisphere	52	49	60	83	107	116	117	115	110	100	86	67	88	34	194	7	240
Equatorial component h_1, $10^{21}\,\mathrm{kg\,m^2\,s^{-1}}$																	
Entire Earth	454	396	306	290	158	−112	−140	−44	26	144	306	406	182	278	7	34	204
Northern Hemisphere													−286				
Southern Hemisphere													468				
Equatorial component h_2, $10^{21}\,\mathrm{kg\,m^2\,s^{-1}}$																	
Entire Earth	−16	−7	81	28	−48	−17	106	42	−97	−86	−148	−9	−14	54	106	55	32
Northern Hemisphere													−144				
Southern Hemisphere													130				

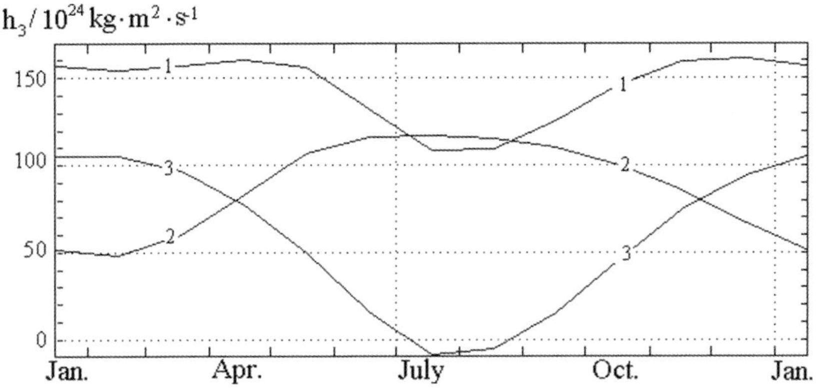

Figure 7.8 Seasonal variations in the angular momentum of zonal winds h_3. Atmosphere as a whole (1); Northern Hemisphere (2); Southern Hemisphere (3).

A comparison of (7.23) with (7.21) shows their good agreement. Slight differences in the annual mean values and phases can possibly be associated with the upper 20- and 10-hPa levels added to the reanalysis data. Actual climate changes could also have some effect.

The annual means of h_3 over the Northern and Southern Hemispheres are 56×10^{24} and $88 \times 10^{24}\,\mathrm{kg\,m^2\,s^{-1}}$, respectively. This means that the annual mean intensity of the zonal circulation in the Northern Hemisphere is 36% less than that in the Southern Hemisphere. However, the annual amplitude of h_3 in the Northern Hemisphere is 67% larger than that in the Southern Hemisphere. Specifically, it is $57 \times 10^{24}\,\mathrm{kg\,m^2\,s^{-1}}$ in the Northern Hemisphere and only 34×10^{24} in the Southern Hemisphere. The cause of this difference is that the Northern Hemisphere embraces many more continental areas than the Southern Hemisphere does. The initial phases φ_1 in the Northern and Southern Hemispheres differ by 178°, that is, the oscillations are opposite in phase and appreciably compensate each other. However, the compensation is not complete because of the difference in the annual amplitudes. The semiannual oscillations in both hemispheres have almost identical initial phases. Therefore, although their amplitudes (in each hemispheres) are much smaller than the annual amplitude, they are added for the entire globe, to become of the same order as the annual amplitudes. The semiannual oscillations are most clearly pronounced in the low-latitude stratosphere.

A remarkable feature of h_3 is that its value is not zero on average but is $+143.9 \times 10^{24}\,\mathrm{kg\,m^2\,s^{-1}}$. This means that the atmosphere as a whole rotates from west to east faster than the Earth. Specifically, while the former makes 71 revolutions, the latter makes 70 revolutions around its axis. This phenomenon is known as the super-rotation of the atmosphere. Its nature is discussed in Chapter 8. Of course, different parts of the atmosphere rotate in a different manner. The atmosphere rotates faster than the Earth in the middle and subtropical latitudes and slower – in the equatorial zone. The period of revolution of the atmosphere relative to the Earth's surface is 58 and 92 days in the Southern and Northern Hemispheres, respectively. In the jet-stream areas, the atmosphere can rotate around the Earth in less than 10 days.

7.4
Estimations of Seasonal Variations in the Earth's Rotation

Given the variations with time in the angular momentum of the atmosphere's zonal wind h_3, it is easy to calculate the variations in the velocity of the Earth's rotation. Indeed, as it follows from (4.26) and (4.33),

$$\frac{dv_3}{dt} = -\frac{d\chi_3}{dt} \tag{7.24}$$

Upon integrating (7.24) from the point of time t_0 to the point of time t we obtain:

$$v_3(t) - v_3(t_0) = -\chi_3(t) + \chi_3(t_0) \tag{7.25}$$

However, as follows from (4.12) and (7.4), $v_3 = \frac{\omega_3 - \Omega}{\Omega}$ and $\chi_3 = \frac{h_3}{C\Omega}$. On this basis, equality (7.23) can be rewritten in the form:

$$\frac{\omega_3(t) - \omega_3(t_0)}{\Omega} = \frac{\delta\omega}{\Omega} \equiv v'_3 = -\frac{h_3(t) - h_3(t_0)}{C\Omega} = -\frac{\delta h_3}{C\Omega} \tag{7.26}$$

Since the seasonal variations in the angular momentum of zonal winds h_3 are described by expression (7.23), the Earth's rotation seasonal variation that is caused by these variations is expressed as:

$$\begin{aligned} v'_3 \cdot 10^{10} &= -15.2 \sin \oplus - 42.1 \cos \oplus + 17.9 \sin 2\oplus + 20.1 \cos 2\oplus \\ &= 44.8 \cos(\oplus - 200^\circ) + 26.9 \cos(2\oplus - 42^\circ) \end{aligned} \tag{7.27}$$

This is close to the approximation of the observed nontidal irregularity of the Earth's rotation. Also, the variations in the Earth's rotational velocity v'_3 can be calculated immediately by the h_3 values presented in the first row of Table 7.2. The seasonal course of the v'_3 values calculated in such a way is presented in Figure 7.9 (curve 2), whereas curve 1 shows the seasonal variations in the angular

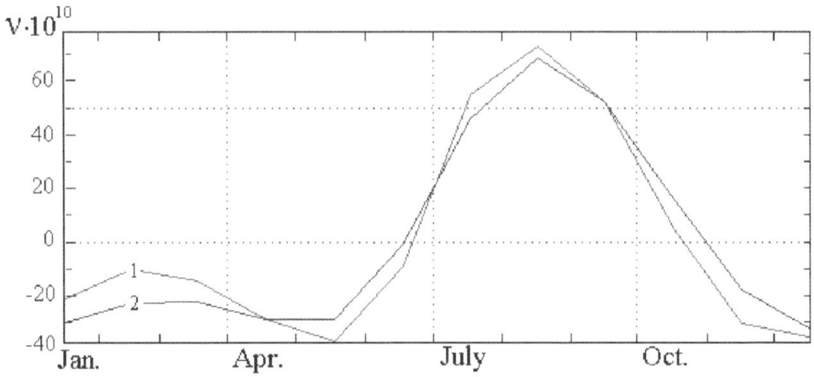

Figure 7.9 Seasonal variations in the velocity of the Earth's rotation. By astronomical observations (1); calculated by the h_3 values of the angular momentum of zonal winds (2).

velocity, which are synchronically observed by astronomical methods and are averaged over 1962–2000. The tidal oscillations with periods $\leq$1 year are eliminated from these data. The two curves show a good agreement. Small discrepancies between them may be due to the errors of the obtained values, so to the effect of other factors (such as the air and water masses redistribution).

In Figure 7.10, the series of the mean monthly values of h_3 for 1948–2000 is synchronously compared with the series of the Earth's angular momentum taken with the opposite sign – $-C\Omega\nu'_3$, which is calculated by the seasonal variations ν'_3 of the Earth's rotational velocity (see formula (7.26)). Both curves closely coincide; the coefficient of correlation $r = +0.84$.

Another conclusive piece of evidence of the functional relationship between the seasonal variations of the Earth' rotation ν'_3 and the variations in the angular momentum of zonal winds is the consistency of temporal variations in the amplitudes of the annual and semiannual harmonics of these processes. We have performed the necessary harmonic analysis, using the mean monthly values of h_3 for calculating the series of deviations of h'_3 from their moving mean annual values h_3 and the amplitudes and phases of the annual, semiannual, and quarterly harmonics. For this purpose, the system of 12 conditional equations with 6 unknowns has been solved, using the fixed sequence of 12 mean monthly values of h'_3:

$$\begin{aligned} h'_{n,i} &= a_n \sin\Theta_i + b_n \cos\Theta_i + c_n \sin 2\Theta_i + d_n \cos 2\Theta_i + e_n \sin 4\Theta_i + f_n \cos 4\Theta_i \\ &= A_n \sin(\Theta_i - \varphi_{1,n}) + B_n \sin 2(\Theta_i - \varphi_{2,n}) + C_n \sin 4(\Theta_i - \varphi_{4,n}) \end{aligned} \tag{7.28}$$

Here, $n = 1, 2, 3, \ldots, N$ is the number of the sequence of 12 mean monthly values of h'_3 ($n = 1$ for January–December of 1962, $n = 2$ for February 1962–January 1963, and so on); $i = 1, 2, 3, \ldots 12$ is the number of a month in the fixed sequence; A, B, C, and φ_1, φ_2, φ_4 are the unknown amplitudes and initial phases of the annual, semiannual, and quarterly harmonics, respectively. The solution of only one such system (7.28) allows one to find the unknown coefficients a, b, c, d, e, f and their root-mean-square errors; further simple calculations provide the required amplitudes A, B, C and their errors. Thus, shifting the window from month to month along the whole series of h'_3 for 1962–2000, we have found the amplitudes of the annual (A), semiannual (B), and quarterly (C) harmonics for each month of the 50-year period under study. Notice that the B and C values characterize the changes in amplitudes only from year to year.

Figures 7.11 and 7.12 show the variations in the amplitudes of the annual and semiannual harmonics of the angular momentum of zonal winds h'_3 for 1962–2000. Thc quarterly amplitudes of harmonic C were small and are not considered here. As is seen from Figure 7.11, the amplitude of the annual harmonic A varies over wide limits (from 13×10^{24} to $34 \times 10^{24}\,\mathrm{kg\,m^2\,s^{-1}}$, the root-mean-square error varying from $\pm 2 \times 10^{24}$ to $\pm 8 \times 10^{24}$). The irregular 6–7-year variations are traced. The high A amplitudes were observed near 1958, 1964, 1970, 1977, 1983, 1988, 1992, 1995, and

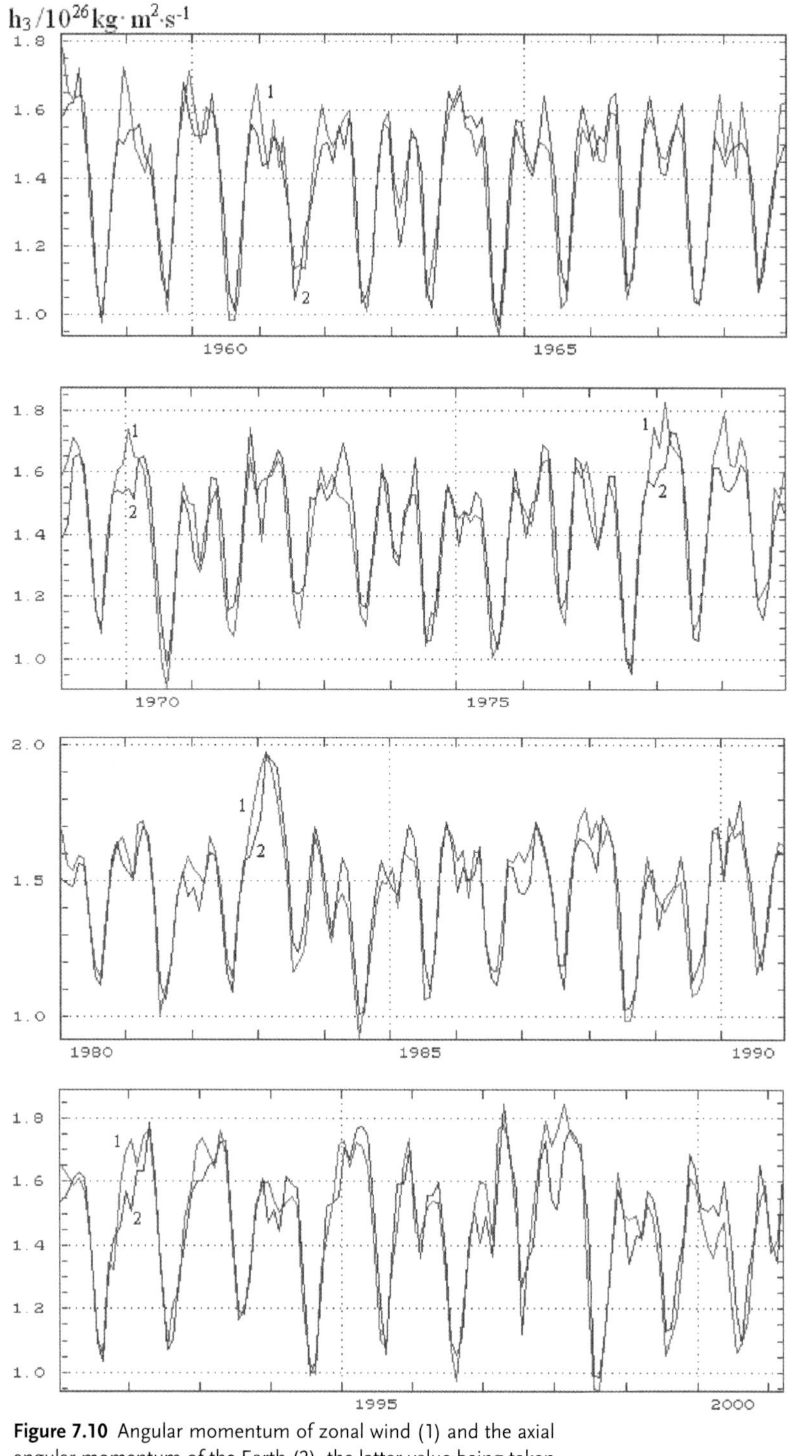

Figure 7.10 Angular momentum of zonal wind (1) and the axial angular momentum of the Earth (2), the latter value being taken with opposite sign.

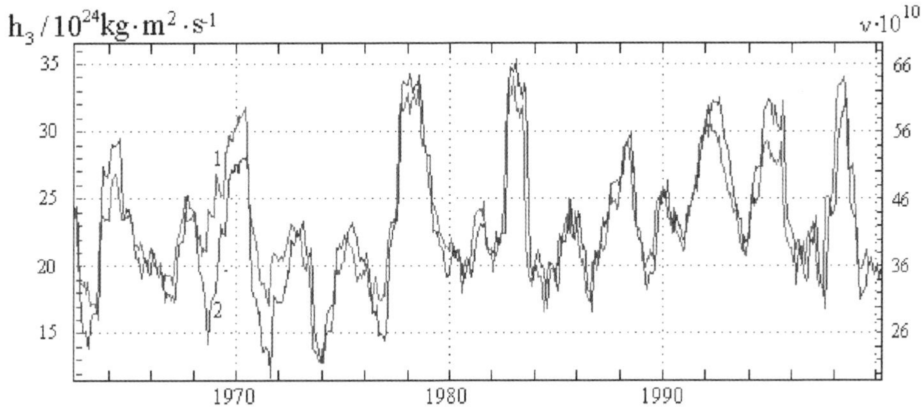

Figure 7.11 Synchronous course of amplitude A of the angular momentum of the annual harmonics of zonal winds h_3 (1) and the Earth's rotation velocity v_3 (2) for 1962–1999.

1998; the low ones – near 1957, 1962, 1966, 1973, 1980, 1986, 1989, 1993, 1996, and 1999.

Amplitude *A* varies inversely to SOI – the index of the Southern Oscillation (see Chapter 9). The coefficient of their correlation is maximum ($r = -0.52$) at the one-month lag of SOI. During the warm phases of ENSO, when SOI < 0, amplitudes *A* increase; during the cold phases of ENSO, when SOI > 0, they decrease.

The amplitude *B* of the semiannual harmonic of the moment h'_3 varies from 4×10^{24} to $22 \times 10^{24}\,\text{kg}\,\text{m}^2\,\text{s}^{-1}$, its root-mean-square error being the same as that for amplitude *A*. One can trace the 2-, 3-year cyclic recurrence in *B* variations. As will be clear from the following, this cyclicity reflects the quasibiennial oscillations (QBO) in the atmospheric circulation. The *B* high amplitudes were observed in 1962, 1966, 1970, 1974, 1977, 1981, 1984, 1985, 1988, 1995, and 1998; the low ones – in 1963,

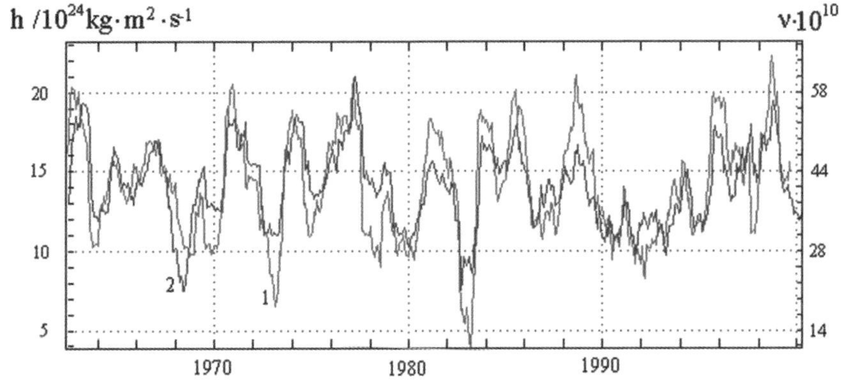

Figure 7.12 Synchronous course of amplitude *B* of the angular momentum semiannual harmonics of zonal winds h_3 (1) and the Earth's rotation velocity v_3 (2) for 1962–1999.

1968, 1969, 1973, 1975, 1978, 1979, 1983, 1986, 1991, 1994, 1997, and 1999. Amplitude A correlates inversely with amplitude B ($r=-0.4$).

The correlation of amplitude B of the semiannual harmonic of the moment h_3' with the index of SOI is in general positive (but is less close than the negative correlation of A with SOI), being the highest ($r=0.4$) at the three-month lag of SOI.

Unlike the amplitudes A and B, the phases of both harmonics vary without any regularity.

The comparison of variations in amplitude A (of the annual harmonic of the angular momentum of zonal winds h_3') with the seasonal variation of the Earth's rotational velocity ν_3' (Figure 7.11, curves 1 and 2, respectively) shows a nearly total coincidence of two curves. Their discrepancies, decreasing with time, reflect a gradual increase in the accuracy and volume of meteorological and astronomical observations. The variations in B amplitudes for h_3' and ν_3' (Figure 7.12, curves 1 and 2, respectively) are also in good agreement. Their comparison indicates that the interannual variations in the parameters of seasonal variations in the Earth's rotational velocity are caused by the instability of the angular momentum of zonal wind; this instability, in its turn, is due to the interannual variability of processes occurring in the climatic system (the El-Nino – Southern Oscillation, QBO and semiannual wind oscillation in the stratosphere, and so on).

Thus, the seasonal variation of the Earth's rotation is mainly due to the seasonal variations in the angular momentum h_3 of atmospheric zonal winds. When the angular momentum h_3 increases, the velocity of the Earth's rotation decreases, and vice versa. In other words, the angular momentum redistributes between the Earth's body and the atmosphere. With this, the total angular momentum of the Earth–atmosphere system remains unchanged. This conclusion is a good example showing that the laws of conservation can be valid not only for experiments in physical laboratories but also for global processes.

As is already noticed in Section 7.3, the angular momentum of zonal winds h_3 is not zero (on average over a year) but makes $144 \times 10^{31}\,\mathrm{kg\,m^2\,s^{-1}}$; the superrotation of the atmosphere is observed. The atmosphere as a whole outruns the Earth; it makes one extra revolution over $2\pi\, n_{33}/h_3 \approx 70$ days. The angular velocity of rotation of the Earth's atmosphere with respect to the Earth's surface is on average $\alpha \approx 1 \times 10^{-6}\,\mathrm{s^{-1}}$; the linear velocity of its rotation at the equator is $6.5\,\mathrm{m\,s^{-1}}$.

Certain researchers use this empirical fact as a basis for the criticism of the existing ideas about the nature of the seasonal variation of the Earth's rotation. The objection is as follows. Since the atmosphere rotates faster than the Earth's body, it must accelerate the Earth's rotation. Because observations do not reveal this acceleration, the conclusion is made that the existing estimates of the effect of winds on the Earth's rotation are erroneous. However, as is seen from Equation 7.26, the constant value of the angular momentum of zonal winds h_3 cannot cause the variation in the Earth's rotation. The latter originates only with a change in the h_3 value. As will be shown in Chapter 8, the constant value h_3 was borrowed by the atmosphere from the Earth at the moment of formation of the zonal circulation. Then, the velocity of the Earth's rotation slowed down by $\delta\nu_3 = 144 \times 10^{24}/513 \times 10^{31} = 28 \times 10^{-9}$ and remains the same at present. In the case of the complete attenuation of zonal circulation, the velocity of the Earth's rotation will take its initial value, that is, it will increase by 28×10^{-9}.

7.5 Equatorial Angular Momentum of Atmospheric Winds

In addition to the polar component h_3, the atmosphere's angular momentum has two equatorial components h_1 and h_2. The component h_1 characterizes the rotation of the atmosphere about the equatorial axis Ox_1 passing through the Earth's center in the direction of the Greenwich Meridian, while h_2 characterizes the rotation of the atmosphere about the equatorial axis Ox_2 passing through the Earth's center in the direction of the 90°E meridian. The values of h_1 and h_2 are either positive or negative for the atmosphere rotating either counterclockwise or clockwise, respectively (when viewed from the end of the Ox_1 and Ox_2 axes, respectively, outside the atmosphere). The equatorial components h_1 and h_2 characterize the meridional motions of the atmosphere.

To analyze temporal variations in the atmospheric equatorial angular momentum $\boldsymbol{h}_e = h_1\boldsymbol{e}_1 + h_2\boldsymbol{e}_2$, we used the series of h_1 and h_2 calculated by David Salstein from the reanalysis starting in 1948 and up to the present time (Salstein *et al.*, 1993; Salstein, 2000; Zhou, Salstein and Chen, 2006).

The monthly normals of the components h_1 and h_2 show that the amplitudes of seasonal variations in the equatorial angular momentum components are very small (Table 7.2). The amplitudes of the diurnal oscillations of h_1 and h_2 are a few tens of times larger than the amplitudes of the seasonal variations.

First, h_1 and h_2 were averaged over each observation time for all days of the year. The resulting normals of h_1 and h_2 are shown in Figure 7.13. It can be seen that the diurnal oscillations of h_1 and h_2 are complicated and amplitude modulated (slowly varying). They have two crests near the summer and winter solstices and two nodes at the beginning of March and October. Their amplitude in June is nearly twice as large as that in December.

Second, we average h_1 and h_2 over each calendar month and analyze the diurnal oscillation of h_1 and h_2 from month to month. It can be seen that the diurnal oscillation of h_1 and h_2 varies widely during the year (Figure 7.14). The largest amplitudes of diurnal oscillations occur in June and December. The variation phases in these months are opposite. In June, the maximum of h_1 is observed at 06:00, while in December at 18:00. The minimums of h_1 in June and December occurs at 18:00 and 06:00, respectively. The maximums of h_2 in June and December occur at 00:00 and 12:00, respectively. The minimum of h_2 in June is observed at 12:00 and in December at 00:00. In other months the amplitudes of the diurnal oscillations of both, h_1 and h_2, are smaller than in June and December. Therefore, their curves lie inside the envelopes for June and December. Only the even months of a year are shown in Figure 7.14 to simplify the plot. The extreme values of the diurnal oscillation of h_2 occur six hours earlier than those of h_1. This corresponds to a phase shift equal to 90°. From October to February, the phases of both components, h_1 and h_2,

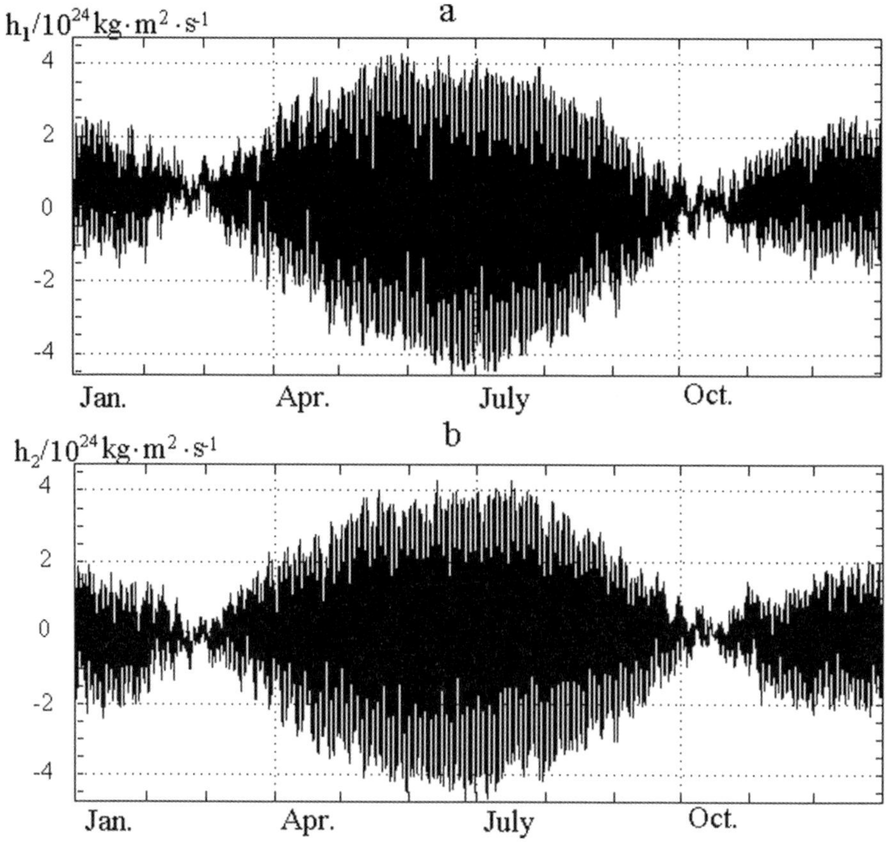

Figure 7.13 Annual course of the equatorial components h_1 (a) and h_2 (b) of the atmospheric angular momentum (the long term means at observation hours).

correspond to those in December, and from March to September – to those in June. This means that the nodes in Figure 7.13 correspond to the changes in the phase of diurnal oscillations. In early March and early October, the components h_1 and h_2 at the nodal points have only a weak semidiurnal oscillations.

The synthesis of the plots of the diurnal oscillations of the equatorial components h_1 and h_2 leads to the conclusion that the vector $\boldsymbol{h}_e = h_1 e_1 + h_2 e_2$, formed by components h_1 and h_2, rotates in the equatorial plane from east to west with a nearly diurnal period (Figure 7.15). Vector $\boldsymbol{h}_e$ describes a trajectory close to an ellipse, one of the foci of which coincides with the Earth's center. In June, vector $\boldsymbol{h}_e$ describes a trajectory corresponding to the outer major ellipse and in December, trajectory corresponding to the inner minor ellipse. The positions of the vector $\boldsymbol{h}_e$ at 00:00, 06:00, 12:00, and 18:00 are indicated next to the ellipses (Figure 7.15). It can be seen that in June vector $\boldsymbol{h}_e$ is constantly deflected about 100° to the west of the meridian at which the Sun is located; in December it is located 80° east of the respective meridian. The positive direction of $\boldsymbol{h}_e$ indicates that the atmosphere rotates counterclockwise in

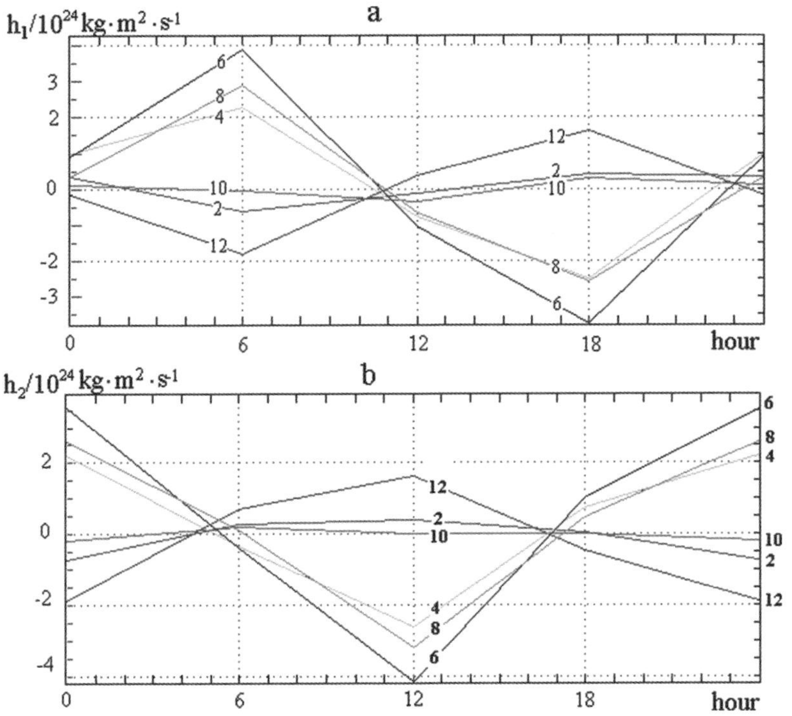

Figure 7.14 Diurnal course of equatorial components h_1 (a) and h_2 (b) of the atmospheric angular momentum in different months. Numerals of lines denote months.

the plane of the meridian perpendicular to the direction of $\boldsymbol{h}_e$. That is, when viewed from the North Pole, in June (more exactly, from March through September), the atmosphere rotates from the daytime to the nighttime side. This rotating wind system continuously follows the Sun. The circulation increases at 06:00 and 18:00, when the Sun is located at the meridians 80°E (over Asia) and 260°E (over America), and decreases at 00:00 and 12:00, when the Sun crosses the meridians 170°E (Pacific Ocean) and 350°E (Atlantic Ocean). Because of this the module of vector $\boldsymbol{h}_e$ has the semidiurnal oscillations.

In December (more exactly, from October through February), the atmosphere rotates in the opposite direction. Yet, the wind system still follows the Sun and the orientation of the ellipse axes remains nearly unchanged. The angular velocity of the atmospheric rotation in June is more than twice as high as that in December.

Therefore, the diurnal oscillations of h_1 and h_2 are modulated in amplitude, frequency and phase. Recall that the parameters of modulated oscillations (amplitude, frequency, and phase) are also the functions of time but vary more slowly than the original oscillation (carrier signal). For example, the carrier frequency of a radio transmitter is of the order of several million hertz, while the frequency of the variation in the amplitude of radio waves is equal to the frequency of sound waves (from 16 to 20 000 hertz).

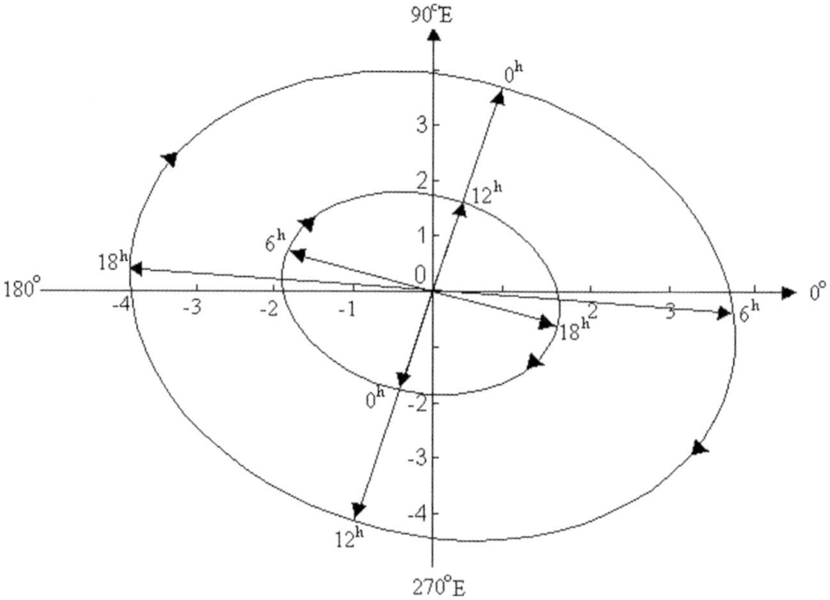

Figure 7.15 Diurnal rotation of the equatorial vector $\boldsymbol{h}_e = h_1\boldsymbol{e}_1 + h_2\boldsymbol{e}_2$ of the atmospheric angular momentum in June (outer ellipse) and December (inner ellipse).

In the case of oscillations in the atmospheric angular momentum, the carrier oscillation is the diurnal one, which is ultimately caused by day and night cycles due to the Earth's rotation. It is modulated by the annual and monthly harmonics and superharmonics due to the solar and lunar forcing, respectively. Signals at these frequencies can be treated as a kind of ultralow frequency sound transmitted by the atmosphere at the diurnal frequency.

Recall that an amplitude-modulated oscillation can be described by the model (Zernov, Karpov and Smirnov, 1972)

$$h = A\left[1+\sum_{i=1}^{N} m_i \cos(\Omega_i t + \Phi_i)\right]\cos(\omega t + \varphi) = A\cos(\omega t + \varphi) + \sum_{i=1}^{N}\frac{Am_i}{2}\cos[(\omega + \Omega_i)t + \varphi + \Phi_i] + \sum_{i=1}^{N}\frac{Am_i}{2}\cos[(\omega - \Omega_i)t + \varphi + \Phi_i] \quad (7.29)$$

Here, h is the angular momentum; A, ω, and φ are the amplitude, circular frequency, and initial phase of the carrier (diurnal) oscillation; m is the modulation depth; Ω and Φ are the frequency and phase of the amplitude modulation of the carrier oscillation; t is time; and i is the index of a harmonic.

Let us consider in more detail the structure of oscillations in model (7.29). The first term on its right-hand side is the carrier oscillation. The second term describes the harmonic components with frequencies $\omega + \Omega_i$. They are called the upper side frequency band. The third term describes the harmonic components with frequencies $\omega - \Omega_i$. They are called the lower side frequency band. The amplitudes of the side components are $Am_i/2$.

In the case of frequency modulation, the oscillations can be described by the model (Zernov, Karpov and Smirnov, 1972)

$$h = A[1 + m\cos(\Omega t + \Delta\Omega \sin \nu t)]\cos \omega t = A[1 + mJ_0(\Delta\Omega)\cos \Omega t + m\sum_{n=1}^{\infty} J_n(\Delta\Omega)\cos(\Omega + n\nu)\, t + m\sum_{n=1}^{\infty}(-1)^n J_n(\Delta\Omega)\cos(\Omega - n\nu)\, t]\cos \omega t \quad (7.30)$$

Here, $\Delta\Omega$ and ν are the amplitude and circular frequency of variations in the modulation frequency Ω, m is the modulation depth, and $J_n(\Delta\Omega)$ is the nth-order Bessel function of argument $\Delta\Omega$. The rest notations are the same as in (7.29). The spectrum of frequency-modulated oscillations (even if the modulation signal is harmonic) consists of an infinite number of the side components arranged symmetrically about ω at distances that are the multiples of ν. The amplitudes A_n of the side components are expressed in terms of the nth-order Bessel function of the first kind: $A_n = Am|J_n(\Delta\Omega)|$.

The power spectrum S of the complex series $h_1 + ih_2$ is calculated using Ch. Bizouard's software program designed for computing the complex fast Fourier transform. As data, we used 61 360 values of h_1 and h_2 measured at 00:06 over 1958–1999. The resulting spectrum S is shown in Figure 7.16, where the vertical axis represents the spectral density on a logarithmic scale in decibels, and the horizontal axis represents the frequencies in the cycles over the mean solar day.

Figure 7.16(a) reveals an intense peak at a frequency of −1 cycle per day (cpd). To the right of it, there is a single peak at −0.9295 cpd, which corresponds to the principal lunar wave O_1 in the expansion of the tidal potential (see Section 5.2). High

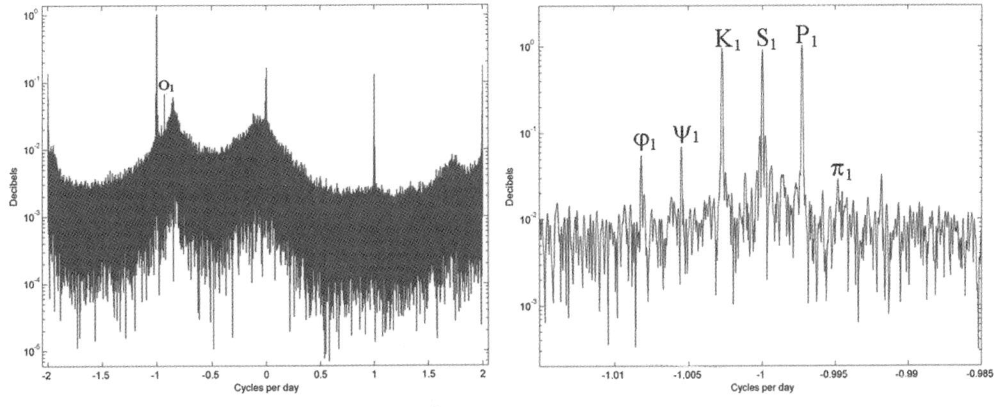

Figure 7.16 The power spectrum of the equatorial components $h_1 + ih_2$ of the atmospheric angular momentum (a) and its retrograde diurnal band (b).

peaks can be seen at low (≈0 cpd) and high frequencies (±2 cpd). The peak at positive frequencies (+ 1 cpd) is many times lower than that at −1 cpd. A remarkable feature is that S has an intense wide maximum at about −0.85 cpd. Minima of the spectral density are observed at about 0.5 and ±1.5 cpd. Interestingly, near the positive frequency + 1.7 cpd there is a weak maximum resembling in shape that observed at −0.85 cpd.

Let us analyze the fine structure of the most interesting regions of S by increasing its resolution. First, we inspect the frequency range from −0.98 to −1.02 cpd, which is shown in Figure 7.16b). Here, the central peak of 41×10^3 at the frequency −1 cpd corresponds to the well-known thermally driven diurnal tidal wave S_1 (Chapman and Lindzen, 1970; Zharov, 1996a, 1997a). The carrier wave S_1 is surrounded by three pairs of side lines arranged symmetrically at equal distances from S_1.

The first pair is composed of the line (on the left) at a frequency of −0.9973 cpd, which is identified with the principal solar wave P_1, and the line (on the right) at −1.0027 cpd, which corresponds to the lunar-solar declination wave K_1. The composition of S_1 with P_1 and K_1 induces the modulation (a slow change) with an annual period of the diurnal amplitude of components h_1 and h_2.

The second pair is composed by the line (on the left) at −0.9945 cpd (identified with the solar elliptic wave π_1) and the line (on the right) at −1.0055 cpd (identified with the solar elliptic wave ψ_1)(Table 5.2). The composition of S_1 with π_1 and ψ_1 gives an amplitude modulation, with a semiannual period, of the diurnal oscillations of components h_1 and h_2.

The third pair is composed of the left line at −0.9918 cpd and the right line at −1.0082 cpd, which is identified with the solar declination wave φ_1. Composition of S_1 with the third wave pair gives an amplitude modulation of diurnal oscillations of h_1 and h_2 with a terannual period (one-third of the year).

In the low-frequency range (≈0 $(\text{day})^{-1}$), we can see a peak of 11×10^3 at the annual frequency and a hardly noticeable peak at the semiannual frequency. No peaks are observed in the frequency range of long-period tides.

In the range of semidiurnal oscillations, there is a peak of 5×10^3 at 1.9973 cpd, which corresponds to the large solar elliptic wave T_2. Another small peak at 1.9945 cpd is not identified with tidal waves.

The most striking detail of the total spectrum of h_1 and h_2 (Figure 7.16a) is a blurred maximum of the spectral density at −0.85 cpd. Its height is indicative of a high power of h_1 and h_2, and the width shows considerable fluctuations of the period. What lies behind this phenomenon and why does the atmospheric circulation produce strong noise in this frequency range?

In papers on the atmospheric tides there is an indication on the existence of the normal antisymmetric mode ψ_1^1 (Yanai wave) with the azimuth wave number $s = 1$ and with a period of 28.94 hours (the frequency −0.83 cpd) (Volland, 1988; Hamilton and Garsia, 1986; Eubanks, 1993; Brzezinski, Bizouard and Petrov, 2002). This mode ψ_1^1 corresponds to the retrograde wave moving in the atmosphere from east to west. From this point of view, the small maximum of a similar shape near the double frequency +1.7 cpd (Figure 7.16a)) can be explained by the presence of the antisymmetric mode ξ_2^1 with a period of 0.6 day.

However, weather forecasters dealing with atmospheric disturbances and waves on an everyday basis observe only the large-scale Rossby waves and the synoptic vortices that move eastward rather than westward. Reflecting on this contradiction leads us to a fundamentally new explanation of the $\boldsymbol{h}_e$ spectrum, which is as follows. The series $h1 + ih2$ is calculated from the meteorological elements observed at fixed times UTC (which are multiples of 6 h). Time is measured relative to fixed stars with instruments located on the Earth's surface. The instrument rotating together with the Earth plays the role of a clock hand, while the stars serve as the clock face. The star time measured is used to calculate the mean solar time UTC. Thus, the meteorological measurements involve both the frequencies f_i of the moving atmospheric disturbances and the frequency ω of the Sun's apparent rotation in the sky. The oscillation frequencies ν_i of the components of $h1 + ih2$ are the difference between f_i and the Sun's circular frequency $\omega = 1$ cpd: $\nu_i = f_i - 1$ cpd. A researcher, who tries to interpret the observations, locates on the Earth. He does not perceive the diurnal rotation of the Earth and analyzes all atmospheric disturbances relative to the fixed objects on the Earth's surface (using synoptic maps). The Sun's apparent rotation is excluded from consideration, and the disturbances have the frequency $f_i = \omega + \nu_i$. If the observed frequency is $\nu = -0.85$ cpd, the frequency of a disturbance wave is $f = 1 - 0.85 = +0.15$ cpd. Accordingly, the disturbance period is about 7 days. These disturbances are not the Yanai waves but rather the Rossby waves propagating eastward together with the synoptic structures. They are especially pronounced in the time–longitude sections of the atmosphere and ocean. As an example, Figure 7.17 shows such a section for the daily 500-hPa height anomalies at latitude 40° S in November 2004. Time (days) is plotted on the vertical axis and the longitudes on the horizontal axis. Positive and negative anomalies are shown by solid and dotted lines, respectively. The extended downward inclined crests (positive anomalies) and troughs (negative anomalies) can be clearly seen in Figure 7.17. The trajectory slope indicates that the anomalies move eastward with time.

How can the complications introduced into the spectra of geophysical characteristics by the Earth's rotation be avoided? For the unknown disturbance waves to be reliably detected in any time series, the diurnal frequency ω has to be eliminated, that is, the oscillations of $h1 + ih2$ have to be demodulated. For this purpose, we used observations made strictly at daily intervals, that is, only the envelope amplitude of the carrier frequency was analyzed. Therefore, the diurnal frequency vanished, while all the frequencies lower than $\frac{1}{2}$ cpd persisted. The spectra of $h1 + ih2$ from 1948 to 2006 were calculated for each fixed observation time separately (that is, with a resolution of 24 hours). All the spectra were found to be similar. For this reason, we present only the spectrum of $h1 + ih2$ at 12:00 UTC (Figure 7.18). Unfortunately, the spectrum of the series measured at a fixed observation time no longer contains the diurnal or within-day component. It can be seen that the spectrum has changed considerably. Specifically, at the negative frequencies, no O_1 or Yanai ψ_1^1 waves are present any longer. They have transformed into the fortnightly (13.7 days) and the quasiweekly (7.8 days) lunar waves, respectively. Let us note that if h_1 and h_2 were daily averaged, we would obtain the spectrum of the white noise (Figure 7.19). In addition, the total

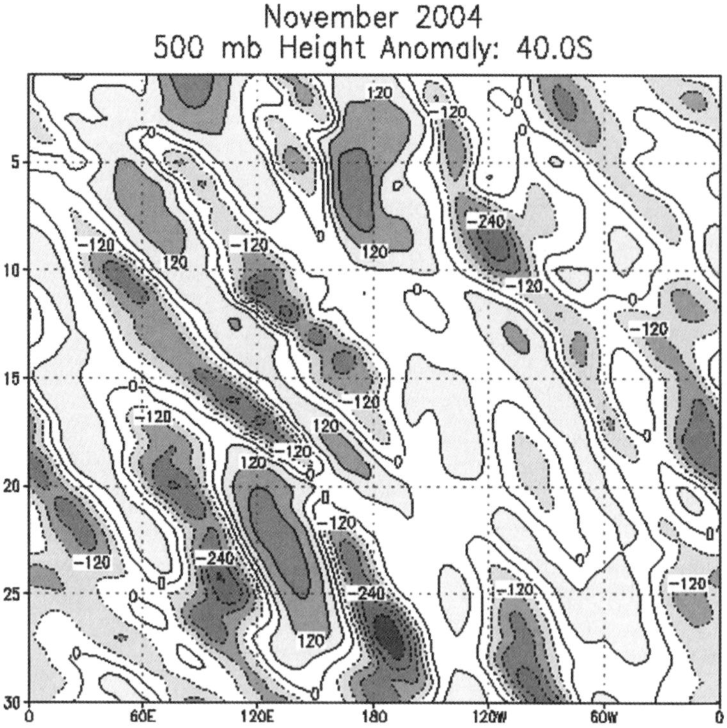

Figure 7.17 Time–longitude section of the 500-hPa height anomalies at 40°S in November 2004 (Climate Diagnostics Bulletin, No. 04/11, CPC).

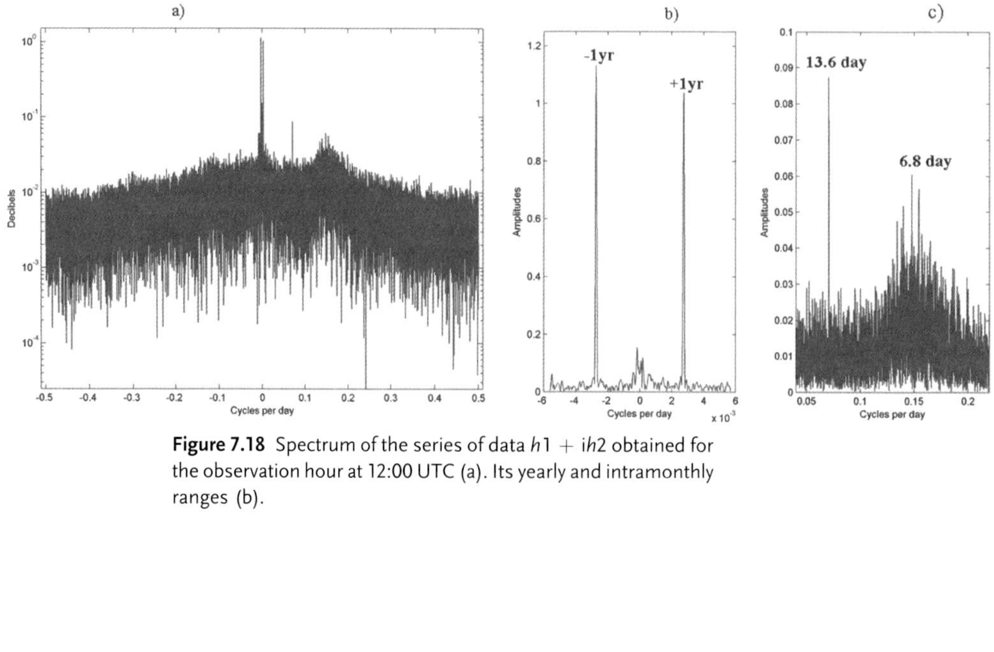

Figure 7.18 Spectrum of the series of data $h1 + ih2$ obtained for the observation hour at 12:00 UTC (a). Its yearly and intramonthly ranges (b).

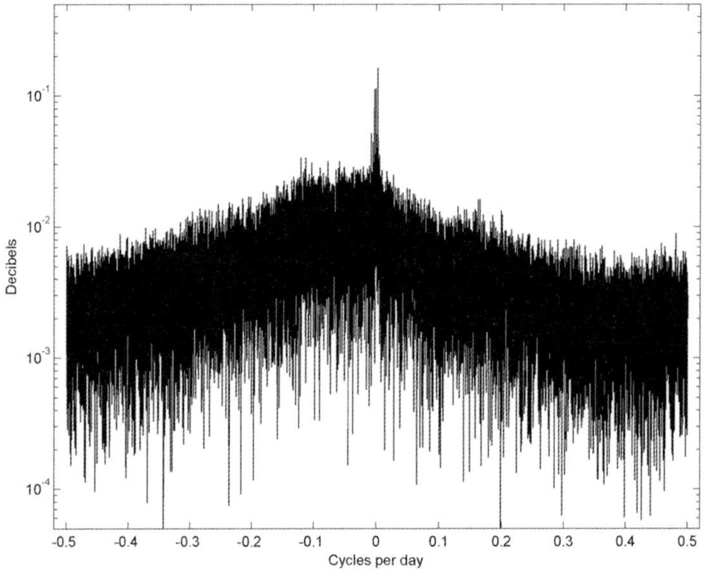

Figure 7.19 Spectrum of the series of the mean daily $h1 + ih2$.

power of spectrum in Figure 7.19 is almost ten times as small as the total power of spectrum in Figure 7.18.

The standing disturbances in vector $h1 + ih2$ can be demodulated by analyzing the series of its magnitudes (Sidorenkov, 2003b). The fact is that the phase of the daily oscillations of component h_1 lags behind the phase h_2 by 6 hours, or 90°. This means that in conformity with model (7.29), we have:

$$h_1 = A(1 + m \cos \Omega t)\cos \omega t \tag{7.31}$$

$$h_2 = A(1 + m \cos \Omega t)\sin \omega t \tag{7.32}$$

By computing the vector $\boldsymbol{h}_e$ magnitude, we obtain:

$$h_e = \sqrt{h_1^2 + h_2^2} = A\,(1 + m \cos \Omega\, t) \tag{7.33}$$

that is the module $\boldsymbol{h}_e$ has a different spectrum from the one the components h_1 and h_2 have. It does not have the carrier and side frequencies, but there is the modulation frequency Ω.

Having done the necessary calculations, we obtained the time series of the vector magnitude $h_e = \sqrt{h_1^2 + h_2^2}$ for 1958–2000 with a 6-hour step. The power spectrum of the obtained series h_e is given in Figure 7.20. We can see three powerful peaks at low frequencies: $-1/365$, $-1/183$ and $-1/122$ cpd. They represent the annual, semi-annual, and terannual frequencies, respectively. They are the basic frequencies of modulation of the diurnal thermal atmospheric tide S_1.

In the total power spectrum of the equatorial components h_1 and h_2, the amplitude modulation of wave S_1 by the annual frequency results in the appearance of the side

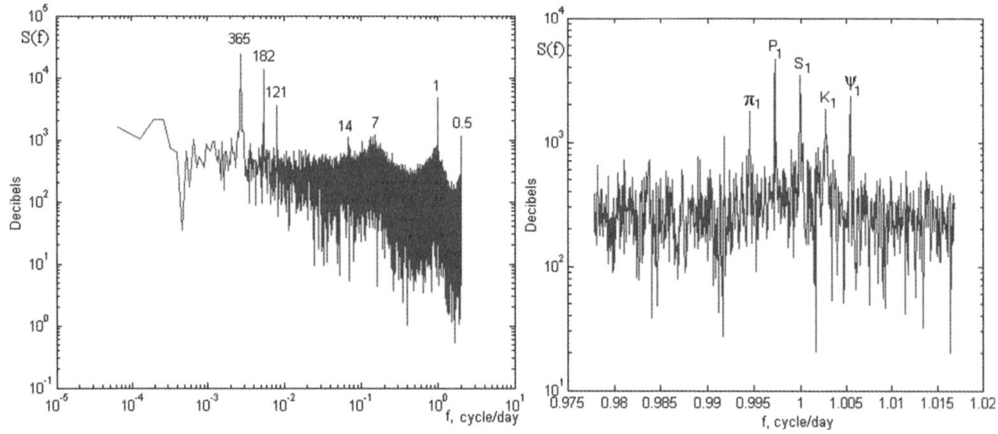

Figure 7.20 Spectrum of the modulus of the equatorial projection of the angular momentum vector $h_e = \sqrt{h_1^2 + h_2^2}$ (a) and its near-diurnal band (b).

peaks at the frequencies of the gravitational tides P_1 and K_1, by the semiannual frequency – at frequencies π_1 and ψ_1, and by the terannual frequency – at the frequencies of -0.9918 cpd and -1.0083 cpd. The latter corresponds to the wave of the gravitational tide φ_1.

Let us consider the modifications that take place in the near diurnal areas (Figure 7.20b)). From comparison of spectrums in Figures 7.16(b) and 7.20(b) we can see that peaks P_1, S_1 and K_1 have decreased more than a hundred-fold. Their energy has transferred into the area of subharmonics of the annual frequency. The energy of waves ψ_1 and -0.9918 cpd has decreased 1.5 times, and the energy of wave π_1 has increased even by 25%. Wave φ_1 has completely disappeared, whereas the peak on frequency $-1/122$ cpd has appeared. The main lunar wave O_1 has also completely disappeared, whereas new peaks formed in the range of frequencies corresponding to half the lunar month (the most powerful peak falls on the frequency of $-1/14.75$ cpd; this frequency corresponds to half the lunar synodical period).

In the spectrum of the vector magnitude $\boldsymbol{h}_e$, the strong maximum at a frequency of -0.85 cpd (Figure 7.16a)) has also disappeared, whereas a broad maximum of about a week's frequency has appeared.

The spectrum of the vector magnitude $\boldsymbol{h}_e$ has no vast maximum at 0.85 cpd (Figure 7.18) but shows a strong maximum in the vicinity of a 7-day frequency. It indicates that the amplitude of the diurnal oscillations of h_1 and h_2 does change with a period varying from 5 to 9 days.

Thus, the equatorial angular momentum $\boldsymbol{h}_e$ rotates with the diurnal period westward in the equatorial plane, so that it is 100° ahead of the Sun's meridian from March to September and 80° behind it from October to February. The length of $\boldsymbol{h}_e$ varies with periods of 0.5, 1, 7, 14, 122, 183, and 365 days. The annual, semiannual, and terannual harmonics result from the variation in the atmospheric circulation due to the Sun's annual motion between the Northern and Southern Tropics. The fortnightly and weekly harmonics result from the effect of lunar tides on the atmospheric circulation. The semidiurnal harmonics are caused by both the gravitational tides and the diurnal variation in the solar light absorption due to the inhomogeneity of the Earth's surface. In the annual range, a peak with a period of 355 rather than 365 days was detected, which definitely indicates its lunar origin, since this is the lunar year! The question arises as to why no lunar components were detected earlier. The answer is obvious. Meteorologists use the averaged data (monthly and annual means), which contain no information on the tidal oscillations. For example, if h_1 and h_2 were daily averaged, we would obtain nearly constants with no amplitude or frequency modulation of the diurnal carrier frequency. The averaging of observations destroys the natural information flow at the diurnal frequency (Figure 7.19).

7.6 Atmospheric Excitation of Nutations

Synthesis of the time variations of the projections h_3 and h_e of the relative atmospheric angular momentum vector $\boldsymbol{h}$ onto the polar axis and the equatorial plane leads to the conclusion that vector $\boldsymbol{h}$ experiences a nutation motion with a near-diurnal period. This is shown graphically in Figure 7.21. Vector $\boldsymbol{h}$ is deflected from the polar axis by an angle $\theta = \text{arctg}(h_e/h_3)$ (in June, up to $\approx 2°$) and rotates from east to west, describing a conical surface. Its vertex is the Earth's center, and the directrix is a curve,

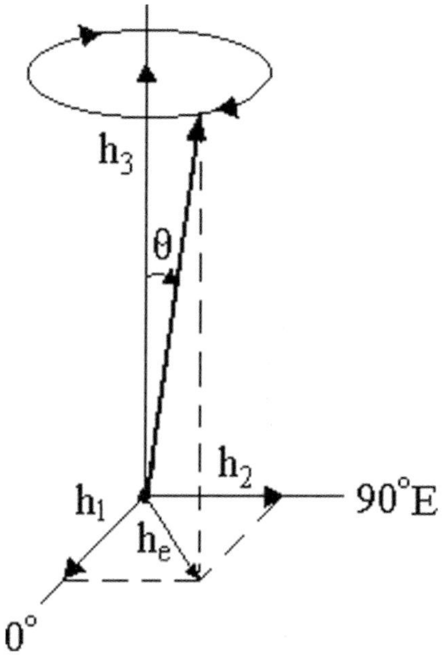

Figure 7.21 Diurnal nutation of the atmospheric angular momentum vector $\boldsymbol{h} = h_1\boldsymbol{e}_1 + h_2\boldsymbol{e}_2 + h_3\boldsymbol{e}_3$.

the projection of which onto the equatorial plane coincides with the trajectory of $\boldsymbol{h}_e = h_1\boldsymbol{e}_1 + h_2\boldsymbol{e}_2$ (see Figure 7.15). The aperture of the cone described by vector $\boldsymbol{h}$ is equal to the magnitude of $|\boldsymbol{h}_e|$. It is clear that all the above features of the vector $\boldsymbol{h}_e$ motion are valid for vector $\boldsymbol{h}$. The amplitude of the retrograde diurnal nutation of $\boldsymbol{h}$ reaches a maximum near the June and December solstices and is close to zero at the beginning of March and in mid-October, when the nutation phase changes by 180°. From March to September, the meridian of vector $\boldsymbol{h}$ is constantly deflected by about 100° to the west and from November to February by 80° to the east of the solar meridian.

In consequence of the law of conservation of the angular momentum of the Earth–atmosphere system, the nutation of atmospheric angular momentum vector $\boldsymbol{h}$ should be accompanied by the reverse nutation of the Earth's angular momentum vector $\boldsymbol{H}$, because in a closed system $\mathrm{D}\boldsymbol{H}/\mathrm{D}t = -\mathrm{D}\boldsymbol{h}/\mathrm{D}t$ (Sidorenkov, 2003b).

The above tidal harmonics in the variations of the vector $\boldsymbol{h}$ components should excite the oscillations of the components of vector $\boldsymbol{H}$, which are opposite in sign and have the same tidal frequencies. However, to the tidal frequencies in the Earth's rotating reference system there correspond definite nutation frequencies in the celestial reference system (Melchior, 1983; Moritz and Mueller, 1987). The nutation frequency $\Delta\sigma_j$ is determined as the difference between the tidal frequency σ_j and the sidereal angular velocity of the Earth's rotation ω_j:

$$\Delta\sigma_j = \sigma_j - \omega_j \tag{7.34}$$

Table 7.3 presents the nutation terms excited in the Earth's rotation by the tidal oscillations of the atmospheric angular momentum vector $\boldsymbol{h}$.

As is seen from Table 7.3, the atmospheric tide P_1 is the equivalent to the prograde semiannual nutation, tide S_1 – to the prograde annual nutation, tide K_1 – to the precessional movement or the offset of the Celestial Ephemeris Pole in space, tides ψ_1 and φ_1 are the retrograde annual and semiannual nutations, respectively. Less significant are the tide π_1 and the tide with a frequency of −0.9918 cpd. They cause

Table 7.3 Tidal oscillations of the atmospheric angular momentum vector $\boldsymbol{h}$ and the respective terms of nutation of the Earth's rotation axis.

Tide	Frequency σ, cpd	Amplitude	Argument of nutation	Period, days	Type and genesis of the nutation
—	−0.8557	16	4s	6.8	Weekly, Moon
O_1	−0.9295	29	2s	13.7	Fortnightly, Moon
—	−0.9918	16	4h	91	Quarterly, Sun
π_1	−0.9945	13	3h	122	Terannual, Sun
P_1	−0.9973	470	2h	183	Semiannual, Sun
S_1	−1	410	h	365	Annual, Sun
K_1	−1.0027	430		∞	Precession, Moon and Sun
ψ_1	−1.0055	37	−h	−365	Retrograde annual, Sun
φ_1	−1.0082	25	−2h	−183	Retrograde semiannual, Sun

the prograde terannual and quarterly nutations of the Earth's spin axis, respectively. The main lunar tide O_1 in the atmosphere excites the prograde fortnightly nutation. The variations in the atmospheric angular momentum in a broad band near the -0.85 cpd frequency can cause the prograde nutation with a period of about 6.8 day.

As is known, the presence of the ellipsoidal liquid core inside the Earth generates an almost diurnal nutation with a frequency of $1/T_{\text{fcn}} = 1.0050663$ cpd. The frequency of tide ψ_1 is very close to this frequency. This produces the resonance that results in a considerable amplification of the effect of this wave on the excitation of nutation (Brzezinski, 1994; Zharov and Gambis, 1996;Bizouard, Brzezinski and Petrov, 1998; Brzezinski, Bizouard and Petrov, 2002). In the frequency domain, this effect is described by the expression (Brzezinski, 1994;Brzezinski, Bizouard and Petrov, 2002):

$$p(\sigma) = \left(a\frac{\sigma_{\text{cw}}}{\sigma_{\text{fcn}} - \sigma} + \frac{\sigma_{\text{cw}}}{\sigma_{\text{cw}} - \sigma}\right)\underline{\chi}(\sigma) \tag{7.35}$$

where $\sigma_{\text{cw}} = 2\pi(1 + i/2Q_c)/T_c$ and $\sigma_{\text{fcn}} = 2\pi(1 + i/2Q_f)/T_{\text{fcn}}$ are the frequencies of the Chandler wobble and of the free core nutation, respectively; $T_c = 433.3$ days and $T_{\text{fcn}} = 0.995$ day are their periods; $Q_c = 170$ and $Q_f = 5722$ are their quality factors for the Chandler wobble and the free core nutation, respectively; a is a dimensionless coefficient expressing the response of the free core nutation mode to the pressure ($a_p = 9509 \times 10^{-5}$) and wind ($a_w = 5489 \times 10^{-7}$) excitation $\chi(\sigma)$ (Brzezinski, Bizouard and Petrov, 2002). As shown in (Gross, 1993), the effect of diurnal tides on the motion of poles can be expressed in the frequency domain as the corrections to the nutation terms described by a simple equation: $n(\sigma) = -p(\sigma)$.

The effect of atmospheric tides on the nutation of the Earth's spin axis was estimated in many papers (Zharov and Gambis, 1996;Bizouard, Brzezinski and Petrov, 1998; Brzezinski, Bizouard and Petrov, 2002). It was found in (Bizouard, Brzezinski and Petrov, 1998) that the total contribution of the atmospheric winds and pressure causes the following nutations of the Earth's spin axis: the prograde annual nutation with an amplitude of 77 μas; the retrograde annual nutation with an amplitude of 53 μas and the prograde semiannual nutation with an amplitude of 45 μas. The constant offset of the pole is estimated in the longitude $\delta\psi \sin \varepsilon_0 = -86$ μas and in the inclination $\delta\varepsilon = 77$ μas.

The analysis of temporal variations in components h_1 and h_2 shows that the atmospheric contribution to the nutation considerably varies with time. Earlier this feature of the atmospheric effect was found (Bizouard, Brzezinski and Petrov, 1998). Therefore, the account of the effect of the atmosphere in the nutation by way of adding the constant atmospheric corrections to nutation terms is inefficient. Permanent calculations of atmospheric corrections using the operative analyses of the atmospheric global circulation models are necessary. Then, these calculations can be used for correcting the astronomical observations in real time.

8
Nature of the Zonal Circulation of the Atmosphere

8.1
Observational Data

The air motion in the atmosphere is irregular. To make sense of this chaos, methods of the temporal and spatial averaging are widely applied to the observed fields of wind, atmospheric pressure, temperature, and other characteristics. The atmospheric characteristics are most variable along the meridian and height; therefore, they are generally averaged either along the meridian or by constructing the meridian–height sections.

The distribution of the zonal (that is, the meridian-averaged) component of wind velocity in the atmosphere is shown [in the meridian section] in Figure 7.1 (Sidorenkov, 1976). It is seen that on average over the year, the westerly winds (positive values) prevail in the atmosphere. The easterly winds prevail only near the equator. The boundaries between these winds (the zero isotachs) lie (at the sea level) at the latitudes of $\pm 30^\circ$. Up to a level of 200 hPa, the zone of easterly winds narrows toward the equator and then, higher up, it widens toward the middle latitudes. In its most narrow place, this zone is limited by the latitudes $\pm 6^\circ$, whereas at a level of 30 hPa it extends from 40°N to 25°S. The maximum velocity of westerly winds is observed at a level of 200 hPa in the subtropics near the latitudes of $\pm 35^\circ$. The wind velocity reaches 20 m/s in the Northern Hemisphere and 29 m/s in the Southern Hemisphere. These averaged velocity maxima of westerly winds reflect the presence of the subtropical jet streams in the upper troposphere. The strongest easterly winds are observed in the equatorial stratosphere: at a level of 30 hPa and 10°N latitude, their velocity reaches 12 m/s. In the near-equatorial troposphere, the velocity of easterly winds does not exceed 4 m/s.

Near the Earth's surface, the trade winds having the eastern component prevail at low latitudes and the westerly winds in moderate and high latitudes. The zonal component of wind velocity changes its sign at the so-called "horse latitudes" (near the latitude of 30° in both hemispheres), where the zones of tropical calms are located. Figure 8.1 illustrates the dependence of the mean annual velocity u of the zonal wind at a level of 850 hPa on the geographical latitude. One can see that in the zone of $40^\circ < \theta < 140^\circ$, the velocity u changes with the latitude from its negative

The Interaction Between Earth's Rotation and Geophysical Processes. Nikolay S. Sidorenkov

ISBN: 978-3-527-40875-7

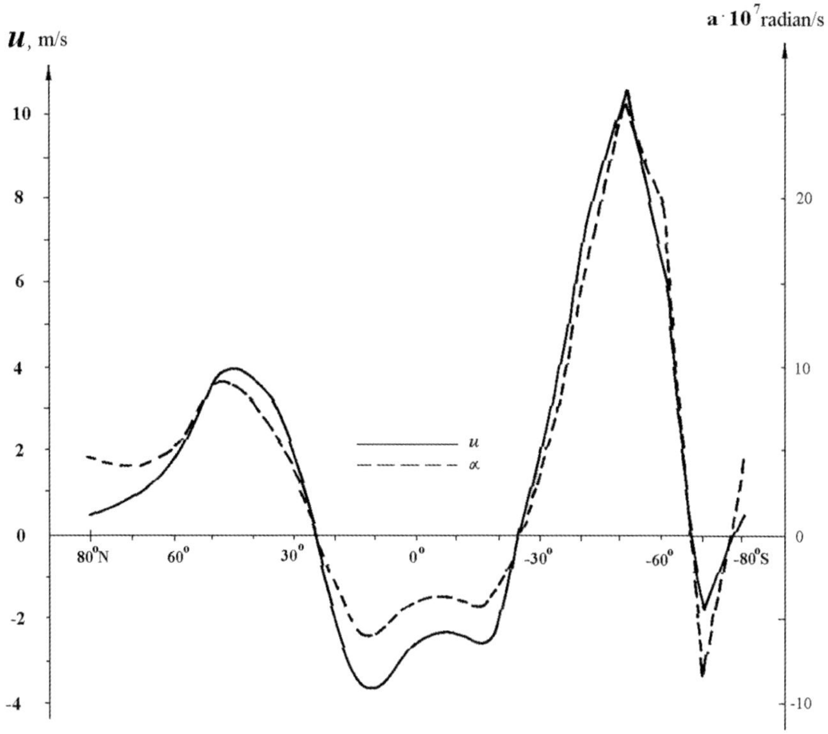

Figure 8.1 Latitudinal changes in the mean annual velocity of zonal wind u and the angular velocity α of the rotation of the atmosphere relative to the Earth's surface, at a level of 850 hPa.

values to maximum positive values. This latitudinal dependence of velocity u is well approximated by the zonal spherical harmonic of the second degree:

$$u \approx c(2/3 - \sin^2\theta) \tag{8.1}$$

where c is a positive constant, θ is the colatitude. The negative values of velocity u near 70°S are associated with the outflow of cold air from the Antarctic and are observed only up to a height of 2 km (see Figure 7.1).

It is found that variations in the intensity of the atmosphere zonal circulation are accompanied with changes in the velocity of the Earth's rotation (Munk and Mcdonald, 1960; Sidorenkov, 1976). The zonal circulation intensifies owing to the inflow of the angular momentum from the Earth and attenuates owing to its outflow to the Earth. It is shown that the angular momentum of zonal winds is not equal to zero; it averages to about $13 \times 10^{25}\,\mathrm{kg\,m^2\,s^{-1}}$, that is, the atmosphere as a whole rotates around the polar axis more rapidly than the Earth rotates; thus, the super-rotation of the atmosphere takes place (Sidorenkov, 1976, 1980b, 1982a, 1991b).

Solar radiation heats most intensely the Earth's surface at low latitudes. Thus, there is an excess of heat influx at low latitudes and its deficiency at high latitudes. It is natural to expect that at such heat regime, the atmospheric pressure near the Earth's

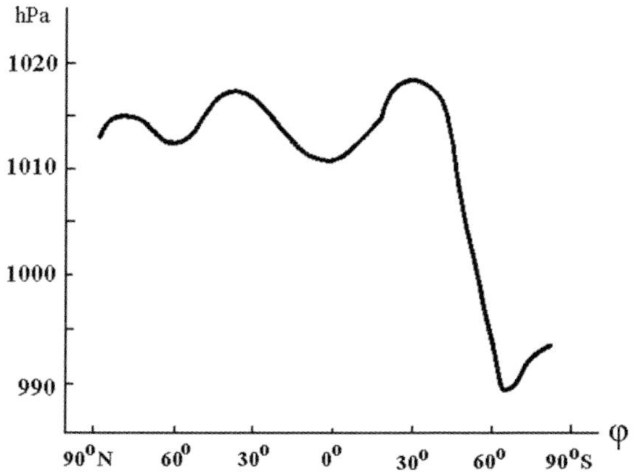

Figure 8.2 Latitudinal changes in the mean atmospheric pressure at sea level.

surface would constantly increase along the meridian from the equator toward the poles. However, actually, the pressure grows only as far as to the "horse latitudes" (35°) and then rapidly drops down to its minima near the polar circles (near the 65° latitude) (Figure 8.2). Thus, the main specific feature of the distribution of the zonal atmospheric pressure near the Earth's surface is clearly defined subtropical maxima (near the 35° latitude); moving away from them, the atmospheric pressure decreases both toward the equator and the poles.

Although the zonal circulation and the field of atmospheric pressure have long been simulated by the numerical models of the atmosphere general circulation, (Dymnikov, Petrov and Lykosov, 1979), there was no adequate physical explanation of the observed peculiarities of zonal circulation and air pressure latitudinal profiles (Lorenz, 1967) until the investigations of Sidorenkov (1980b, 1982a, 1991b), in which they were successively explained. The conception of N.S. Sidorenkov is given below.

8.2 Translational–Rotational Motion of Geophysical Continua

The motion of a rigid body is described either by two vectorial equations (those of the momentum and angular momentum balances) or by six scalar equations (those of the balances of three projections of the momentum and three projections of the angular momentum of a body). The equation of the momentum balance describes the translational motion of a body (when all the points of a body move precisely as its center of inertia does). The equation of the angular momentum balance describes the rotation of a body around its center of inertia. In continuum mechanics, the motion of a continuum is described using solely the equation of the momentum balance. In this case, a theorem of the reciprocity of tangential stresses (or the symmetry of the stress tensor) is proved, and then it is assumed that the equation of the angular momentum balance is identically satisfied when accounting for the equation of the momentum balance alone (Landau and Lifshitz, 1959; Massey, 1983).

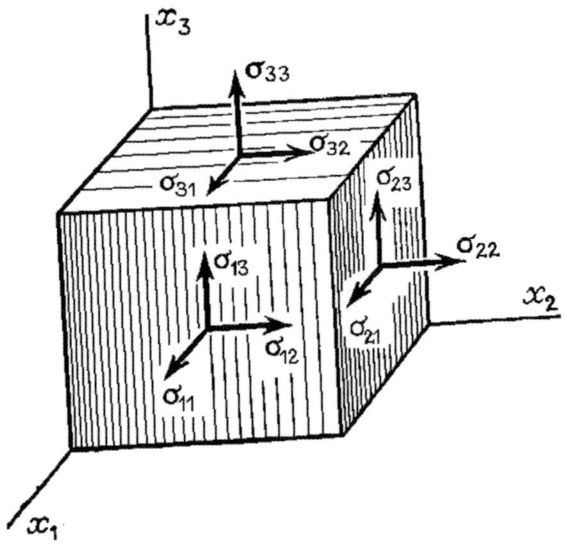

Figure 8.3 Components of the stress tensor.

What would happen if the stress tensor is really symmetrical $\sigma_{ij} = \sigma_{ji}$ (Figure 8.3)? In that case, the moments of forces of tangential stresses acting on the physical particle (the microvolume) under consideration would compensate each other and the particle would not be able to rotate (Sidorenkov, 2008a). In the isotropic space (Figure 8.4), such as where the geophysical and astrophysical envelopes are, the particle would move only translationally over the spherical surface of an equal potential of the gravitational field (equipotential surface). From here on we consider the geophysical continua. If particles move in a translational way relative to the inertial cosmic space, they must rotate relative to the Earth's surface. Each particle would rotate around its individual axis that passes through the individual center of inertia of this particle. Hence, the orientation of particles relative to each other would

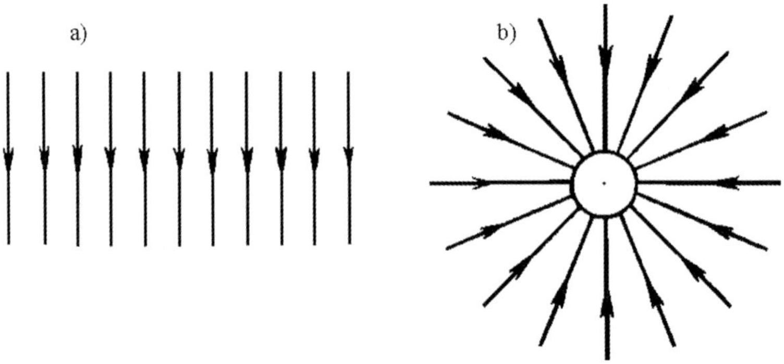

Figure 8.4 Homogeneous (a) and isotropic (b) spaces.

constantly vary. Also, there should be friction between particles, and the results of mixing and friction would depend on the size of particles. For example, the atmosphere and hydrosphere contain compound molecules, clusters of water, ice crystals, and so on. If the sizes of particles are smaller than those of the above formations, the latter would behave as though they parted. At the translational motion along the equipotential surface, these parts of formations will rotate as autonomous bodies (independently of one another). As a result, the order of arrangement of molecules would be disturbed, and a mash of atoms and fragments of formations will form in place of clusters and crystals.

In nature, there is neither any mixing of orientations of particles nor friction between them. Therefore, we can state that the model of motion proposed in the modern continuum mechanics is inadequate.

Let us consider the second contradiction. The model of translational motion proposed in the modern continuum mechanics yields different results depending on the choice of the particle size. Figures 8.5a and 8.6a illustrate the translational motion of a particle in the case of its dividing into two and four parts, respectively. As is seen from these figures, these parts rotate individually relative to the Earth's surface. The results of their motion differ, whereas by their nature they must be similar at any fragmentation of the original particle. However, this takes place only in the case of the translational–rotational motion of particles. The orientation of the chosen parts of particles relative to the Earth's center must remain unchanged. They always as though "face" the Earth's center (Figures 8.5b and 8.6b) and, hence, rotate relative to the inertial cosmic space. In other words, each particle has its own spin, its own angular momentum, which is equal to the product of the moment of inertia of the particle and the angular velocity of its rotation. Note that at rest relative to the

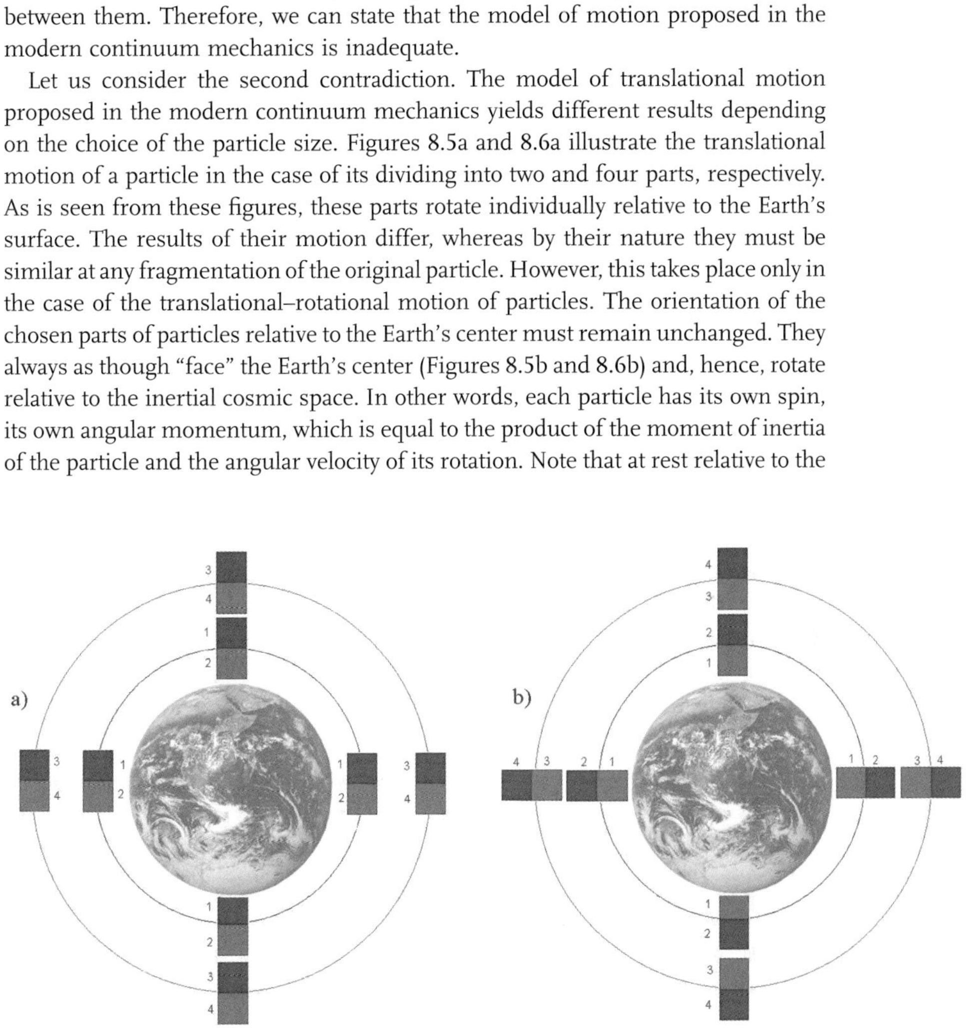

Figure 8.5 Translational (a) and translational–rotational (b) motions of two particle's parts over the surfaces of the fixed geopotential level.

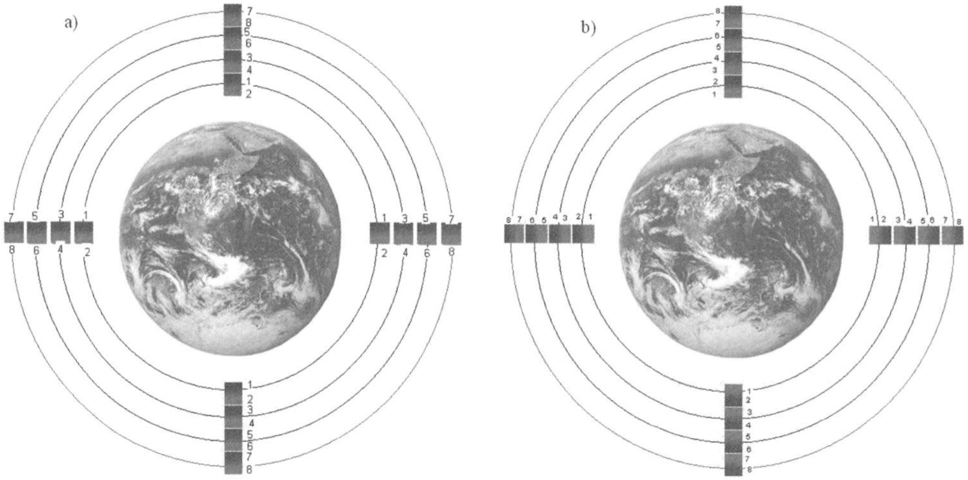

Figure 8.6 Translational (a) and translational–rotational (b) motions of four particle's parts over the surfaces of the fixed geopotential level.

rotating Earth, all elementary particles rotate around their own axes that pass through their centers of inertia, parallel to the axis of the Earth's rotation. The angular velocity of rotation is the same for all particles and coincides with the angular velocity of the Earth's rotation. In the case of the relative motion over the equipotential surface, particles acquire an additional rotation with the angular velocity $\alpha = u/r$, where u is the linear velocity of the translational motion of the particle, and r is the radius of curvature of the trajectory of its movement. In this case, the angular velocities of revolution of the particle over the equipotential surface and rotation of the particle around its center of inertia are equal.

The cause of the proper rotation of particles can be understood from the example of the motion of a balloon executing a round-the-world voyage. The balloon's gondola is always situated at the bottom and the envelope with gas is at the top. Hence, the balloon is subject to the effect of the gravity force moment, which makes the balloon turn following the changes in the direction of the vertical. At any deviation of the balloon axis from the local vertical, there arise a couple of gravity forces relative to the balloon inertia center, which turns the balloon to fit the position of the local vertical. During one round-the-world flight, the balloon will turn around its inertia center once. In other words, the force of gravity makes the bodies moving in the horizontal direction turn to follow the changes in the direction of the vertical. It is just in this way that all bodies on the Earth move, all the Earth's continua of geospheres (atmosphere, hydrosphere, mantle, and core).

In addition, the orbital revolutions of the major natural satellites of the planets are in most cases synchronous with their spin rotations, so that the satellite's angular velocities of orbital revolution and spin rotation appear to be equal. Due to this, the satellite always presents the same hemisphere to an attracting planet. Evidently, the gravitational field tends to make the bodies of the system rotate in such a way as they would rotate if the system was solid. It is just in a solid body that each physical particle revolves around the common axis of rotation and around its proper axis that goes through the mass center of the particle, parallel to the common axis of rotation.

Thus, all bodies and all particles of continua in the isotropic space, which the space around the Earth and any gravitating celestial bodies is, move in a translational–rotational way. This motion can be adequately described only under the condition of using all equations of motion (the equations of the momentum and angular momentum balances). The description of the continuum motion that is accepted in the modern mechanics contains an error caused by the neglect of the fact that all the particles moving in the isotropic space have their proper rotations (spins). Therefore, it is impossible to omit the equation of the angular momentum balance. In this chapter we demonstrate the possibility to successively solve the problems that are unsolvable in classic hydrodynamics, using the balance equation for the atmospheric angular momentum. New fundamental results in the theory of turbulence and mechanics of continuum were obtained by Nikolaevskiy (2003), who always used the equations of the angular momentum balance.

8.3 Genesis of the Zonal Circulation

As is mentioned above, the year-averaged heat balance of the equatorial and tropical areas is positive and that of the polar areas is negative, that is, the atmosphere is warmed at low latitudes and is cooled at high latitudes. The density of the warm air is lower than that of the cold air. Therefore, there is an air-density gradient between the equator and the poles. Such nonuniform horizontal distribution of air density initiates convective movements in the field of the gravity force. Under the effect of the Archimedean forces, the atmospheric air moves to eliminate the density gradient, whereas the sources of heat and cold restore this gradient. As a result, the convective motion in the atmosphere is constantly maintained. In the context of thermodynamics, a heat engine is operating in the atmosphere, the equatorial zone playing the role of a heater and the polar area – the role of a cooler.

Observations show that the convective movements in the atmosphere occur in the form of a disordered turbulent motion of large air masses along the meridian rather than in the form of cells enclosed between the equator and poles. Some air masses that are formed at low latitudes break out far toward the pole, others that originated at high latitudes penetrate to the equator. Characteristic horizontal dimensions L of air masses – the macroturbulent formations – are thousands of kilometers. The vertical extension (thickness) of air masses is hundreds of times smaller L than (approximately 10–20 km).

It is clear that the macroturbulent horizontal mixing of the atmosphere levels out the distinctions in the distributions of all the substances, whose horizontal gradient in the meridian direction differs from zero. This may be density, temperature, moisture, admixtures, energy, and so on. To understand the process of formation of the zonal circulation, it is sufficient to analyze the transport of the axial component of the angular momentum l. Jeffreys (1926) was the first who pointed out the importance of the law of conservation of the angular momentum l for the atmospheric circulation.

Let us observe the atmospheric motion from the inertial (for example, the heliocentric) reference frame. In the absence of heating the atmosphere is at rest with respect to the Earth's surface but rotates together with the Earth relative to the axis of its diurnal rotation; the specific (per unit mass) absolute angular momentum l in it is distributed according to the law

$$l = \Omega R^2 \sin^2\theta \quad (8.2)$$

where R is the geocentric distance approximately equal to the Earth's radius; Ω is the angular velocity of the Earth's diurnal rotation.

The l value is directed along the instant axis of the Earth's rotation, which does not virtually change its position (Sidorenkov, 1973). Hence, the l value is considered as the scalar one. The absolute angular momentum steadily decreases from the equator (where its specific value is maximum and equal to ΩR^2) to the poles (where $l = 0$). The mean value of $\bar{l}$ in the atmosphere is obviously equal to the absolute angular momentum of the entire atmosphere divided to its mass

$$\bar{l} = \int_W l\rho \, \mathrm{d}V \Big/ \int_W \rho \, \mathrm{d}V = n_{33}\,\Omega/M \approx 0.674\Omega\, R_0^2 \quad (8.3)$$

Here, W is the volume of the atmosphere, n_{33} and M are the moment of inertia and the mass of the atmosphere. When estimating $\bar{l}$ in (8.3), it is taken into account that, according to Table 6.1, $n_{33} = 0.1413 \times 10^{33}\,\mathrm{kg\,m^2}$; $M = 5.159 \times 10^{18}\,\mathrm{kg}$; $R_0 = 6.371 \times 10^6\,\mathrm{m}$. Note that in the case of a uniform spherical shell, $\bar{l} = 2/3\,\Omega R^2$. The latitude φ at which $l = \bar{l}$ is easily to be found from the condition $\sin^2\theta = 0.674$; it is equal $\pm 35°$ ($\theta_1 = 55°$ and $\theta_2 = 125°$). At low latitudes ($|\varphi| < 35°$) $l > \bar{l}$; at moderate and high latitudes, ($|\varphi| > 35°$) $l < \bar{l}$).

Due to the macroturbulent horizontal mixing, the distribution of the absolute angular momentum along the latitude levels off: it decreases at low latitudes, where the l value is high ($l > \bar{l}$); and increases at moderate and high latitudes, where l is small ($l < \bar{l}$). This process is similar to the well-known decrease in the temperature contrast between the heated and cooled areas due to air mixing: the air temperature drops in warm areas and rises in cold areas. An observer from the heliocentric reference frame will see that as the distribution of the absolute angular momentum levels off, the differential rotation of the atmosphere starts. It is relatively slow at low latitudes and rapid at high latitudes. As for the Earth, it rotates from west to east as a perfectly rigid body, with the angular velocity being equal in all its points. Therefore, the decrease in the absolute angular momentum of the adjacent air layer manifests itself in the lag of its rotation relative to the rotation of the Earth's surface and, consequently, in the initiation of easterly winds. The increase in the absolute angular momentum manifests itself in a more rapid air rotation (as compared with the rotation of the Earth's surface) and, consequently, in the initiation of westerly winds.

The zonal movement of atmospheric air relative to the Earth's surface (which, in this case, serves as the reference frame) can be characterized either by the relative angular velocity α or the wind velocity u. The latter is related to the α value by the expression $u = \alpha R \sin\theta$. As is known, the velocity of zonal winds in the Earth's atmosphere is of the order of magnitude of 10 m/s. Consequently, $\alpha \approx 10^{-6}\,\text{s}^{-1}$, or almost a hundred times smaller than the angular velocity Ω (see Figure 8.1).

With account for the zonal relative movements, the distribution of the specific absolute angular momentum in the atmosphere is described by the formula

$$l = (\Omega + \alpha) R^2 \sin^2\theta \tag{8.4}$$

With mixing, the l value in the zone of 35°N–35°S decreases; the α and u velocities in this zone have negative values, that is, the easterly winds prevail there. To the north and to the south of this zone, the l value increases; therefore, the α and u values are positive and the westerly winds prevail there. At the latitude of 35°, the absolute angular momentum is equal to its mean value for the entire atmosphere; therefore, the angular momentum does not change with mixing here and winds do not initiate. To illustrate this qualitative reasoning, let us consider an idealized example.

Let the whole atmosphere be instantly mixed up in such a way that the specific absolute angular momentum l in it become everywhere the same, that is, $l = \bar{l}$ (the absolute angular momentum of the whole atmosphere remained unchanged). As a result, the zonal movements should be observed, whose velocities can be found from the equality $l = \bar{l}$, or $(\Omega + \alpha) R^2 \sin^2\theta = 0.674\,\Omega R^2$. Hence, it follows

$$\alpha = \Omega(0.674 - \sin^2\theta)/\sin^2\theta \tag{8.5}$$

It is easy to see that in the zone from $\theta_1 = 55°$ to $\theta_2 = 125°$ ($|\varphi| < 35°$), $\alpha < 0$, that is, the easterly winds will be observed. The maximum velocity of easterly winds will be observed at the equator, where $\alpha = -0.326\,\Omega$. Within the zones from $\theta < 55°$ and $\theta > 125°$ ($|\varphi| > 35°$), $\alpha > 0$, that is, the westerly winds should be observed, whose velocity interruptedly grows with the increasing latitude. The change in the sign of winds occurs at the 35° latitude.

Thus, the steadily existent zones of westerly winds and easterly winds near the Earth's surface are caused by the process of leveling off the distribution of the absolute angular momentum in the meridian direction. This leveling results from the macroturbulent horizontal mixing of the atmosphere, the original cause of mixing being different heating of the troposphere in the equatorial and polar zones.

Notice that according to the concepts existing in geophysical hydrodynamics, the macroturbulence sharpens the gradients and transfers the energy from small scales to large scales (Lorenz, 1967; Starr, 1968). However, these effects are only apparent. They are associated with the noninertial reference systems that are used in the geophysical hydrodynamics. In the inertial reference frame chosen above, an observer records not sharpening but leveling off gradients in the distributions of all substances and characteristics, including the specific values of momentum q, angular momentum l, and energy E. This can be seen in Figures 8.7 and 8.8, which represent schematically the latitudinal changes in the angular momentum and energy in the actual atmosphere and the atmosphere at rest. This is also easy to understand from

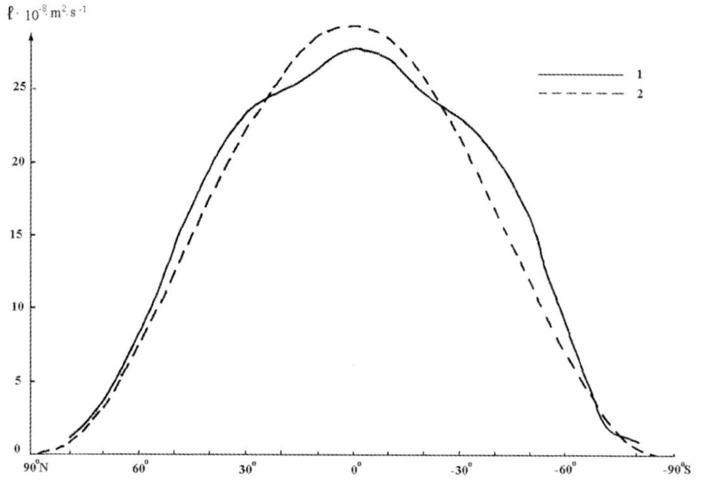

Figure 8.7 Scheme of latitudinal changes in the specific angular momentum l in the real atmosphere (1) and in the atmosphere at rest relative to the Earth's surface (2).

the expressions for the q, l, and E values represented in the inertial reference frame with the origin in the Earth's center:

$$\begin{aligned} q &= (\Omega+\alpha)R\sin\theta = \Omega R\sin\theta + u \\ l &= (\Omega+\alpha)R^2\sin^2\theta = \Omega R^2\sin^2\theta + uR\sin\theta \\ E &= \frac{1}{2}(\Omega+\alpha)^2 R^2\sin^2\theta \approx \frac{\Omega^2 R^2}{2}\sin^2\theta + u\Omega r\sin\theta \end{aligned} \tag{8.6}$$

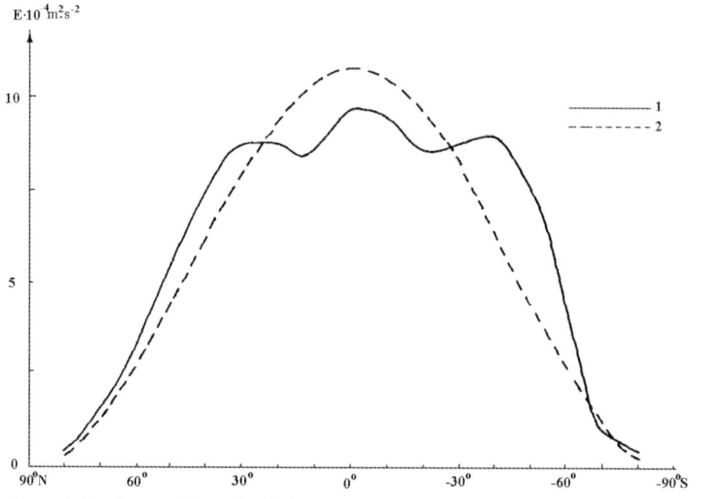

Figure 8.8 Scheme of latitudinal changes in the specific energy E in the real atmosphere (1) and in the atmosphere at rest relative the Earth's surface (2).

The first terms in the right-hand sides of these expressions are the momentum q_r, angular momentum l_r, and energy E_r in the case of the atmosphere at rest relative to the Earth's surface ($u = 0$). The values of these terms are greater than their mean values at low latitudes and smaller than those at high latitudes. The last terms of the equations describe the additions arising at the expense of the relative movements (winds). The results showing that in the surface layer of the atmosphere, in the zone of $|\varphi| < 35°$ $u < 0$ (the easterly winds) and in the zones of $|\varphi| > 35°$ $u > 0$ (the westerly winds) are the observation facts (see Figure 8.1). One can see that the values of q, l, and E in the zone of $|\varphi| < 35°$ are smaller and in the zones of $|\varphi| > 35°$ larger than the values of q_r, l_r, and E_r. To put it another way, the macroturbulence decreases the values of q_r, l_r, and E_r in the areas where these values are large (at low latitudes) and increases wherever they are small (at moderate and high latitudes). The distributions of the values of momentum, angular momentum, and energy of the absolute motion are leveled off, as it occurs at the classic processes of transfer.

This brings up the question: why does the macroturbulence not quench but initiate the relative air motion – the zonal circulation of the atmosphere (see Figure 7.1)? In order to answer this question, let us recall that any system occurs in the equilibrium state when all its parts either are at rest or move uniformly and rectilinearly with the same velocity $V = \text{const}$.

The atmosphere that is at rest relative to the Earth's surface but rotates together with the Earth with the angular velocity $\Omega = \text{const}$ is far from the equilibrium state, (if we consider it from the above point of view). Its "equilibrium" is due to the gravity force. If we eliminate this force, the air particles will "slide off" the Earth at a tangent to parallels, forming nonequilibrium plane-parallel flows with large transverse (along the meridian) gradients of velocity. Therefore, the state at which the atmosphere rotates as a solid body with the angular velocity $\Omega = \text{const}$ would be more properly called the constrained equilibrium state. The latter differs substantially from the state of the uniform rectilinear movement of the system with the velocity $V = \text{const}$, which can be called the free equilibrium state.

The macroturbulence mixing of the atmosphere makes it free (to some extent) from the ties of the gravity force and tends to transform the atmosphere from the constrained equilibrium state ($\Omega = \text{const}$) into the free equilibrium state ($V = \text{const}$). The more intense is the meridian mixing of the atmosphere, the stronger is this effect. As the meridian mixing attenuates, the atmosphere returns into the constrained equilibrium state with $\Omega = \text{const}$.

The vertical extension of the atmosphere is very small. Therefore, given the vertical mixing of the atmosphere, the distinction between the constrained ($\Omega = \text{const}$) and free ($V = \text{const}$) equilibrium states is so slight that it is generally overlooked.

The q, l, and E become equal due to the redistribution of their quantities along the meridian. They increase at moderate and high latitudes mainly at the expense of the inflow of the momentum, angular momentum, and energy, respectively, from low latitudes and, as will be shown below, from the Earth. These characteristics (including energy) are transported by eddies, which gives an impression that all the energy of zonal circulation is taken from eddies. Actually, the sources of energy require special consideration. Here, we only state that with the redistribution of the momentum and kinetic energy their integral values remain approximately constant; only the integral value of the angular momentum is kept strictly constant. The redistribution of this parameter is considered below.

The way the macroturbulent formations transport the angular momentum can be understood with the help of the following idealized example. Let us consider the balance of the angular momentum fluxes through some fixed latitude. Let the conic surface of the parallel of a constant colatitude θ be crossed by the air masses that move chaotically southward within some segments of the latitudinal circle and northward within other segments. Let us also assume that these air masses have formed at a distance of $\Delta\theta_N$ to the north and $\Delta\theta_S$ to the south of the given parallel θ. Then the air masses moving from the south will transport the absolute angular momentum $M_S{\cdot}l\,(\theta + \Delta\theta_S)$ through the parallel θ northward, and the air masses moving from the north will transport the angular momentum $M_N{\cdot}l\,(\theta - \Delta\theta_N)$ through the parallel θ southward. The flux directed to the south is assumed to be positive and the flux directed to the north is negative. In this case, the balance of fluxes Q will have the form

$$Q = M_N l\,(\theta - \Delta\theta_N) - M_S l\,(\theta + \Delta\theta_S) \tag{8.7}$$

Since the systematic inflow or outflow of air mass in individual parts of the atmosphere does not occur on average over a year, then $M_N = M_S = M$. Expanding the l values in Equation 8.7 into series up to the first degree of $\Delta\theta$, we obtain the expression for the total flux:

$$\begin{aligned} Q &= M\left\{\left[l(\theta) - \frac{dl}{d\theta}\Delta\theta_N\right] - \left[l(\theta) + \frac{dl}{d\theta}\Delta\theta_S\right]\right\} \\ &= -M\frac{dl}{d\theta}(\Delta\theta_N + \Delta\theta_S) \approx -M\Omega R^2 \sin 2\theta(\Delta\theta_N + \Delta\theta_S) \end{aligned} \tag{8.8}$$

One can see that the value of the angular momentum flux Q is negative in the Northern Hemisphere ($0^\circ < \theta < 90^\circ$) and positive in the Southern Hemisphere ($90^\circ < \theta < 180^\circ$). Hence, in both hemispheres the angular momentum fluxes are directed from the equator to the poles. The empirical investigations are in good agreement with these conclusions (Lorenz, 1967; Starr, 1968).

8.4
Nature of the Atmosphere Superrotation

The angular momentum horizontal flux from the equator to the poles exists constantly, because the macroturbulent mixing in the atmosphere occurs continuously. As a result, the angular momentum would systematically decrease at low latitudes and increase at high latitudes, which should manifest itself in the uninterruptedly increasing velocity of easterly and westerly winds. This is not actually observed, because with the appearance of relative motions (winds), the forces of friction of atmospheric air against the Earth's surface originate instantly. This entails the angular momentum exchange with the Earth and hence, the angular momentum fluxes start.

The angular momentum vertical flux is obviously proportional to the gradient of the specific absolute angular momentum dl/dR, which is described by the expression

$$\frac{dl}{dR} = (\Omega + \alpha)2R\sin^2\theta + \frac{d\alpha}{dR}R^2\sin^2\theta \approx \frac{d\alpha}{dR}R^2\sin^2\theta \quad (8.9)$$

Here, it is taken into account that since $\Omega = 2\pi/1$ day, $\alpha = 2\pi/30$ day, $R \approx R_0 = 6371$ km, $d\alpha/dR \approx \alpha/1$ km, then the first term in the right part of expression (8.9) is almost 200 times smaller than the second term. In the zone of easterly winds, the gradients $d\alpha/dR < 0$ and $dl/dR < 0$; therefore the angular momentum flux is directed upward – from the Earth to the atmosphere. In the zone of westerly winds, the gradients $d\alpha/dR > 0$ and $dl/dR > 0$; consequently, the angular momentum flux is directed downward, from the atmosphere to the Earth.

The vertical fluxes of the angular momentum accelerate the slowly rotating parts of the atmosphere and slow down its rapidly rotating parts; in other words, they attenuate the easterly and westerly winds. On the contrary, as is mentioned above, the horizontal fluxes of the angular momentum strengthen the zonal winds. The higher the wind velocity, the larger is the gradient dl/dR; consequently, the more intense is the vertical flux of the angular momentum and the more significant is the attenuation of winds. Thus, the zonal winds continue to strengthen until the gradient $d\alpha/dR$ (more precisely dl/dR) becomes so large that the slowing-down effect of the micro-turbulent viscosity will compensate the accelerating effect of the horizontal flux of the angular momentum. Only then is the equality of the angular momentum inflow and outflow attained and the stationary state established (within the part of the atmosphere under consideration). These conclusions are in good agreement with the empirical investigations of the angular momentum fluxes in the atmosphere (Lorenz, 1967; Starr, 1968).

The investigations of the angular momentum conservation in the atmosphere was initiated by Jeffreys (1926) and resumed in the 1950s. Many studies are dedicated to this problem, the reviews of these works being made by Lorenz (1967) and Starr (1968). In the majority of these investigations the vertical and horizontal fluxes of the angular momentum are assessed using the data of the wind aloft observations. It is shown that the angular momentum is transported in the atmosphere from the equator to the poles, mainly by the eddy fluxes (the macroturbulence); only in the tropics an essential part of this transport is due to the mean circulation within the Hadley cell. The meridian flux reaches its maximum intensity at 30°N, where it equals $50 \times 10^{18}\,\text{kg}\,\text{m}^2\,\text{s}^{-1}$ in winter and $13 \times 10^{18}\,\text{kg}\,\text{m}^2\,\text{s}^{-1}$ in summer. This angular momentum flux, which is directed northward through the 30°N parallel, is equal to the total inflow of the angular momentum from the Earth's surface to the atmosphere in the zone of easterly winds of the Northern Hemisphere. The upward flow of the angular momentum in the low latitudes is completely compensated by its downward flow at moderate and high latitudes.

Direct calculations of the angular momentum of the atmosphere have shown (Sidorenkov, 1976) that the zonal circulation cannot be simply explained by the redistribution of the angular momentum between different parts of the atmosphere. The angular momentum of the easterly winds is 4–8 times smaller than that of the westerly winds. Therefore, the value of the angular momentum of zonal winds in the entire atmosphere h is not equal to zero; its year-averaged value is $128 \times 10^{24}\,\mathrm{kg\,m^2\,s^{-1}}$. Thus, the atmosphere as a whole rotates around the polar axis more rapidly than the Earth, that is, the so-called superrotation of the atmosphere takes place.

To understand the causes of the angular momentum accumulation in the atmosphere, let us consider the following circumstance. The angular momentum inflow and outflow occur in the atmosphere under different geometric conditions. Similar vertical flows at different latitudes are attained at dissimilar vertical gradients of the relative angular velocity $\mathrm{d}\alpha/\mathrm{d}R$. Indeed, let us compare the values of $\mathrm{d}\alpha/\mathrm{d}R$ in the zones of the easterly and westerly winds, at which the equality of the angular momentum vertical fluxes is achieved. Since the balance of fluxes suggests the equality of gradients of the specific absolute angular momentum $\mathrm{d}l/\mathrm{d}R$ in the zones of easterly and westerly winds, then, with account for (8.5), we obtain

$$\begin{aligned} &(\mathrm{d}\alpha/\mathrm{d}R)_{\mathrm{E}} R^2 \sin^2\theta_{\mathrm{E}} = (\mathrm{d}\alpha/\mathrm{d}R)_{\mathrm{W}} R^2 \sin^2\theta_{\mathrm{W}} \\ &\text{or} \\ &(\mathrm{d}\alpha/\mathrm{d}R)_{\mathrm{E}} : (\mathrm{d}\alpha/\mathrm{d}R)_{\mathrm{W}} = \sin^2\theta_{\mathrm{W}}/\sin^2\theta_{\mathrm{E}} \end{aligned} \tag{8.10}$$

Thus, the vertical gradient of the relative angular velocity should be much greater in the zone of westerly winds than in the zone of easterly winds. For example, to achieve the equality of the angular momentum vertical fluxes, the gradient $\mathrm{d}\alpha/\mathrm{d}R$ at the 60° latitude should exceed 4 times the relevant gradient at the equator. This increase in the gradient is achieved at the expense of wind velocity; therefore the westerly winds in the atmosphere should be much stronger than the easterly winds. In other words, the equality of the angular momentum vertical fluxes can be achieved only when the angular momentum of westerly winds exceeds that of easterly winds.

In order to understand from where the excess of the angular momentum of westerly winds comes, let us trace how the zonal circulation originates and is maintained. With the development of macroturbulent mixing, the angular momentum flux from the equator to the poles originates (Figure 8.9) and gives rise to the easterly winds at low latitudes and the westerly winds at high latitudes (at the expense of redistribution of the angular momentum in the atmosphere between the low and high latitudes). The microturbulent viscosity tends to weaken the wind: it gives rise to the vertical flows of the angular momentum from the Earth to the atmosphere in the zone of easterly winds and from the atmosphere to the Earth in the zone of westerly winds (see Figure 8.9).

At the initial stage, the effect of macroturbulent viscosity exceeds the effect of microturbulent friction, and the velocity of zonal winds increases. When a certain wind velocity is achieved, the angular momentum inflow from the Earth into the atmosphere in the tropical zone compensates completely its losses through the meridian transport to the poles. At the same time, the angular momentum that inflows to moderate and high latitudes has no time to flow down to the Earth (because of the insufficiently high velocity of westerly winds) and accumulates. The angular momentum is being "pumped" from the Earth into the atmosphere through the zone of easterly winds.

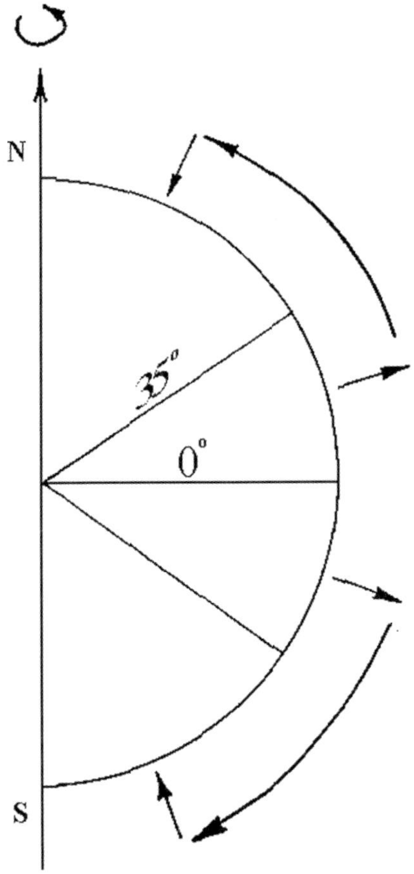

Figure 8.9 Scheme of the flows of the angular momentum in the atmosphere.

The angular momentum of the atmosphere increases, whereas the angular momentum of the Earth decreases, which manifests itself in its rotation that slows down. The angular momentum is being "pumped" into the atmosphere until the velocity of westerly winds becomes so high that the outflow of the angular momentum from the atmosphere to the Earth in the zone of westerly winds counterbalances completely its inflow from the Earth in the zone of easterly winds. Only then is the stationary state established: the turbulence will pump over the angular momentum along the meridian from the Earth's surface in the zone of easterly winds to the Earth's surface in the zone of westerly winds, without the angular momentum losses or accumulation in the atmosphere; the intensity of zonal circulation will be maintained unchanged; the angular momentum of the Earth, which was accumulated earlier, will be confined in the atmosphere; the velocity of the Earth's rotation will be constant but smaller than it was at the initial moment when the atmosphere was at rest relative to the Earth.

When the equator–pole difference of temperatures decreases, the macroturbulence attenuates and the meridian flow of the angular momentum weakens. The angular momentum vertical fluxes adapt to the altered meridian flow with different rates: rapidly in the zone of easterly winds and slowly in the zone of westerly winds. As a result, the outflow of the angular momentum to the Earth exceeds its inflow. The angular momentum accumulated in the atmosphere begins to decrease. It flows down to the Earth, and the velocity of the Earth's rotation increases. This process lasts until the westerly winds weaken to the degree at which the angular momentum outflow to the Earth becomes equal to its inflow in the zone of easterly winds and the stationary state re-establishes. However, the atmosphere will hold a smaller value of the angular momentum, which was taken from the Earth during the start of zonal circulation.

When the equator–pole temperature contrast increases, the process develops similarly to its development at the initiation of zonal circulation, which is described above.

8.5 Theory of the Zonal Atmospheric Circulation

Now that we have a clear idea of the processes of redistribution of the angular momentum in the atmosphere, let us deduce the differential equation for the distribution of the mean annual angular momentum in the atmosphere and assess the angular momentum of the zonal circulation, which is adopted by the atmosphere from the Earth. Let us follow the works of the author (Sidorenkov, 1982a, 1991b) and consider an elementary latitudinal zone in the atmosphere, of length $2\pi R \sin\theta$, width $R\,d\theta$, and thickness (height) dR; its volume is $2\pi R^2 \sin\theta\, d\theta\, dR$, mass $\rho\, 2\pi R^2 \sin\theta\, d\theta\, dR$, and angular momentum $l\rho\, 2\pi R^2 \sin\theta\, d\theta\, dR$ (here, ρ is the air density).

It is known that the turbulent mixing is responsible for the transport of substances, the flow of any scalar quantity being proportional to its gradient. In our case (l is a scalar quantity), the flow of the angular momentum is equal to $k\nabla l$, where k is the coefficient of turbulence viscosity. Consequently, at the expense of macroturbulence, the angular momentum $2\pi R \sin\theta_e\, dR\, k_\theta \frac{\partial \rho\, l_e}{R\,\partial\theta}$ inflows into the elementary latitudinal zone through the lateral vertical face presented to the equator and the angular momentum $2\pi R \sin\theta_p\, dR\, k_\theta \frac{\partial \rho\, l_p}{R\,\partial\theta}$ outflows through the lateral face presented to the pole. As a result, the angular momentum will be accumulated (or lost) in this latitudinal zone, or, which is the same, the moment of force (torque) will affect this zone:

$$-\frac{\partial}{R\,\partial\theta}\left(2\pi R \sin\theta\, dR\, k_\theta \frac{\partial \rho\, l}{R\,\partial\theta}\right) R\,\partial\theta$$

where k_θ is the coefficient of macroturbulence viscosity.

The microturbulence transports the angular momentum through the horizontal faces. The angular momentum $2\pi R_d^2 \sin\theta\, d\theta\, k_R \partial\rho\, l_d/dR$ flows through the lower face and $2\pi R_u^2 \sin\theta\, d\theta\, k_R\, \partial\rho l_u/dR$ through the upper face. The angular momentum losses (accumulation) due to this transport are

$$\frac{\partial}{\partial R}\left(2\pi R^2 \sin\theta\, d\theta\, k_R \frac{\partial \rho\, l}{\partial R}\right) dR$$

where k_R is the coefficient of microturbulence viscosity.

The change in the angular momentum within the zone under consideration, which is equal to $\partial(2\pi R^2 \sin\theta\, d\theta\, dR\rho l)/dt$, is determined from the balance of the angular momentum accumulation and losses, or from the sum of the moments of forces:

$$-\frac{\partial}{\partial t}(2\pi R^2 \sin\theta\, d\theta\, dR\, \rho l) = -\frac{\partial}{\partial \theta}\left(2\pi R \sin\theta\, dR\, k_\theta \frac{\partial \rho\, l}{R\, \partial \theta}\right) d\theta - \frac{\partial}{\partial R}\left(2\pi R^2 \sin\theta\, d\theta\, k_R \frac{\partial \rho\, l}{\partial R}\right) dR \tag{8.11}$$

The air density in the considered elementary zone can be assumed to be constant. After canceling the similar constant values, we obtain the differential equation of the second order in partial derivatives for the distribution of the mean annual specific absolute angular momentum in the atmosphere. It describes the mean annual zonal circulation of the atmosphere, because the distribution of value l in the atmosphere at rest is known:

$$\frac{\partial l}{\partial t} = \frac{1}{R^2 \sin\theta} \frac{\partial}{\partial \theta}\left(k_\theta \sin\theta \frac{\partial l}{\partial \theta}\right) + \frac{1}{R^2}\frac{\partial}{\partial R}\left(R^2\, k_R \frac{\partial l}{\partial R}\right) \tag{8.12}$$

We can derive Equation 8.12 by other means, using the hypothesis of closure from the semiempirical theory of turbulence. Below, we demonstrate this possibility, following (Sidorenkov, 1980b).

Let us choose the stationary Cartesian reference frame with the origin in the Earth's center of masses and the axes oriented relative to the "immobile stars". Let one axis be directed along the axis of the Earth's rotation and two other axes located in the equatorial plane.

Let us write the equation of motion of a unit air volume for this inertial reference frame:

$$\rho \frac{d\mathbf{V}_a}{dt} = -\nabla P - \frac{fM\rho}{R^3}\mathbf{R} + \mathbf{F} \tag{8.13}$$

where $\mathbf{V}_a = [\boldsymbol{\Omega} \times \boldsymbol{R}] + \mathbf{V}_0$ is the absolute motion velocity; $\mathbf{V}_0$ is the wind velocity; $\boldsymbol{\Omega}$ is the angular velocity of the Earth's rotation; $\mathbf{R}$ is the geocentric radius vector of the considered volume; ρ is the air density; f is the gravitation constant; M is the Earth's mass; $\nabla \boldsymbol{P}$ is the gradient of atmospheric pressure P; $\mathbf{F}$ is the friction force related to the unit volume; t is the time.

Let us multiply each term of Equation 8.13 vectorially from the left by the geocentric radius vector $\mathbf{R}$ and take into account that the friction force in the free atmosphere is negligibly small. Then, we obtain the equation for the absolute angular momentum of the unit air volume in the atmosphere:

$$\mathbf{R} \times \rho \frac{d\mathbf{V}_a}{dt} = \rho \frac{d}{dt}[\mathbf{R} \times \mathbf{V}_a] = [\nabla \mathbf{P} \times \mathbf{R}] \tag{8.14}$$

The moment of the Earth's gravity force is equal to zero, because this force is directed along the radius $\mathbf{R}$. Projecting the vector equation 8.14 onto the axis of the

Earth's rotation, we obtain:

$$\rho\frac{dl}{dt} = -\frac{\partial P}{\partial\lambda} \tag{8.15}$$

where $l = \Omega R^2 \sin^2\theta + u\,R\sin\theta = (\Omega + \alpha)R^2\sin^2\theta$ is the projection of the absolute angular momentum of the unit air mass onto the axis of the Earth's rotation; $u = \alpha R\sin\theta$ is the velocity of the zonal wind (the positive direction is eastward); α is the angular velocity of the zonal air motion relative to the Earth's surface; θ is the polar angle, or the colatitude φ up to $\pi/2$; λ is the longitude that is read off eastward.

Notice that the projections of the absolute angular momentum onto the axes located in the equatorial plane are negligibly small as compared with l, because the angle of deviation of the instant vector $\mathbf{\Omega}$ from the long-term mean does not exceed 10^{-6} rad (Sidorenkov, 1973) and $\mathbf{V}_0 \ll \mathbf{V}_\mathrm{a}$.

The left part of Equation 8.15 can be written out (with the help of the equation of continuity) in the form

$$\rho\frac{dl}{dt} = \frac{d\rho l}{dt} - l\frac{d\rho}{dt} = \frac{\partial\rho\, l}{\partial t} + \nabla\cdot\rho\, l\boldsymbol{V}_\mathrm{a} - l\left(\frac{d\rho}{dt} + \rho\nabla\cdot\boldsymbol{V}_\mathrm{a}\right) = \frac{\partial\rho l}{\partial t} + \nabla\cdot\rho\, l\boldsymbol{V}_\mathrm{a} \tag{8.16}$$

Let us write down the divergence $\nabla\cdot\rho l\mathbf{V}_\mathrm{a}$ of the flow of the projection of the absolute angular momentum in the spherical reference frame and substitute (8.16) into Equation 8.15. Then we obtain

$$\frac{\partial\rho l}{\partial t} = -\frac{1}{R^2}\frac{\partial}{\partial R}(R^2\rho l v_R) - \frac{1}{R\sin\theta}\frac{\partial}{\partial\theta}(\sin\theta\,\rho l v_\theta) - \frac{1}{R\sin\theta}\frac{\partial}{\partial\lambda}(\rho l v_\lambda) - \frac{\partial P}{\partial\lambda} \tag{8.17}$$

Since our interest is the zonal circulation of the atmosphere, we average Equation 8.17 over the longitude, that is, we integrate all its terms over λ from 0 to 2π. Doing so, we take into account that the integrals of two last components are equal to zero, because the discontinuities of pressure P and velocity u on the mountain ridges can be neglected. Introducing the conventional designations $\bar{x} = \frac{1}{2\pi}\int_0^{2\pi} x d\lambda$, we have:

$$\frac{\partial\overline{\rho l}}{\partial t} = -\frac{1}{R^2}\frac{\partial}{\partial R}\left[R^2(\overline{\rho l}\,\overline{v_R} + \overline{\rho\, l'v'_R})\right] - \frac{1}{R\sin\theta}\frac{\partial}{\partial\theta}\left[\sin\theta\,(\overline{\rho l}\,\overline{v_\theta} + \overline{\rho l'v'_\theta})\right] \tag{8.18}$$

The components of the form of $\overline{\rho l}\,\overline{v_i}$ and $\overline{\rho\, l'\, v'_i}$ in Equation 8.18 reflect the transport of the projection of the absolute angular momentum by the ordered (streamline) circulation and by eddies (the turbulence), respectively. The investigations on the components of the angular momentum balance in the atmosphere, the results of which are generalized in monographs (Lorenz, 1967; Starr, 1968), have shown that the angular momentum transport is mainly due to the turbulence. The role of the ordered (streamline) circulation in the atmosphere is relatively small. Relying on these empirical data, we can neglect the terms of the form of $\overline{\rho l}\,\overline{v_i}$ (as opposed to those of $\overline{\rho\, l'\, v'_i}$).

As is known, the intuitive phenomenological relationships that connect the flows of different physical parameters with their gradients are widely used in physics.

In particular, in the semiempirical theory of turbulence, the flows of the momentum, which are described by Reynolds stresses, are considered to be proportional to the gradient of the mean wind velocity. According to these hypotheses, the flows of angular momentum in the vertical and meridian directions can be expressed by the following approximate equations, respectively:

$$\overline{\rho l' v'_R} \approx -A_R \frac{\partial \bar{l}}{\partial R} \quad \text{and} \quad \overline{\rho l' v'_\theta} \approx -A_\theta \frac{\partial \bar{l}}{R\,\partial \theta} \tag{8.19}$$

where A_R and A_θ are the interchange coefficients of the vertical and meridian turbulent mixing, respectively.

Notice that the classic hydrodynamics, which are based on the theorem of the symmetry of the stress tensor, does not allow approximations (8.19). However, the criterion of truth is observations rather than attractive theorems. Numerous empirical studies have shown that in the real atmosphere, there is a well-pronounced transport of the angular momentum in the vertical (due to the microturbulence mixing) and meridian (due to the macroturbulence mixing) directions but the compensating transport along the horizontal and the parallel circle is absent (Lorenz, 1967; Matveev, 1965; Starr, 1968), that is, according to the data of observations, the stress tensor is asymmetric. This observational fact and arguments in Section 8.2 provide the basis for approximations $\overline{\rho\, l' v'_R} \approx -A_R \frac{\partial \bar{l}}{\partial R}$ and $\overline{\rho\, l' v'_\theta} \approx -A_\theta \frac{\partial \bar{l}}{R\,\partial \theta}$.

Substituting (8.19) into (8.18), we obtain the differential equation of the second order in partial derivatives for the absolute angular momentum projection onto the axis of the Earth's rotation:

$$\frac{\partial \rho \bar{l}}{\partial t} = \frac{1}{R^2 \sin\theta} \cdot \frac{\partial}{\partial \theta}\left(\sin\theta\, A_\theta \frac{\partial \bar{l}}{\partial \theta}\right) + \frac{1}{R^2}\frac{\partial}{\partial R}\left(R^2 A_R \frac{\partial \bar{l}}{\partial R}\right) \tag{8.20}$$

Since $k_\theta = A_\theta/\rho$, and $k_R = A_R/\rho$, Equation 8.20 coincides with Equation 8.12.

Equations 8.12 and 8.20 show that the rate of changes in the projection of the absolute angular momentum in the unit volume is determined by the sum of convergences of the angular momentum, or, which is the same, of the torque arising due to the angular momentum macroturbulent transport (the first term) and microturbulent transport (the second term in the right part of formulas (8.12) and (8.20). Equation 8.12 allows us to calculate the zonal circulation of the atmosphere, because the distribution of l_r value in the atmosphere at rest (but rotating together with the Earth) is known.

Let us try to find the solution to Equation 8.12. For this purpose, let us choose the boundary conditions and make simplifications. An obvious boundary condition is the condition of the "adhesion" of air to the Earth's surface:

$$\alpha = 0,\ l = \Omega R_0^2 \sin^2\theta \quad \text{for} \quad R = R_0 \tag{8.21}$$

where R_0 is the Earth's radius. At the upper boundary of the atmosphere, where $R = R_\infty$, the vertical flow of the angular momentum approaches zero; therefore, we may assume:

$$\partial l/\partial R = \partial \alpha/\partial R = 0 \quad \text{for} \quad R = R_\infty \tag{8.22}$$

The coefficients of macro- and microturbulence viscosity depend on the latitude and height. As a first, very rough approximation, we may assume

$$k_\theta = \text{const} \tag{8.23}$$

$$k_R = k_z \sin\theta \tag{8.24}$$

where $k_z = \text{const}$. We cannot assume the coefficient k_R to be the latitude independent, because, as is shown in the theory of the atmosphere boundary layer, this coefficient is proportional to the wind velocity $u = \alpha R \sin\theta$.

Let us consider the stationary (the time-independent) zonal circulation of the atmosphere. In this case

$$\frac{\partial}{\partial R}\left(k_R R^2 \frac{\partial l}{\partial R}\right) = -\frac{1}{\sin\theta}\frac{\partial}{\partial\theta}\left(k_\theta \sin\theta \frac{\partial l}{\partial\theta}\right) \tag{8.25}$$

Substituting the coefficients k_θ and k_R from (8.23) and (8.24) into (8.25) and taking into account that $\partial l/\partial R \approx R^2 \sin^2\theta\, \partial\alpha/\partial R$ and $\partial l/\partial R \approx \Omega R^2 \sin 2\theta$ we obtain:

$$\frac{\partial}{\partial R}\left(R^4 \frac{\partial\alpha}{\partial R}\right) = -6\Omega \frac{k_\theta}{k_z} \cdot \frac{2/3 - \sin^2\theta}{\sin^3\theta} R^2 \tag{8.26}$$

Let us introduce the designations for the constant value $2\Omega k_\theta/k_z = G$ and for the function of latitude $(2/3 - \sin^2\theta)/\sin^3\theta = F(\theta)$. Integrating Equation 8.26, we obtain:

$$\alpha = -GF(\theta)\ln R - C_1(\theta)/3R^3 + C_2(\theta) \tag{8.27}$$

The unknown functions of latitude, $C_1(\theta)$ and $C_2(\theta)$, are easily to find from conditions (8.22) and (8.21), respectively:

$$C_1(\theta) = GF(\theta) R_\infty^3 \tag{8.28}$$

$$C_2(\theta) = GF(\theta)(\ln R_0 + R_\infty^3/3R_0^3) \tag{8.29}$$

Substituting these functions into expression (8.27), we obtain the final expression for the relative angular velocity of the atmosphere rotation:

$$\alpha = GF(\theta) f(R) \equiv 2\Omega \frac{k_\theta}{k_R} \cdot \frac{2/3 - \sin^2\theta}{\sin^3\theta}\left[\frac{R_\infty^3}{3}\left(\frac{1}{R_0^3} - \frac{1}{R^3}\right) - \ln\frac{R}{R_0}\right] \tag{8.30}$$

Formula (8.30) shows that in both hemispheres the relative angular velocity (and, consequently, the wind velocity) is negative at latitudes $< 35^\circ$ and positive at latitudes 35°, the sign changing at 35°N and 35°S. The increase in the velocity with height follows a complicated law. Formula (8.30), which describes the velocity of zonal wind, virtually coincides with formula (8.1) obtained from empirical data. This validates our theory.

Let us assess the angular momentum of the atmosphere zonal circulation obtained in the above way. As a first approximation, let us consider the uniform atmosphere with constant density $\rho = \rho_0 = 1.29\,\text{kg}\,\text{m}^{-3}$, so that its height is equal to $R_\infty - R_0 \approx 8\,\text{km}$:

$$h = \int_W \alpha R^2 \sin^2\theta \rho \, dV = 2\pi G \rho_0 \int_{R_0}^{R_\infty} R^4 f(R) dR \int_0^\pi F(\theta) \sin^3\theta \, d\theta$$

$$= \frac{1}{3}\pi^2 G \rho_0 R_0^5 \int_0^{x_\infty} (1+x^4) \left[\frac{(1+x_\infty)^3}{3} \left(1 - \frac{1}{(1+x)^3} \right) - \ln(1+x) \right] dx$$

$$\approx \frac{8}{9}\pi^2 \Omega \frac{k_\theta}{k_z} \rho_0 R_0^2 (R_\infty - R_0)^3 = 17 \times 10^{21} \frac{k_\theta}{k_z} \text{kg}\,\text{m}^2\,\text{s}^{-1} \qquad (8.31)$$

Here, $(R - R_0)/R_0 = x$; $(R_\infty - R_0)/R_0 = x_\infty$ and it is assumed that $R_0 = 6.37 \times 10^6\,\text{m}$; $\Omega = 7.29 \times 10^{-5}\,\text{s}^{-1}$.

According to empirical data, the values of the coefficients of macroturbulence viscosity are within the limits of 10^6–$10^7\,\text{m}^2\,\text{s}^{-1}$ (Panchev, 1967) and those of microturbulence viscosity are within 1–10 $\text{m}^2\,\text{s}^{-1}$ (Matveev, 1965). Thus, $k_\theta/k_z \approx 10^6$ and, consequently, the angular momentum $h \approx 17 \times 10^{27}\,\text{kg}\,\text{m}^2\,\text{s}^{-1}$, or approximately 140 times larger than its observed value. This is likely to be associated with the fact that the values of the coefficient of microturbulence viscosity k_z that are used in meteorology do not coincide with its values for the entire atmosphere (because they are usually calculated from the local rather than planetary data). Note that I.A. Kibel and N.E. Kochin arrived at similar conclusions. When calculating the zonal circulation from the given distribution of temperature in the atmosphere, they obtained plausible results only at the coefficients k_z of the order of magnitude of $10^3\,\text{m}^2\,\text{s}^{-1}$ (Izekov and Kochin, 1937).

However, we did not pose the problem to determine the precise value of h. We had only to show that the angular momentum of zonal winds differs from zero and is positive; hence it is adopted from the Earth. This is substantiated by expression (8.31).

Thus, the qualitative semiempirical theory developed above adequately explains the main features of the zonal circulation observed in the troposphere. A more plausible pattern of the atmosphere zonal circulation can be obtained by way of significant refining the coefficients of turbulence viscosity (k_θ and k_R) specified in the calculations.

The theory of zonal circulation formulated above is valid for any case of mixing, including the molecular viscosity. In the latter case, it is only sufficient to replace the coefficients of turbulent viscosity by those of molecular viscosity. Recall that the coefficient of molecular viscosity in the atmosphere is approximately equal to $1.5 \times 10^{-5}\,\text{m}^2\,\text{s}^{-1}$ and the coefficient of horizontal macroturbulent viscosity is 10^6–$10^7\,\text{m}^2\,\text{s}^{-1}$, that is, the first coefficient is by about 11 orders of magnitude smaller than the second coefficient. Also, in the case of molecular mixing, the horizontal and vertical coefficients of viscosity are almost equal, whereas in the case of turbulent mixing the ratio k_θ/k_R is 10^6–10^4. Thus, the zonal circulation that originates due to the molecular viscosity is so small that it is rather difficult to find.

8.6 Nature of the Subtropical Maxima of Atmospheric Pressure

Let us take the spherical reference frame that is rigidly bound with the Earth and rotates with it with the constant angular velocity Ω. The equation for the relative motion of the unit air mass in this system has the form

$$dV/dt = -\nabla P/\rho - 2(\Omega \times V) + g + F \tag{8.32}$$

Here, P is the atmospheric pressure; **g** is the acceleration of gravity; F is the friction force.

Let us consider the zonal circulation of air in the free atmosphere, where the friction force F can be neglected. At the zonal circulation, air moves parallel to the latitudinal circles. Therefore, the wind velocity can be presented in the form of $V = dr/dt = \alpha \times r$ and the relative acceleration in the form

$$dV/dt = (\boldsymbol{\alpha} \times dr/dt) = \boldsymbol{\alpha} \times (\boldsymbol{\alpha} \times r) = -\alpha^2 r \tag{8.33}$$

Here, α is the vector of the angular velocity of the air rotation relative to the Earth's surface, r is the radius vector, which is perpendicular to the axis of the Earth's rotation and directed from it toward the air particle under consideration. One can see that in the case of zonal circulation, the relative acceleration is simply the centripetal acceleration.

The expression for the Coriolis force can be transformed in the same way:

$$2(\Omega \times V) = 2(\Omega \times (\boldsymbol{\alpha} \times r)) = -2\Omega \alpha r \tag{8.34}$$

It is taken into account that in expressions (8.33) and (8.34) $\boldsymbol{\alpha} \cdot r = \Omega \cdot r = 0$. Substituting expressions (8.33) and (8.34) in (8.32), we obtain the equation for the zonal circulation of air without friction:

$$-\nabla P/\rho + (2\Omega\alpha + \alpha^2)r + g = 0 \tag{8.35}$$

Let $\mathbf{e}_\lambda$, $\mathbf{e}_\theta$ and $\mathbf{e}_R$ be the orthogonal unit vectors, which are directed eastward at a tangent to the latitudinal circle, northward at a tangent to the meridian, and vertically upward. Recall that the direction of the vertical on the Earth's surface is specified by the summarized vector **g** of Newton's gravitation forces and the centripetal force, which acts on the unit mass rotating together with the Earth at the angular velocity Ω.

Let us write out Equation 8.35 in the projections onto the direction $\mathbf{e}_\theta$. For this, we multiply the scalar value of each term by the unit vector $\mathbf{e}_\theta$:

$$(-1/\rho R)(\partial P/\partial\theta) + (2\Omega\alpha + \alpha^2)R\sin\theta\cos\theta = 0 \tag{8.36}$$

Here it is taken into account that $\mathbf{g}\cdot\mathbf{e}_\theta = g\mathbf{e}_R\cdot\mathbf{e}_\theta = 0$; $\mathbf{r}\cdot\mathbf{e}_\theta = R\sin\theta\cos\theta$. Since the angular velocity of the Earth's rotation Ω is several tens of times more than the angular velocity of the atmosphere rotation α from west to east relative to the Earth's surface, then the term containing α^2 in Equation 8.36 can be neglected. Substituting expression (8.30) for α into (8.36) and dividing the variables of integration, we obtain:

$$dP = 2\rho\,\Omega\alpha R^2 \sin\theta\cos\theta\,d\theta = 4\rho\Omega^2 R^2 \frac{k_\theta}{k_z} f(R) \frac{2/3 - \sin^2\theta}{\sin^2\theta} \cos\theta\,d\theta \tag{8.37}$$

Integrating Equation 8.37 under the condition that at the equator $P(90^\circ) = P_e$ we have:

$$P = P_e + 4\rho\Omega^2 R^2 \frac{k_\theta}{k_z} f(R) \left[\frac{-3\sin^2\theta + 5\sin\theta - 2}{3\sin\theta} \right] \qquad (8.38)$$

The function of latitude (square brackets in (8.38)) is equal to zero at the equator. When moving toward the poles, this function increases and reaches its maxima at the colatitudes 54.7° and 125.3° in the Northern and Southern Hemispheres, respectively; then it decreases as far as the poles. Thus, formula (8.38) shows that the atmospheric pressure has the subtropical maximum at the 35.3° latitude and it decreases when moving toward the equator and the poles.

The subtropical maxima originate under the effect of the Coriolis force horizontal components, which displace the atmospheric air from the belt of westerly winds toward the lower latitudes and from the belt of easterly winds in the opposite direction, toward the higher latitudes. The "horse latitude" (35°) is the latitude toward which the air is pumped from these two wind belts. The atmospheric pressure in the subtropical high-pressure zone grows under the effect of the Coriolis force horizontal components for so long as they are counterbalanced by the horizontal pressure gradient. If we had not cancelled the term containing α^2 from (8.36), that is, if we had accounted for the horizontal component of the centrifugal force of the relative rotation of air, then an additional, very small, term would have been added:

$$4\rho\Omega^2 R^2 \frac{k_\theta^2}{k_z^2} f^2(R) \left[\frac{6\sin^2\theta - 5\sin^4\theta + 9\sin^4\theta \ln|\sin\theta| - 1}{9\sin^4\theta} \right] \qquad (8.39)$$

The value of this term progressively increases from $-\infty$ at the poles to 0 at the equator; that is, the subtropical high-pressure zone can be slightly displaced toward the equator at the expense of this term. This is clear, because the centrifugal force of the relative air motion initiates the displacement of air masses toward the equator; due to this, some additional growth of the equatorial bulging of the atmosphere is observed.

8.7 Mechanism of Seasonal Variation

It has been known even since the last century that the trade winds near the Earth's surface cross the equator and penetrate into the summer hemisphere as far as the intertropical convergence zone (ITCZ). The latter migrates along the meridian, depending on the season of the year. The mean latitude of the ITCZ location varies from 5° S in January to 15° N in July. It was shown in the 1960s that in the upper layers of the equatorial zone there existed the compensating air flows from the summer hemisphere into the winter hemisphere (Kidson, Vincent and Newell, 1969; Rao, 1964). The direction of these flows coincides with the air circulation in the Hadley cell of the winter hemisphere. It is since considered that the Hadley cell of the winter hemisphere extends into the summer hemisphere, following the Sun. The boundary, to which the Hadley cell penetrates into the summer hemisphere, is the ITCZ. The latter separates the circulation of the winter hemisphere from the circulation of the summer hemisphere. Therefore, it is assumed in many studies that the meridian flows of the mass, moisture, pollutants, and other substances do not penetrate through the ITCZ and the air masses lying to the north and to the south of the ITCZ, respectively, are dynamically isolated from each other.

Below, it will be shown that this concept is not quite consistent with the real pattern of atmospheric circulation. In December–February and in June–August, there exists the global cell of atmospheric circulation between the polar areas of the Northern and Southern Hemispheres. It affects the processes of the mass, moisture, and pollutants transport not only in the tropical zone but also in all other zones of the Earth and is responsible for complicated seasonal variations in the angular momentum and indices of atmospheric circulation, concentrations of gaseous and ionic admixtures, and so on.

The initial cause of atmospheric circulation is the horizontally nonuniform heating of the atmosphere. From the positions of hydrodynamics, the generation of kinetic energy associated with this heating can be considered as the phenomenon of a heating engine.

It is well known that during the whole year, the air temperature in the troposphere decreases from the equator to the poles. This uneven heating up of the troposphere gives rise to the heating engine of the first type, the HEFT (Shuleykin, 1968). The main consequence of the HEFT operation is the zonal air circulation in the troposphere (westerly winds at moderate and high latitudes and easterly winds at low latitudes) and the meridian circulation (the direct Hadley cell at the subtropical latitudes and the reverse and direct Ferrel cells at moderate and polar latitudes, respectively).

It is also known that the air is heated differently over oceans and continents, in winter and summer. This temperature contrast gives rise to the heating engine of the second type. The main consequence of its operation is the monsoonal circulation and the redistribution of air masses between continents and oceans.

Apart from these two universally known heating engines, there is the interhemispheric heating engine (IHHE), which was first described in Sidorenkov (1975). The IHHE exists due to the temperature contrast between the winter and summer hemispheres. This contrast is easily seen from the analysis of the latitudinal change in air temperature in the lower layer of the atmosphere (the layer thickness is >20 km) rather than in the troposphere. Let us consider, for example (following Sidorenkov, 1988), the latitudinal changes in the zonal temperature in the layer between the 1000 and 30 hPa levels ($T = 9.738 \cdot H$, where H is the thickness between the 1000 and 30 hPa in km) (Figure 8.10). The initial data are taken from the climatic atlases (Zastavenko, 1972, 1975). It is seen from Figure 8.10 that unlike in the troposphere, the temperature gradient in the layer between the 1000 and 30 hPa levels does not change its direction at the equator. In January, T progressively decreases from the South Pole toward the North Pole. The temperature difference between the poles is 21 °C. In July, on the contrary, T decreases from the North Pole toward the South Pole, the temperature differing by 34 °C.

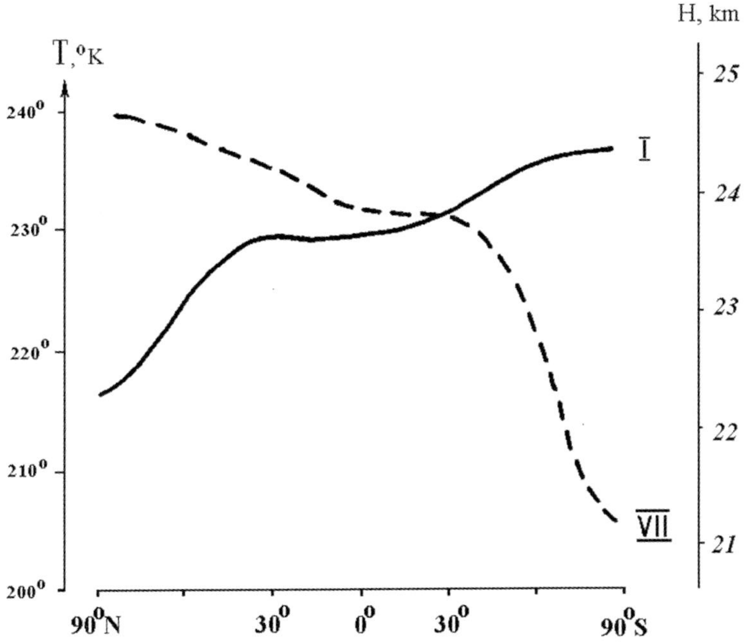

Figure 8.10 Latitudinal course of the zonal temperature T or the thickness H between the 1000 and 30 hPa geopotential levels in January (I) and July (VII). The H is measured in hPa kilometers.

Thus, if we bear in mind the mean temperature in the layer between the 1000 and 30 hPa levels, then the warmest atmosphere is not over the equator and not even over the tropic but over the pole of the summer hemisphere; the coldest atmosphere is over the pole of the winter hemisphere.

This latitudinal distribution of temperature should give rise to the cell of air circulation between the polar areas of the summer and winter hemispheres. This circulation is directed along the meridian from the summer hemisphere to the winter hemisphere at the top of the layer under consideration and in the opposite direction at its bottom. In order to check this conclusion, we used the meridian components v_θ of the wind velocity vector for January and July (Guterman, 1975, 1978) and averaged them over the longitude.

As a result, we had the meridian sections of velocity v in January (Figure 8.11) and July (Figure 8.12), which were similar to those obtained in (Guterman and Khanevskaya, 1972). The Hadley and Ferrel cells in these sections are strongly distorted by some other circulation superimposed on these cells.

To distinguish this circulation from the summary pattern, we averaged the velocities v over January and over July: $\bar{v} = \frac{1}{2}\{v(I) + v(VII)\}$ (that is, we singled out the Hadley and Ferrel cells); then we calculated the deviations (anomalies) $v' = v - \bar{v}$ of the v values from the mean velocity $\bar{v}$ for January and July, respectively (that is, we singled out the superimposed interhemispheric circulation).

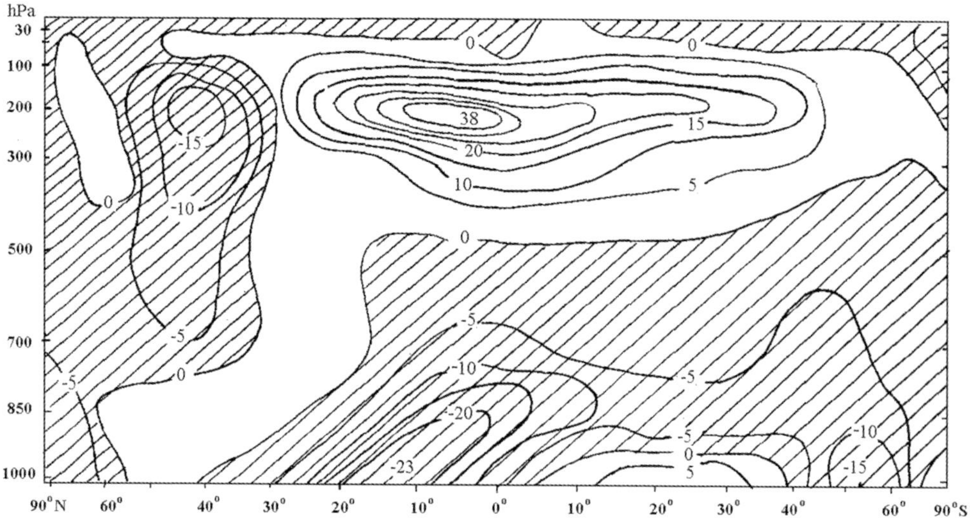

Figure 8.11 Height–latitude section of the longitude-averaged meridian wind in January (positive direction is from the South to the North; the isotachs are digitalized in 0.1 m/s).

Figure 8.13 demonstrates the distribution of v' with latitude and height in January. The meridian section of v' in July is similar to that in January but the circulation has the opposite direction. As was earlier presumed, in the lower layer of the atmosphere (up to 450 hPa) the northerly winds are prevalent in January and the southerly

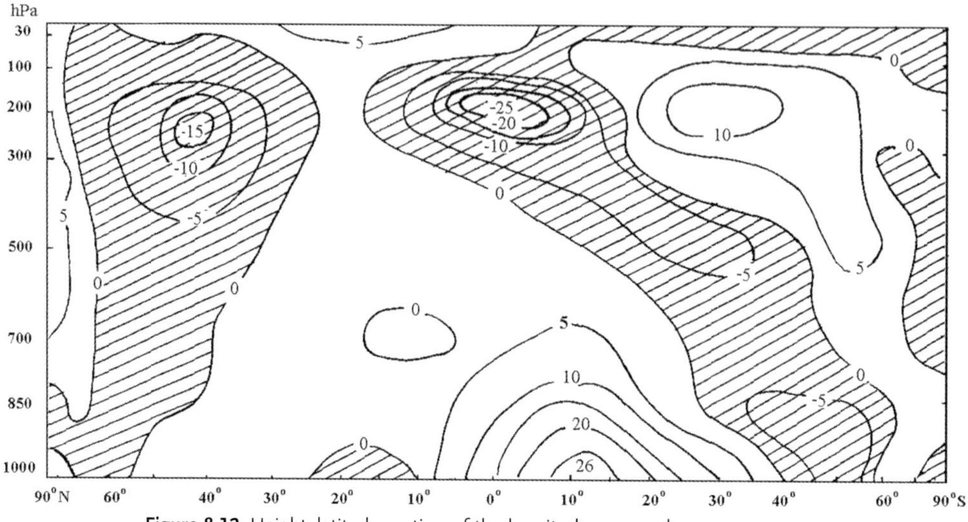

Figure 8.12 Height–latitude section of the longitude-averaged meridian wind in July (positive direction is from the South to the North; the isotachs are digitalized in 0.1 m/s).

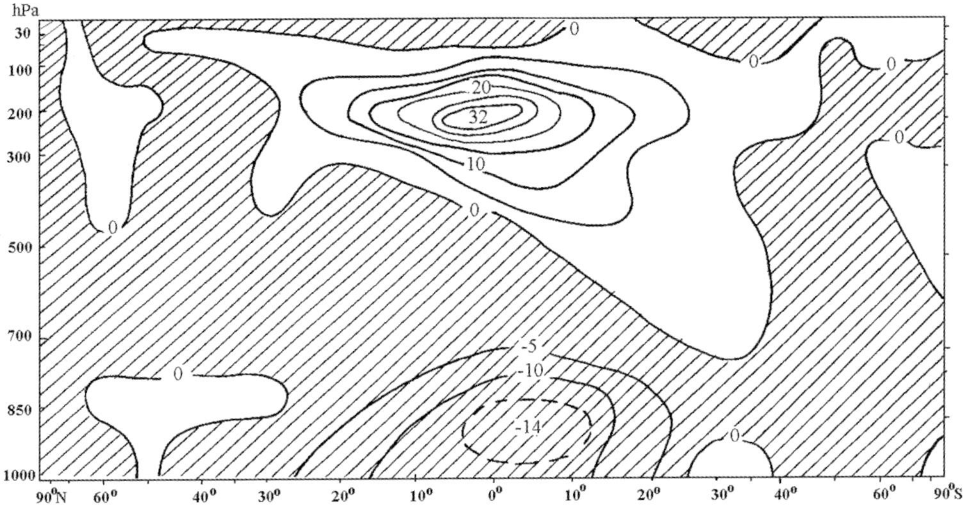

Figure 8.13 Height–latitude section of the longitude-averaged anomalies of the meridian wind in January (positive direction is from the South to the North; the isotachs are digitalized in 0.1 m/s).

winds – in July. In the upper layer (450–50 hPa), on the contrary, the southerly winds are prevalent in January and the northerly winds in July. This interhemisphere circulation is particularly distinct at low latitudes, where wind velocities reach 1.4 and 3.2 m/s at the bottom and top of the atmosphere, respectively. As one would expect, the circulation weakens when moving from the equator toward the poles. At moderate latitudes, there are slight deviations from the interhemisphere circulation, which are due to an incomplete elimination of the Hadley and (especially) Ferrel cells when calculating v'.

In January, the velocity $\bar{v}'$ averaged over the latitude with the account for the spherical geometry increases from $-0.45\ \mathrm{m\,s^{-1}}$ (near the Earth's surface) to $0.76\ \mathrm{m\,s^{-1}}$ (at the 200 hPa level), the sign changing at the 450 hPa level. In July, the velocities $\bar{v}'$ have the opposite sign: below 450 hPa the southerly winds prevail and above 450 hPa the northerly winds prevail.

Thus, the empirical data confirm the existence of the global cell of the interhemisphere circulation (IHC) between the polar areas of the summer and winter hemispheres. Note that the velocity of the meridian wind drastically changes along parallels. The most intense transport of the air and the inherent in it substances occurs through few narrow meridian "tubes" – the jet streams located near the upper boundary of the planetary boundary layer of the atmosphere. In some sectors, the direction of transport can be opposite. However, the velocity of transport averaged over the whole perimeter of the parallel has the direction and value characteristic of the given height and season. Note also that since the mean velocity $\bar{v}'$ varies within ± 0.45 m/s, then the mean displacement of particles along the meridian near the Earth's surface over half a year is a mere 4500 km, or 41° of latitude.

The larger the difference between temperatures of the summer and winter hemispheres, the more intense is the IHC. Consequently, the velocity v' in the lower layer of the atmosphere should change with time:

$$v' \sim T_N - T_S \approx \Pi + E \cos \Omega (t - t_N) \tag{8.40}$$

Here, T_N and T_S are the integral temperatures of the Northern and Southern Hemisphere, respectively, which vary with the yearly period:

$$\begin{aligned} T_N &= \bar{T}_N + A_N \cos \Omega (t - t_N) \\ T_S &= \bar{T}_S + A_S \cos \Omega (t - t_S) \end{aligned} \tag{8.41}$$

where $\bar{T}_N$ and $\bar{T}_S$ are the mean annual values of $\bar{T}_N$ and $\bar{T}_S$. A_N, A_S, Ωt_N, Ωt_S are the amplitudes and the initial phases of annual variations of T_N and T_S; Ω is the circular frequency; t is the time; the constant $\Pi = \bar{T}_N - \bar{T}_S$ is the mean annual superheating (excess of heating up the Northern Hemisphere as compared with the heating up the Southern Hemisphere); $E = A_N + A_S$. is the amplitude. When deriving (8.40), it is taken into account that the annual variations in T_N and T_S are nearly opposite in phase, that is, $\Omega(t_N - t_S) \approx \pi$.

On average over a year, the Northern Hemisphere is warmer than the Southern Hemisphere ($\Pi > 0$). Therefore, the southerly (positive) wind in the lower layers of the troposphere is observed longer (from April through November) than the northerly (negative) wind (from December through March).

A good indicator of the atmospheric heating-engine operation and the circulation cells initiated by it is the mass transport (Shuleykin, 1968). The existence of the IHC suggests a deficiency of air mass in the summer hemisphere and its excess in the winter hemisphere. Hence, some portion of air mass should overflow from the Northern Hemisphere into the Southern Hemisphere from January through June and return back from July through January. Indeed, careful calculations show that from January to July, the air mass decreases by 4×10^{15} kg in the Northern Hemisphere and increases by the same value in the Southern Hemisphere (Sidorenkov and Stekhnovsky, 1971). The respective increments of the hemisphere-averaged atmospheric pressure are 1.6 hPa.

The IHC should transport pollutants from one hemisphere into another. The direction and velocity of this transport depend on the season and the atmospheric layer containing these pollutants. Unfortunately, there are no purposeful observations of such transport. However, there are numerous data on the transport of the nuclear explosion products from the hemisphere where the explosions took place in another hemisphere (Malakhov, 1971). Some of these data are given in Table 8.1. One may see that the meridian velocity of the radioactive decay products transport is 2–5 m/s. This is much higher than the velocity of diffusive mixing but is close to the wind velocity v'.

Thus, the data on the transport of the nuclear explosion products substantiate the existence of the IHC in the atmosphere. This is also evidenced by the pesticides found in snow samples in the Antarctica, the pesticides being in use only in some countries of the Northern Hemisphere in the postwar years (Wolff and Peel, 1985).

Table 8.1 Examples of a rapid interhemisphere transport of the nuclear explosion products.

Date and latitude of explosion	Date and latitude of detection of the explosion products	Velocity of meridian transport (m/s)
13 February 1960	27 February 1960	5.0
26°N	35°S	
1 April 1960	1–3 May 1960	2.9–2.6
26°N	41°S	
16 October 1964	December 1964	2.3–1.4
40°N	41°S	
14 May 1965	from 3 June to 3 July 1965	5.1–2.2
40°N	41°S	
9 May 1966	13–20 June 1966	2.4–2.0
40°N	26°–34°N	
July 1966	August 1966	2.0–1.2
23°S	23°N	
July 1968	September 1968	1.5–1.1
22°S	50°N	

Since about 90% of atmospheric moisture is contained in the lower 5-km layer and the wind in this layer flows from the winter hemisphere into the summer hemisphere, then the IHC should carry away the atmospheric moisture from the winter hemisphere into the summer hemisphere. Due to this moisture transfer, precipitation should be more abundant in summer (at all latitudes) than in winter. Unfortunately, it is very difficult to verify this proposition, because the summer increase in evaporation also contributes to atmospheric precipitation. Indirect evidence of the atmospheric moisture transfer from the winter hemisphere into the summer hemisphere is the seasonal course of precipitation on the mountain slopes that are extended along the parallel and present a barrier on the route of the moisture flow. An example is the southern slopes of the Himalayas in India. In June–September, when the air moisture is directed from the Southern Hemisphere into the Northern Hemisphere (toward the mountain slopes), precipitation amounts to 11 000 mm; in December–March, when the air moisture is transferred in the opposite direction, precipitation scarcely falls there, as one would expect.

The IHC intensity changes with time. The more intense the IHC, the larger amount of air moisture is transferred from the winter hemisphere into the summer hemisphere and the more abundant is precipitation in the summer hemisphere (and the less abundant in the winter hemisphere). When this circulation is weak, the distribution of precipitation anomalies is reversed: the precipitation deficiency takes place in the summer hemisphere and the precipitation excess in the winter hemisphere. It is clear that the latitude-averaged moisture flux q along the meridian is proportional to the velocity v' below the 450 hPa surface and changes with time in the same way as the velocity v' does:

$$q \sim v' \sim \Pi + E\cos(t - t_{\mathrm{N}}) \tag{8.42}$$

On this basis, we can conclude that from November through April the air moisture is transferred from the Northern Hemisphere into the Southern Hemisphere ($q < 0$) and from May through October from the Southern Hemisphere into the Northern Hemisphere ($q > 0$). When integrating expression (8.42) over time for the annual period P, the second component is equal to zero; then the annual amount of moisture Q that comes from the Southern Hemisphere into the Northern Hemisphere is

$$Q \sim \Pi \cdot P \tag{8.43}$$

Since $\Pi > 0$, the positive (from the Southern Hemisphere into the Northern Hemisphere) moisture flux exceeds the negative flux, and $Q > 0$. As superheating Π grows, the moisture flux Q from the Southern Hemisphere into the Northern Hemisphere increases. As superheating weakens, the moisture flux Q weakens too. As will be shown below, the value of the superheating Π can be determined from the analysis of seasonal variations in the angular momentum of the atmosphere. The deviations of the moisture amount Q from its normal value can be judged from the anomalies of Π. The characteristic time of the change in superheating Π is significant (several months). Thus, having determined the value of Π, we can assess the anomaly of Q and the expected hemisphere-averaged anomaly of precipitation within the next few months.

The seasonal variations in the angular momentum of zonal winds h and in the indices of the atmosphere zonal circulation are described by a complicated curve (see Figure 8.14b), which has two maxima (in November and April) and two minima (in August and January). The minimum in August is much more pronounced than that in January. These variations are due to the IHC seasonal changes. This dependence can be demonstrated with the help of thermodynamic analysis of the atmospheric heat engines. Let us perform this analysis in the most simplified form, that the physical essence of the problem was not complicated by cumbersome mathematical operations. A more rigorous demonstration can be found in the author's publications (Sidorenkov, 1975b, Sidorenkov, 1978).

Any atmospheric heat engine transforms the heat energy into the kinetic energy of wind. The rate of the kinetic energy generation is called the engine capacity $\Re$. The angular momentum h is directly proportional to the engine capacity: $h \sim \Re$. The capacity $\Re$ is in its turn proportional to the amount of solar energy that comes to the heat engine in a unit time (the solar radiation power W): $\Re \sim \eta W$, where η is the efficiency coefficient, which is directly proportional to the difference between the temperatures of the heater and the cooler ($\eta \sim T_H - T_C$). Thus, the whole chain of relationships has the form:

$$h \sim \Re \sim \eta W \sim (T_H - T_C) W \tag{8.44}$$

There are several heat engines in the atmosphere (Shuleykin, 1968). The heat engines of the second type effect but on the angular momentum of the meridian winds, that is, on the equatorial components of the vector of the atmosphere angular momentum. As for the angular momentum h of zonal winds, which we are interested in, it is only affected by the heat engines that have the meridian components of

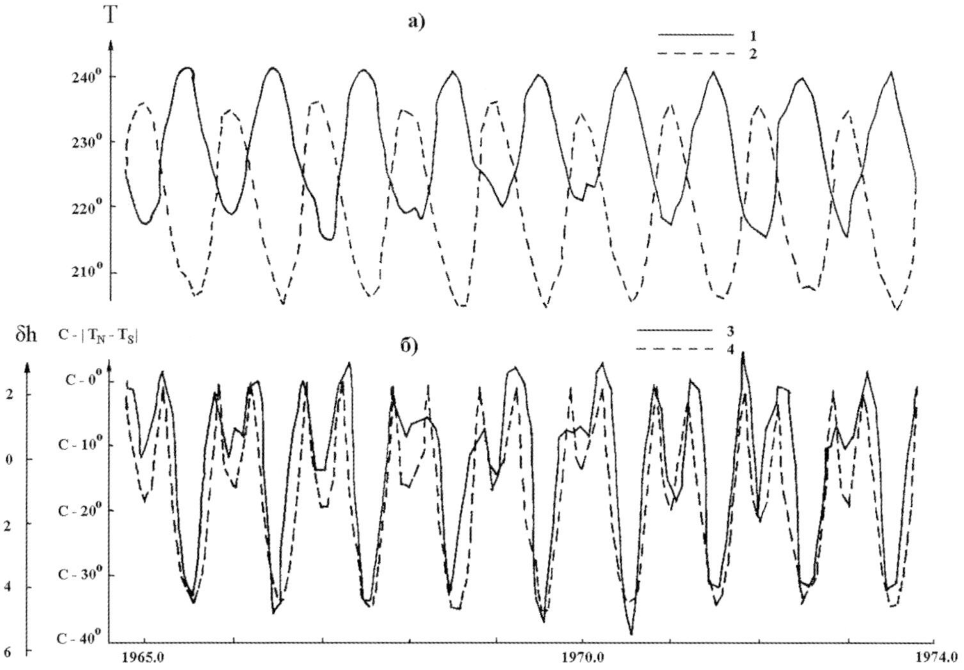

Figure 8.14 Seasonal variations: (top) in the air temperature T of the layer between the 1000 and 50 hPa levels at the North (curve 1) and South (curve 2) poles and also (bottom) of the angular momentum of zonal winds δh (curve 3) and the modulus of the difference of temperatures of the layer between the 1000– and 50 hPa levels at the North and South poles $C - |T_N - T_S|$ (curve 4).

temperature gradients. These are the heat engine of the first type (FTHE) and the interhemisphere heat engine (IHHE) (Sidorenkov, 1975b).

There are two FTHEs. One of them operates in the Northern Hemisphere and another in the Southern Hemisphere. Their heater is the equatorial zone and the coolers are the northern and southern polar caps, respectively. The capacities of these engines change with time according the following laws:

$$\Re_{\mathrm{N}} \sim \frac{1}{2}(T_{\mathrm{E}} - T^*_{\mathrm{N}})W; \quad \Re_{\mathrm{S}} \sim \frac{1}{2}(T_{\mathrm{E}} - T^*_{\mathrm{S}})W \tag{8.45}$$

Here, T_{E}, T^*_{N} and T^*_{S} are the mean absolute air temperatures at the equator and the northern and southern polar caps, respectively; W is the total solar radiation coming to the entire atmosphere in a unit time.

In the case of IHHE, the heater is the atmosphere of the summer hemisphere and the cooler is the atmosphere of the winter hemisphere. Therefore, its capacity changes in a more complicated way:

$$\Re_M \sim \left\{ \begin{array}{l} (|T_{\mathrm{N}} - T_{\mathrm{S}}|)W \cdot \cdot \text{when} \cdot \cdot T_{\mathrm{N}} > T_{\mathrm{S}} \\ (|T_{\mathrm{S}} - T_{\mathrm{N}}|)W \cdot \cdot \text{when} \cdot \cdot T_{\mathrm{S}} > T_{\mathrm{N}} \end{array} \right\} = |T_{\mathrm{N}} - T_{\mathrm{S}}|W \tag{8.46}$$

where T_N and T_S are the integral temperatures of the Northern and Southern Hemispheres, respectively. When transforming the right side of (8.46), it is taken into account that the expression in brackets is equivalent to the modulus of the difference between the above temperatures.

The angular momentum h of the entire atmosphere depends on the combined contribution of the FTHEs and IHHEs. It is known that the FTHEs redistribute the angular momentum between the low and moderate latitudes and adopt it from the Earth (Sidorenkov, 2002a, 2002b). The IHHEs do not adopt the angular momentum from the Earth; they only redistribute it between the winter and summer hemispheres (the positive angular momentum is accumulated in the winter hemisphere and the negative momentum in the summer hemisphere). The angular momentum redistribution between particular parts of the atmosphere does not affect the value of h for the entire atmosphere. Therefore, at first sight it seems that the IHHE does not affect the value of h. However, this is not so. The point is that the part of the atmosphere that participates in the IHHE operation is excluded from the FTHE operation, that is, the IHHE hinders the FTHE operation. Thus, the contribution of the IHHE to the combined capacity and corresponding to it angular momentum h should be negative.

Thus, having summed up the contributions of all heat engines, we obtain:

$$h \sim [1/2(T_E - T_N^*) + 1/2(T_E - T_S^*) - |T_N - T_S|]W \tag{8.47}$$

Let us consider temporal changes in all the values of expression (8.47). The change in the solar radiation power W that comes from the Sun over the year is due to the eccentricity of the Earth's orbit. The value of W in January is approximately by 7% larger than that in July. The change in the expression taken in square brackets is much more significant. For example, it is smaller by about 40% in July than in December. Therefore, hereafter we will neglect the variations in W and will consider the variations in the quantity taken in square brackets.

It is known that the air temperature at the equator is nearly constant in the course of the year, and the air temperatures over the polar caps and the hemispheres vary with the yearly period. The initial phases of variations in T_N^* and T_S^* and also in T_N and T_S differ by about 180°; the amplitudes of variations can be approximately equal, that is:

$$\begin{aligned} T_N^* &= \bar{T}_N^* + A^* \cos\Omega(t - t_0^*) \\ T_S^* &= \bar{T}_S^* - A^* \cos\Omega(t - t_0^*) \end{aligned} \tag{8.48}$$

where $\bar{T}_N^*$ and $\bar{T}_S^*$ are the mean annual air temperatures over the northern and southern polar caps, respectively; A^* is the amplitude; Ωt_0^* is the initial phase of the annual variations in the air temperature over the northern polar cap. Substituting (8.48) and (8.41) into (8.47), we obtain:

$$h \sim C - |T_N - T_S| = C - |\Pi + E\cos\Omega(t - t_0)| \tag{8.49}$$

where $C = T_E - 1/2(\bar{T}_N^* + \bar{T}_S^*)$ is a constant; the values Π, E and Ωt_0, enter into formula (8.40).

Figure 8.14 clearly demonstrates the origin of the seasonal variations in the angular momentum h. Here, the integral temperatures T_N and T_S are the mean monthly temperatures of the layer between the 1000 and 50 hPa levels in the centers of the circumpolar eddies, which are located close to the geographic poles. The center of the circumpolar eddy of the Southern Hemisphere coincides in most cases with the South Pole. Therefore, we used the data immediately from the Amundsen–Scott station. The layer between the 1000 and 50 hPa levels temperature variations in the Northern (T_N) and Southern (T_S) Hemispheres for 1965–1974 are shown in Figure 8.14a (curves 1 and 2, respectively). One can see that these temperatures vary with the yearly period and have the phase difference close to 180°. The superheating of the Northern Hemisphere (as compared with the heating of the Southern Hemisphere $\bar{T}_N > \bar{T}_S$) is also well pronounced. Figure 8.14b shows the changes in the value of $C - |T_N - T_S|$ (curve 4), which is proportional to the angular momentum h. One can see that this curve is the graph of function $C - |\Pi + E \cos \Omega(t - t_0)|$ and its form is very close to that of curve 3 showing the observed seasonal variations in the angular momentum h of zonal winds (Sidorenkov, 1988). The time periods of extremes in both curves coincide to an accuracy of one month. Twice in a year, in April and November, when the air temperatures in the Northern and Southern Hemispheres become equal, the angular momentum h attains its maximum values. In July and January the modules of the temperature differences $|T_N - T_S|$ are maximal, and the h value becomes minimal. The value $|T_N - T_S|$ is greater in July than in January (because of the superheating Π of the Northern Hemisphere as compared with the heating of the Southern Hemisphere). Hence, the minimum of h is much more pronounced in July than in January. The greater (less) the superheating Π, the more (less) significant are these distinctions. At $\Pi = 0$ the minima of h in July and January would be equal.

It is seen from formula (8.49) that for the minima in July (M) and January (m) it holds true (Figure 8.15):

$$M = E + \Pi; \; m = E - \Pi \tag{8.50}$$

Consequently, given the values of M and m, one can find the relative values of the amplitude E and the superheating Π:

$$E = 1/2(M + m); \; \Pi = 1/2(M - m) \tag{8.51}$$

The value of Π is used to forecast the anomaly of the hemisphere-averaged amount of precipitation Q by formula (8.43). Notice also that in the periods when the amplitude E is constant, it is possible to forecast the value of the succeeding minimum by the observed value of the current minimum:

$$M = 2E - m, \quad \text{or} \quad m = 2E - M. \tag{8.52}$$

This circumstance allows one to make prognostic conclusions on the intensity of the zonal and meridian circulations during the coming season (summer or winter).

The operation of FTHE provides a good explanation for the diurnal variations in the angular momentum h of zonal winds. The diurnal course of h has two maxima (at 0 and 12 h UTC) and two minima (at 6 and 18 h). At 0 and 12 h the Sun occurs over

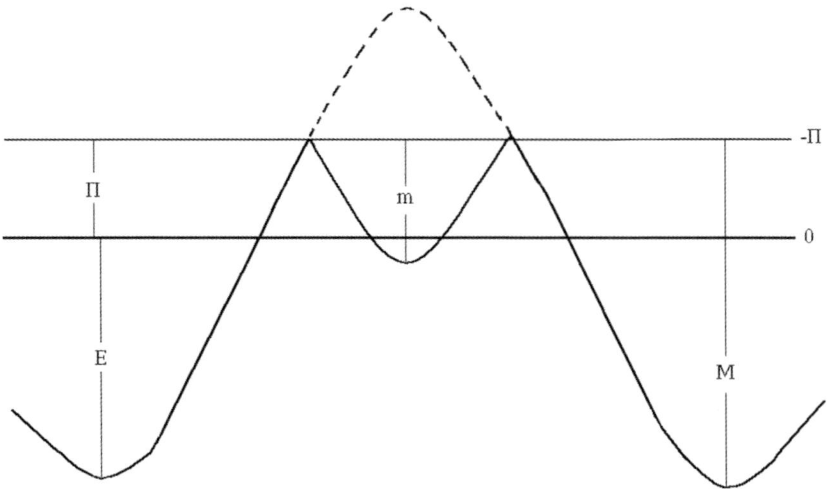

Figure 8.15 Scheme of forming two minima of the angular momentum *h* in an annual course.

the Pacific and Atlantic oceans, respectively; and at 6 and 18 h the Sun is over the Asian and American continents, respectively. The water vapor content of the atmosphere is greater over oceans than over continents. Due to this difference, the solar radiation is more intensely absorbed by the atmosphere over oceans than over continents. Therefore, the efficiency coefficient η of the first-type heat engine has the diurnal course with maxima at 0 and 12 h and minima at 6 and 18 h. In accordance with (8.44), the angular momentum h, which is held in the atmosphere by the FTHE, is proportional to η; therefore, it features similar diurnal course. The 18-h minimum is slightly pronounced, because the effect of the American continent is small as compared with that of the Asian continent.

The IHC is observed not only in the troposphere and stratosphere but also in the thermosphere, where it significantly affects the formation of the so-called "semiannual variations" in the concentrations of certain ions and molecules. It is not surprising that these variations are similar to the seasonal variations shown in Figure 8.14b and are described by formulas similar to (8.49). The mechanism of the origin of these variations can be represented (as a first, rough approximation) as follows.

The concentration r of the ion under consideration depends on the latitude φ. As a rule, on average over a year, r increases or decreases from the equator to the poles, having a maximum or minimum, respectively, at the equator. The measurements of r are performed at this or that fixed station. At the initial moment of time $t = 0$, the concentration $r(0) = f(\varphi_0)$, where φ_0 is the latitude of the station. The interhemisphere circulation transports the atmospheric air (together with the "frozen in" ions) over the station with the velocity $v' = R\dot{\varphi}$, where R is the Earth's radius and $\dot{\varphi}$ is the angular velocity. As a result, the concentration r at the time moment t will be the same as it is at the latitude $\varphi_0 + \Delta\varphi$:

$$r(t) = f(\varphi_0 + \Delta\varphi), \quad \text{where} \quad \Delta\varphi = 1/R\int_0^t v' dt$$

Let the maximum air displacement along the meridian be equal to $\Delta\Phi$. Then, at the low latitudes, where $\varphi_0 < \Delta\Phi$, the atmospheric air will first come from the inherent hemisphere and then from another hemisphere. Since $f(\varphi)$ has the extreme at the equator, then $r(t)$ forms one extreme within the period of this unidirectional displacement. After the sign of v' changes, the air returns and the above changes follow in the reverse sequence. Hence, at the latitudes $\varphi_0 < \Delta\Phi$, two minima and two maxima of $r(t)$ are observed in the course of a year. At $\varphi_0 > \Delta\Phi$, the air from another hemisphere does not reach the station; therefore, $r(t)$ has one maximum and one minimum during a year.

Thus, the IHC transforms the spatial variations in the concentration $f(\varphi)$ into the temporal variations $r(t)$ and thereby it forms the seasonal variations in the concentration under consideration.

The air mass that participates in the IHC depends on the Sun's declination δ. At $\delta = 90°$, the Sun would occur at the zenith over the North Pole. It would be a continuous day in the Northern Hemisphere and a continuous night in the Southern Hemisphere. In this case, the whole mass of the atmosphere would be involved in the IHC. A similar situation would also arise at $\delta = -90°$. At $\delta = 0°$ both hemispheres would be equally illuminated, and the IHC would be absent: the whole mass of the atmosphere would participate in the FTHE operation.

Since the Sun's actual declination at the moments of solstices can attain $\pm 23.5°$, then some part of the atmosphere becomes involved into the IHC and the remaining, larger part participates in the FTHE operation. These parts can be assessed by the data on the seasonal changes in the angular momentum h. During the transitional seasons, when the whole atmospheric mass M_a takes part in the FTHE operation, $h = 145 \times 10^{24}\,\mathrm{kg\,m^2\,s^{-1}}$; and in August, when some part of the atmospheric mass is involved into the IHC, $h = 91 \times 10^{24}\,\mathrm{kg\,m^2\,s^{-1}}$ (Sidorenkov, 1978). Since h is proportional to the mass M that participates in the FTHE operation, we obtain that in August $M = \frac{91}{145} M_a = 0.63 M_a$, where M_a is the whole atmospheric mass; the remaining 37% of the atmosphere mass is involved in the IHC. Note that if we take the daily values of h instead of the mean monthly values, then $M \approx 0.56\, M_a$.

Thus, the IHC is one of the most important circulation cells in the atmosphere. It involves up to 40% of the atmosphere mass. The IHC is responsible for the meridian transport of atmospheric air and moisture, products of nuclear explosions, and pollutants (including their transport through the ITCZ). The observed complicated seasonal and diurnal variations in the angular momentum of zonal winds and the indices of atmospheric circulation, in the concentrations of ions and molecules are also associated with the IHC. The list of processes and phenomena that are due to the IHC will be extended with further investigations.

8.8 Conclusions

In the absence of solar radiation, the atmosphere would be at rest relative to the Earth's surface but it would rotate together with the Earth as a solid body from the west to the east around the polar axis. The nonuniform heating of the atmosphere by solar radiation arises the macroturbulent air mixing that levels off the distribution of the angular momentum in the atmosphere. At high and moderate latitudes, the angular momentum increases, the air in its rotation around the polar axis begins to outrun the Earth's surface, and the westerly winds originate. At low latitudes, on the contrary, the angular momentum decreases, the air lags behind the Earth's rotation, and the easterly winds originate.

The emergence of winds entails the appearance of friction forces, which tend to weaken the wind. The momentum of friction forces provides the inflow of the angular momentum from the Earth to the atmosphere in the zones of easterly winds and its outflow from the atmosphere to the Earth in the zones of westerly winds. The zones of westerly winds are located closer to the Earth's rotation axis than the zones of easterly winds. Because of this, at the same wind velocity in the zones, the momentum of friction forces of the easterly winds exceeds that of the westerly winds, that is, the inflow of the angular momentum from the Earth to the atmosphere exceeds its outflow. Due to this, the velocity of westerly winds increases, until the absolute values of the angular momentum inflow and outflow become equal. This stationary state is only attained when there is some angular momentum, which is taken from the Earth and accumulated in the atmosphere. Ultimately, the velocity of westerly winds in the atmosphere significantly exceeds the velocity of easterly winds; on the whole, the atmosphere rotates around the polar axis more rapidly than the Earth does, and the superrotation of the atmosphere is observed.

The Coriolis force exerts the decisive effect on the formation of the atmospheric pressure field. The horizontal component of the Coriolis force displaces the air from the zone of westerly winds toward the equator and from the zone of easterly winds toward the poles. As a result, an excess of air mass (the pressure maximum) forms near the latitude where the direction of zonal winds changes the sign (at roughly 35°). This maximum is called the subtropical high-pressure zone. The pressure in the subtropical high-pressure zone grows until the meridian component of the force of the pressure gradient counterbalances the horizontal component of the Coriolis force.

The thermodynamic analysis shows that the value of the angular momentum h of zonal winds in the atmosphere remains constant due to the operation of the atmospheric heat engines of the first type. As for the seasonal variations in the value of h and, as a consequence, the seasonal nonuniformity of the Earth's rotation, they are due to the operation of the interhemisphere heat engine. The seasonal variations in the above characteristics are described by the expression similar to (8.51) or by the annual harmonic taken over the module (with the shifted zero point), rather than by the sum of the annual and semiannual harmonics. The intensity of the interhemisphere heat engine operation changes from year to year, entailing changes in the parameters of seasonal variations in the angular momentum h and the velocity of the Earth's rotation.

9
Interannual Oscillations of the Earth–Ocean–Atmosphere System

Despite the centenary history of studies of the Chandler wobble (CW), its nature remains obscure. The mechanism of its excitation is still also under discussion. Among the causes of the Chandler wobble, meteorological and seismic processes are indicated most frequently. However, most estimates of the atmospheric effect suggest that it is small. Widely different views have been expressed on the influence of earthquakes on the excitation of the Chandler wobble. The most comprehensive overview of studies concerning this subject can be found in (Yatskiv *et al.*, 1976; Lambeck, 1980).

In the author's opinion, the negative conclusion about the role of meteorological processes in the excitation of the Chandler wobble was drawn for two reasons. The first is that the meteorological observation network in the Southern Hemisphere is very rare, while it is in this hemisphere that the Earth is pushed by consistent air mass oscillations between the Pacific and Indian oceans (El Niño–Southern Oscillation (ENSO)) (Sidorenkov, 1997, 2000a, 2000b, 2000e, 2001, 2002a, 2002b). The other reason is that estimates are traditionally based on the theory of linear oscillations, which is inadequate in this case. Moreover, the excitation of the wobble is usually estimated near the principal resonance, that is, near the Chandler frequency (Munk and Macdonald, 1960; Wilson and Haubrich, 1976; Vondrak and Pejovic, 1988; Wilson, 2000). However, the oscillations in the Earth–ocean–atmosphere system are nonlinear and the Chandler wobble is excited not at the fundamental resonance frequency but primarily at combination frequencies of the wobble (periods of 2.4, 3.6, 4.8, and 6 years) (Sidorenkov, 1997, 2000a, 2000b, 2000e, 2001).

Below, we analyze the power spectra of oceanic and atmospheric characteristics and series of effective excitation functions of the atmospheric angular momentum. The results confirm the presence of Chandler superharmonics and suggest that the data involve subharmonics of the fundamental period of the Earth's forced nutation (18.6 years). Slow waves in the ocean and the atmosphere are discovered that generate these cycles.

Based on these findings, we conclude that the Earth, the ocean, and the atmosphere exhibit consistent oscillations and influence each other, that is the Earth–ocean–atmosphere system exhibits coherent oscillations initiated by tides. These oscillations

The Interaction Between Earth's Rotation and Geophysical Processes. Nikolay S. Sidorenkov

ISBN: 978-3-527-40875-7

are manifested as the polar motion, El Niño and La Niña events in the ocean, and the Southern Oscillation and the quasibiennial oscillation in the atmosphere. All of them are studied as independent phenomena in different areas of Earth sciences. In contrast, we analyze these phenomena together. First, let us describe them in more detail.

9.1 El Niño–Southern Oscillation

Among the planetary-scale phenomena observed in the Earth–ocean–atmosphere system, ENSO has attracted much interest over the past years. A huge number of publications are available on this phenomenon and its effects. It was shown that the ENSO is related to many of the most noticeable interannual oscillations in atmospheric meteorological elements, oceanic hydrological characteristics, the Earth's rotation, and the polar wobble.

The episodes of long-lived surface-water warming in the central and eastern Pacific Ocean and a variety of accompanying processes are called the El Niño phenomenon. They have been known since the dawn of time. The term "Southern Oscillation" (SO) was introduced by G.T. Walker in 1920 (Walker, 1924). It means air-mass oscillations in the Southern Hemisphere subtropical zone between the Pacific and Indian oceans with a characteristic time of several years. A rise (fall) of the atmospheric pressure in the central and eastern tropical Pacific is accompanied by a fall (rise) of the pressure in the tropical Indian Ocean and near Australia and Indonesia (Figure 9.1).

Inspection of Figure 9.1 shows that the correlation coefficients between 30°N and 35°S are positive in the Eastern Hemisphere and are negative in the Western Hemisphere. The most pronounced negative correlation in pressures (with a correlation coefficient of $r \approx -0.8$) is observed between two regions: Indonesia–Australia and the Pacific Community islands. In other words, there are two oppositely signed SO centers of action: Australian–Indonesian and South Pacific, both located in the tropical Southern Hemisphere (hence, the name "Southern Oscillation").

Since Walker's studies (Walker, 1924), several indices have been proposed to characterize the SO (Ropelewski and Jones, 1987; Trenberth, 1976; Troup, 1965; Wright, 1989; Sidorenkov, 1991c). Usually, they make use of the sea level pressure (SLP) at a single station or a combination of several stations located in the western and eastern Pacific. One of the most justified indices is based on the SLP at Tahiti and Darwin stations, which are located near the oppositely signed SO centers of action. Since the variances of the monthly mean SLP at Tahiti and Darwin are different, the pressure anomalies at these stations have to be normalized so that the South Pacific and Australian– Indonesian centers of action are equally represented.

The index can be calculated in several ways. We follow the technique used at the Climate Analysis Center of the US National Meteorological Center (Sidorenkov, 1991c; Ropelevski, 1987). It can be described as follows. A time series of differences δ_{ym} between normalized SLP anomalies at these stations is calculated from time series of monthly mean SLP at Tahiti and Darwin:

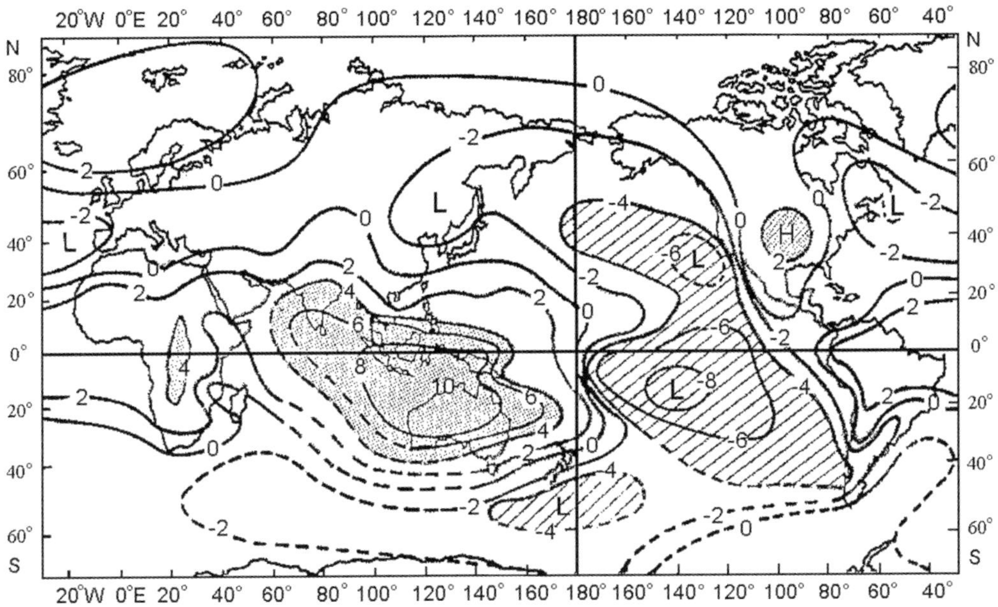

Figure 9.1 Coefficients (multiplied by ten) of correlation between the mean annual sea-level pressure at the station of Darwin, Australia, and at other meteorological stations of the world (according to Trenberth, 1976). Note the high negative correlations over Tahiti – showing the basis for the choice of the Tahiti – Darwin pressure difference as a Southern Oscillation Index (SOI).

$$\delta_{ym} = \left[\frac{P_{ym} - \bar{P}_m}{\varepsilon}\right]_{Ta} - \left[\frac{P_{ym} - \bar{P}_m}{\varepsilon}\right]_{Da} \tag{9.1}$$

where P_{ym} is the actual pressure; $\bar{P}_m$ is the long-term average value (normal); ε is the standard deviation calculated from all pressure anomalies over the period 1951–1980; and y and m denote the year and the month, respectively. The normals $\bar{P}_m$ are calculated from the monthly mean data over 1951–1980. Next, the SOI index is determined as

$$\mathrm{SOI} = \delta_{ym}/\sigma \tag{9.2}$$

where σ is the standard deviation of all the differences δ_{ym} over 1951–1980.

A continuous homogeneous series of monthly mean SOI values starting in 1935 was presented in (Sidorenkov, 1991c, 2000c). Although observations at Tahiti and Darwin were made even earlier, they were unavailable for a long time since they were dispersed over various archives. Due to the careful efforts undertaken by scientists, continuous series of atmospheric pressure from 1866 to 1934 were restored for these stations (Allan *et al.*, 1991; Ropelevski, 1987). As a result, a continuous series of monthly mean SOI indices from 1866 to the present is now available.

Wright (Wright, 1989) calculated three-month mean W indices for ENSO starting in 1851 by using SLP at several (up to eight) stations. To eliminate the inhomogeneities and to fill the gaps in the pressure series used, he employed ENSO indices based on sea surface temperature (SST), precipitation, and air temperature in the central and eastern Pacific. Later, he replaced the W index by the DT index calculated as the difference between the pressure anomalies at Darwin and Tahiti. The DT index was easy to calculate starting in 1935. By using simultaneous DT and W series over 1935–1974, Wright derived a regression equation for DT on W and then calculated the DT indices over the period 1851–1934 from the W indices over this period. At present, a series of DT is available for 150 years.

It has been found that higher (lower) SOI values are accompanied by lower (higher) SST in the eastern and central Pacific (Rasmusson and Carpenter, 1982; Philander, 1990; Sidorenkov, 1991c). For this reason, two extreme phases have been distinguished in ENSO: a warm phase (El Niño) when SOI < 0 and a cold phase (La Niña) when SOI > 0. In El Niño events, the sea level in the eastern Pacific is approximately 50 cm higher than in the western Pacific. An opposite situation occurs in La Niña events (Rasmusson and Carpenter, 1982; Philander, 1990), that is interannual sea-level oscillations with an amplitude of about 50 cm are observed between the eastern and western tropical Pacific.

Thermal oscillations in the ocean are characterized by SST averaged over most representative regions. Series of monthly mean SST indices over various regions, such as *Nino* in the equatorial Pacific, *N.Atl* and *S.Atl* in the Atlantic, *Tropics* in three oceans, and so on, are known to be regularly computed at the US National Centers for Environmental Prediction (NCEP) (Climate Diagnostics Bulletin). Unfortunately, these series are available only from 1950 to the present. Data reconstructions over the *Nino3* and *Nino4* regions from 1903 to the present were calculated at the UK Hadley Center (Parker, Folland, and Jackson, 1995).

Six-hour series of excitation function components χ_i for the atmospheric angular momentum (AAM) from 1958 to 2001 are available at present. They were derived from the NCEP/NCAR reanalysis data at the Subbureau for Atmospheric Angular Momentum (Kalnay *et al.*, 1996; Salstein and Rosen, 1997) (see Section 7.3).

At first glance, the fluctuations in long-term SOI curves look like a simple (Poisson) process. However, a spectral analysis of long-term series of SOI (from 1866 to 1996) and of DT (from 1851 to 1996) has shown (Sidorenkov, 1997, 2000a, 2000c, 2000e, 2001) that the SOI spectral density reaches its maximum values in the period range of 2–7 years (Figure 9.2). For periods shorter than one year, the spectral density is very low (varies from 0 to 1). In both curves, spectral density peaks are exhibited by the components with periods of about 6, 3.6, 2.8, and 2.4 years. The DT index also has a peak at 11.2 years, which corresponds to a weak peak at 13.2 years in SOI. A major feature of the dominant periods in the ENSO indices is that all of them are (to a greater or lesser extent) multiples of the forced nutation period (18.6 years), and the free nutation period, or the Chandler wobble period (1.2 years) (with the superharmonic wave numbers $n_k = 2, 3, 5$).

A spectral analysis of SOI and SST and their anomalies over 1903–1998 and 1950–1998 was performed in (Sidorenkov, 2000a, 2000b, 2000e, 2001). The resulting

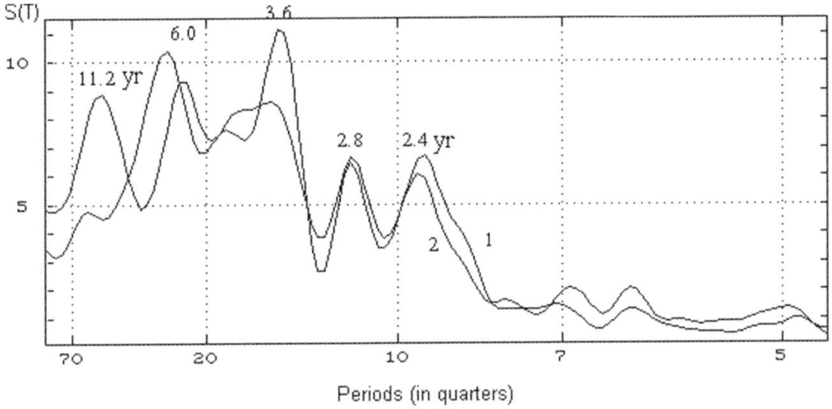

Figure 9.2 Power spectra of the SOI (1) and DT (2) indices. The oscillation periods (in quarters of a year) and the spectral density *S* (*T*) are plotted on the *X*-axis and *Y*-axis, respectively.

spectra for *Nino3*, *Nino4*, and SOI from 1950 to 1998 are illustrated in Figure 9.3. It can be seen that the 3.6- and 2.4-year periods still persist, but the 6-year period in SOI is no longer observed, while a 4.8-year period appears in *Nino3* and *Nino4*.

The periodograms and power spectra of all daily mean AAM excitation components χ_i^P and χ_i^W over 41 years were calculated in (Sidorenkov, 2000a, 2000b, 2000e, 2001). Table 9.1 presents the spectral analysis results for all the series. The third column lists the most prominent cycles in decreasing order of spectral power.

Although the series of ENSO indices we analyze cover a rather short time interval, the cycles determined generally coincide with those derived from 114- and 145-year series (Schneider and Schonwiese, 1989; Sidorenkov, 1997, 2000e, 2001). The difference is that the latter series additionally exhibit a 6-year cycle, while a 4.9-year cycle is observed in the present indices.

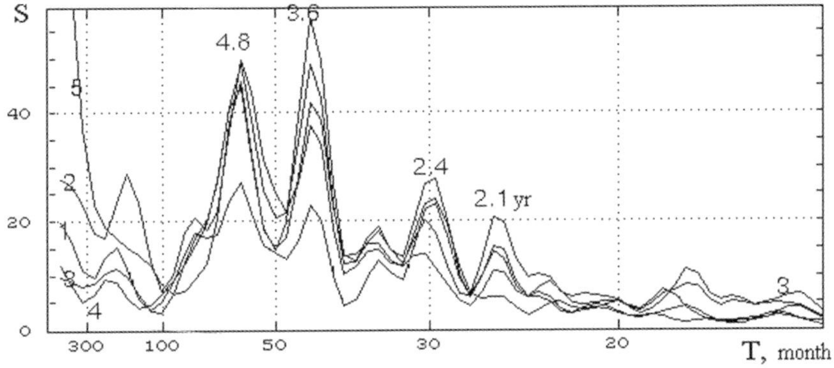

Figure 9.3 Power spectra of the oscillations of the SOI (1), *Nino4* (2), *Nino3.4* (3), *Nino3* (4), and *Tropics* (5) indices. The oscillation periods (in months) and the spectral density *S* (in arbitrary units) are plotted on the *X*-axis and *Y*-axis, respectively.

Table 9.1 Dominant cycles in characteristics of the ENSO, QBO, SST, and the atmospheric angular momentum excitation components.

Index	Region borders	Periods of dominant cycles, years
SOI		4.9; 2.4; 3.6; 2.1
DT		5.8; 3.6; 2.8; 12; 2.4
QBO	5°N – 5°S 0° – 360°	2.4; 2.8; 2.0; 4.8; 1.2
Nino4	5°N – 5°S 160°E – 150°W	4.9; 3.6; 12; 2.4
Nino3.4	5°N – 5°S 170°W – 120°W	4.9; 3.6; 2.4; 2.1
Nino3	5°N – 5°S 150°W – 90°W	3.6; 4.9; 2.5; 2.1
Nino1 + 2	0° – 10°S 90°W – 80°W	3.6; 4.9; 2.9: 2.1
Tropics	10°N – 10°S 0° – 360°	4.9; 3.6; 2.8; 2.4; 2.1
N.Atl	5°N – 20°N 60°W – 30°W	9.3; 3.6; 2.5; 2.1; 5.2
S.Atl	0° – 20°S 30°W – 10°E	12; 5.2; 2.3; 3.5
χ_1^{W}	Globe	2.7; 3.7; 1.86; 5.9; 2.4
χ_2^{W}	Globe	2.7; 3.1; 2.1; 1.8
χ_3^{W}	Globe	2.4; 3.7; 5.1
χ_1^{Pib}	Globe	2.6; 3.7; 4.6; 1.7
χ_2^{Pib}	Globe	2.4; 2.7; 5.9
χ_3^{Pib}	Globe	5.1; 4.1; 2.3; 2.7
18.6 years	subharmonics	6.2; 4.7; 3.7; 3.1; 2.7; 2.3; 2.1; 1.86

A spectral analysis of χ_i^{W} (wind component) and χ_i^{Pib} (atmospheric pressure component corrected for the inverse barometer effect) suggests the presence of forced-nutation subharmonics (4.7, 3.7, 3.1, 2.7, 2.3, 2.1, and 1.86 years). Some of them are close to the Chandler superharmonics of periods 2.4, 3.6, 4.8, and 6 years, that is the analysis of the AAM components does not contradict the results derived from the ENSO indices. However, the AAM series are too short to provide reliable results for long-term cycles.

Thus, all the studies suggest that the ENSO spectra contain components that are not one-year multiples but rather multiples of about the Chandler period (1.2 years) and the forced-nutation fundamental period (18.6 years). The superharmonic periods of the latter are given in the last line of Table 9.1.

9.2
Quasibiennial Oscillation of the Atmospheric Circulation

Among the numerous nonseasonal oscillations in the atmospheric circulation, the quasibiennial one stands out as the most stable and significant phenomenon. The quasibiennial oscillation (QBO) of the atmosphere was discovered in the early 1960s in the study of the equatorial stratospheric circulation. It was found that the direction of the equatorial zonal wind reverses with a period of about 26 months in the atmospheric layer between the altitude of 18 and 35 km. More specifically, westerlies persist at a fixed height for about 10 months, then easterlies blow for about 16 months, and then a new cycle begins (Reed, 1964).

The QBO of the equatorial zonal wind is explained by the interaction of Kelvin waves and mixed Rossby-gravity waves with the zonal wind in the equatorial stratosphere (Lindzen and Holton, 1968; Holton and Lindzen, 1972; Holton, 1975). The nature of Kelvin and mixed Rossby-gravity waves is not clear. In the author's view, mixed Rossby-gravity waves are manifestations of the interaction between diurnal and zonal tides in the atmosphere. They account for an intense wide peak at a frequency of about 0.85 $(\text{day})^{-1}$ in the spectrum of the atmospheric angular momentum (see Section 7.5). It is believed that Kelvin waves propagate into the stratosphere, where they meet a westerly shear zone and are absorbed at the height where their phase velocity coincides with the wind velocity. As a result, the westerlies at this height strengthen, while the absorption of new waves is reduced. Since wave absorption proceeds continuously, the westerly zone gradually descends toward the tropopause at a velocity of about 1 km/month (Figure 9.4). When the westerly zone stretches down to the tropopause, the Kelvin waves have low frequencies due to the Doppler shift, while the mixed Rossby-gravity waves have high frequencies. That is why the latter propagate upward. At the height of semiannual oscillations ($\approx$35 km), they can meet an easterly shear zone, where they are absorbed. As a result, the easterly velocity increases and the easterly zone descends permanently from 35 km to the tropopause, where the cycle is completed. At the same time, Kelvin-wave absorption begins at the height of semiannual oscillations and a new cycle starts.

In this model, the QBO period for wind depends only on the intensity of atmospheric waves and on the distance between the equatorial tropopause and the height of semiannual oscillations in the stratosphere.

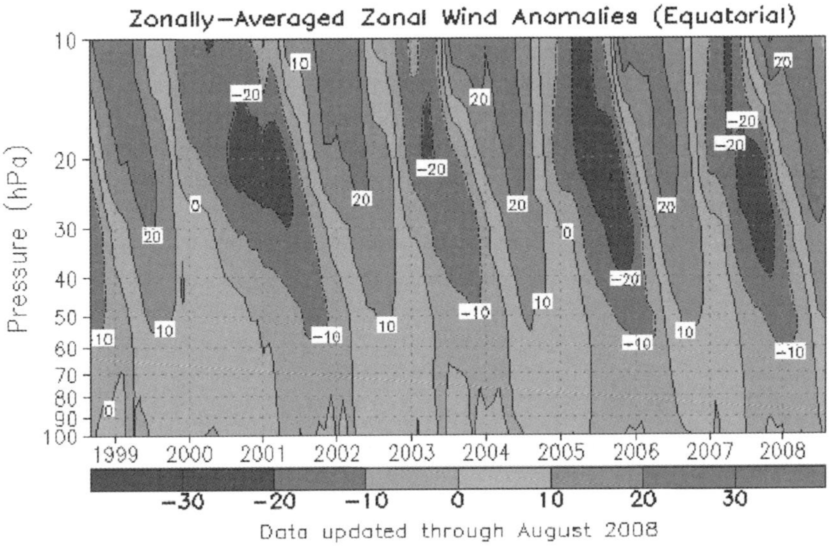

Figure 9.4 Equatorial time–height section of anomalous zonally averaged zonal wind ($m\,s^{-1}$) (CDAS/Reanalysis). Contour interval is 10 $m\,s^{-1}$. Anomalies are departures from the 1979–1995 base period monthly means. (Climate Diagnostic Bulletin, 2008).

A large number of indices (Climate Diagnostics Bulletin) are available that quantify the QBO in equatorial winds. Chuchkalov (1972) suggested the index Π defined as the mass flux through an equatorial meridional cross section with a width of 1 m and a height of 19 to 31 km (i.e. from 66.63 to 10.07 hPa). The wind at five pressure levels (70, 50, 30, 20, and 10 hPa) is required for computing the index Π:

$$\Pi = \iint \rho \boldsymbol{V} \cdot \mathrm{d}\boldsymbol{S} = 1 \cdot \int_{Z_1}^{Z_2} \rho\, u \mathrm{d}z = \frac{1}{g} \int_{P_2}^{P_1} u \mathrm{d}p = \frac{1}{g} \sum_i u_i \Delta P_i \tag{9.3}$$

where $\boldsymbol{V}$ is the wind velocity vector, u is the average zonal wind velocity at the level P ($u > 0$ for westerlies and $u < 0$ for easterlies), $\mathrm{d}\boldsymbol{S}$ is the elementary area of 1 m in width and dz in height, ρ is the air density; g is the acceleration due to gravity, P is the atmospheric pressure, ΔP_i is the pressure increment in the ith layer, $Z_1 = 19$ km, $Z_2 = 31$ km, $P_1 = 66.63$ hPa, and $P_2 = 10.07$ hPa.

My experience gained from dealing with Π has shown that this index is difficult to interpret. For this reason, Π is recalculated to obtain the average zonal wind velocity $\bar{u}$ in the layer of 19 to 31 km. The Π and $\bar{u}$ indices are related by the formula

$$\bar{u} = \frac{\frac{1}{g} \int u \mathrm{d}P}{\frac{1}{g} \int \mathrm{d}P} = \frac{\frac{1}{g} \sum_{i=1}^{5} u_i \Delta P_i}{\frac{1}{g} \sum_{i=1}^{5} \Delta P_i} = \frac{\Pi}{577.14} \tag{9.4}$$

where the notation is as before.

Starting in the 1960s, the QBO in the equatorial stratospheric wind was regularly monitored by the Hydrometeorological Center of the USSR via radiosonde and rocket measurements from research vessels at the equator (Chuchkalov, 1989). Simultaneous observations at different equatorial points differ little from each other, so the wind QBO can be characterized by measurements at a single point close to the equator. This means that, in the equatorial stratosphere, the Earth is surrounded by a jet stream in which the wind velocity varies only with height, latitude, and time.

Since the 1990s, the QBO index has been monitored using 00:00 and 12:00 UTC daily observations from all accessible upper-air stations located near the equator (in the ±5° latitude band). At present, there is a 55-year monthly time series (from 1954 to 2008) of the average zonal wind velocity $\bar{u}$ in the equatorial stratosphere (Sidorenkov, 2000c). This series is presented in Appendix D.

Time variations in the average zonal wind velocity in the equatorial stratosphere are illustrated in Figure 9.5. Over the period 1954–2008, 24 cycles of $\bar{u}$ occurred in the series. The period varied from 21 months in 1972–1973 to about 36 months in 1964–1966. The length of a single cycle averaged over the entire time interval was 28.1 months. The zonal wind velocity varied from −22.5 m/s in July, 1984, to + 18 m/s in January, 1983. The mean velocity $\bar{u}$ over 55 years was −3.8 m/s, and the (standard) rms deviation was ±9.3 m/s.

Figure 9.6 shows the periodogram of $\bar{u}$ calculated from its time series over 1954–1999 (altogether 552 monthly mean values). It can be seen that the Fourier

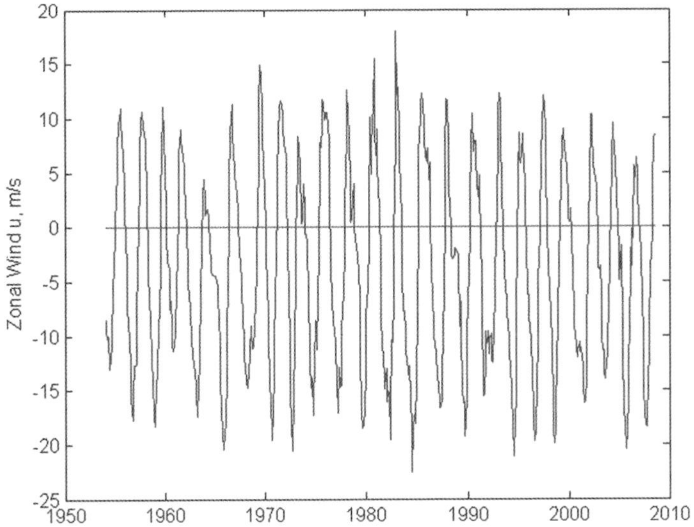

Figure 9.5 Temporal course of the zonal wind average velocity $\bar{u}$ in the equatorial stratosphere (layer of 19–31 km) in 1954–2008.

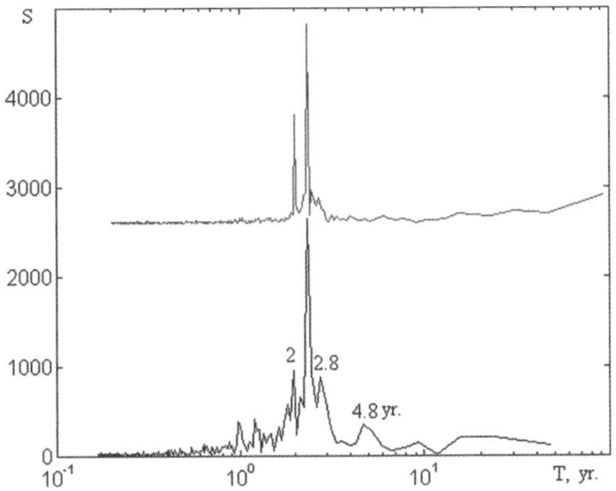

Figure 9.6 Power spectra of the pole coordinate x (top) and the QBO indices $\bar{u}$ (bottom). To demonstrate the curves' similarity, the pole's curve was transformed as follows: $T = 2T_0$ and $S - 30S_0 + 2600$, where T_0 and S_0 are the actual values of the periods T and spectral densities S, respectively.

spectrum is dominated by the harmonic with a period of about 2.33 years (28 months). It is this oscillation that is called QBO. The QBO amplitude varies from 8 to 13 m/s. Another noticeable harmonic is that with a period of about 2.0 years (24 months), but its power is more than 5 times lower than that of the 28-month harmonic. Weak peaks can also be seen at periods of 1.0, 1.2, 2.8, and 4.8 years. The addition of these harmonics explains the amplitude modulation of $\bar{u}$ exhibited in

Figure 9.5. Specifically, the sum of 28 and 24-month cycles is responsible for 14-year beats in the QBO amplitude.

Next, we calculated the periodograms of the North Pole's coordinates x and y from their time series over the period of 1954 to 1999 (altogether 920 values at 0.05-year intervals) (IERS Annual Report, 2000). They turned out to be nearly identical. For this reason, we present only the periodogram of x (Figure 9.6). The component with a Chandler period of 1.18 years dominates the other harmonics. Additionally, a well-pronounced annual harmonic is observed that is more than 3 times less intense than the Chandler one.

A surprising feature is that the spectrum of $\bar{u}$ is similar, within a factor of 2, to that of the pole's coordinates x and y. If the horizontal-axis scale in the spectrum of the pole's coordinates is doubled as shown in Figure 9.6, then all the details in the spectrum of $\bar{u}$ coincide with those in the polar motion spectrum; that is the oscillation in the polar motion is reflected as the doubled-period QBO in the atmosphere. In the equatorial stratosphere, the duration of all the Earth's polar motion cycles is doubled. Presently, one believes that the QBO is driven by momentum transfer from the Kelvin waves and the mixed Rossby gravitation waves to the mean zonal flow (Holton and Lindzen, 1972; Holton, 1975). However, these waves are the manifestation of the tidal movements in the atmosphere (Sidorenkov, 2002a, 2002b). They cause a powerful wide peak near the 0.85 $(\text{day})^{-1}$ frequency in the spectrum of the NCEP/NCAR reanalysis Atmospheric Angular Momentum (see Section 7.5). As early as in 1960, Carl Eckart showed that the Rossby waves are tidal waves (Eckart, 1960, page 279). These facts testify that the Chandler wobble of the poles and the QBO cyclicity of the stratospheric winds are likely to have a common mechanism of excitation that is due to the lunar–solar tides.

9.3
Multiyear Waves

Starting in 1979, I constructed time–longitude sections of equatorial SST and found that the SST anomalies move eastward at a velocity of about 0.25 m/s (Sidorenkov, 1991c). It was also found that slow waves propagate along the equator in the ocean. They travel around the Earth in about 4.8 years. Their velocities are much lower than those of normal equatorial Kelvin waves (the latter travel through the Pacific Ocean in one to three months) (Efimov *et al.*, 1985; Gill, 1982). Analyzing the motion of SST anomalies, I predicted their occurrence in the Pacific equatorial belt (Sidorenkov, 1991c) and realized that the Atlantic El Niño is delayed relative to the Pacific one because several months is required for a wave to travel from the American coast to Africa.

A time–longitude section of equatorial SST anomalies is shown in Figure 9.7. The longitudes from 0° to 360° run along the horizontal axis, while the vertical axis represents months and years from 1979 to 1990. The contour interval is 0.5°. The positive anomalies ($> +0.5°$) are shaded, while the negative ones ($< -0.5°$) are shown in white. No data are available in the longitude ranges of Africa and South

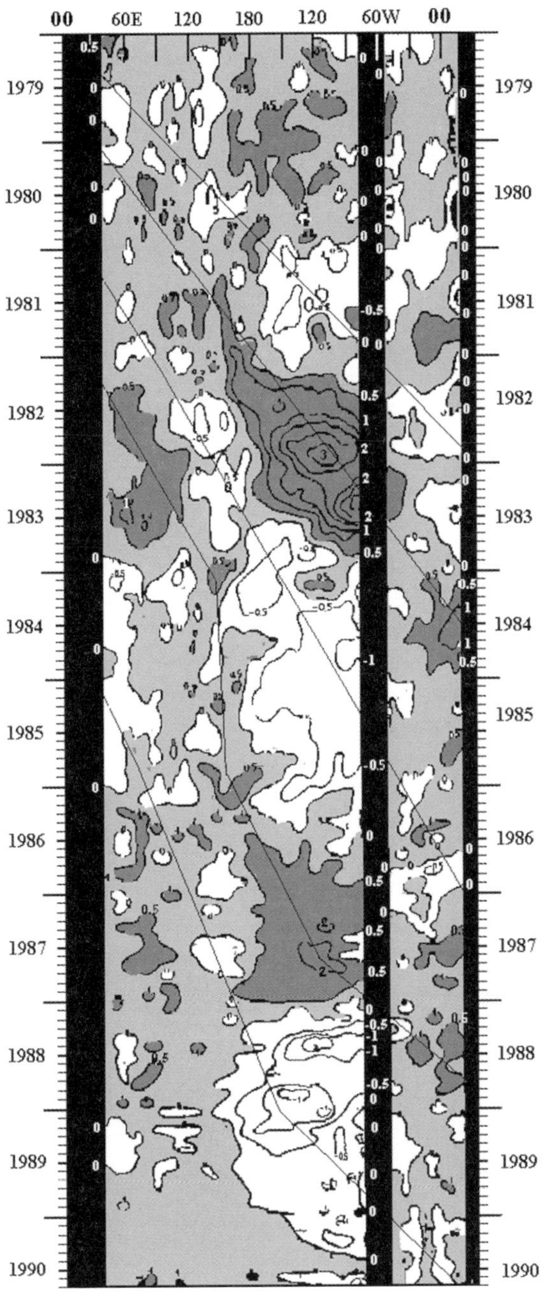

Figure 9.7 Time–longitude section of the sea surface temperature anomalous in the zone of 5° N–5° S from 1979 to 1990. The interval between isolines is 0.5 °C. Africa and South America are blackened. The positive anomalies (> + 0.5°) are shaded, the negative ones (< −0.5°) are light. Thin lines indicate the location of ridges and troughs of the many-year waves (Sidorenkov, 2002a, 2002b).

America, so they are depicted in black. The centers of dominant anomalies are joined by light straight lines. It can be clearly seen that these lines are inclined from left to right. Therefore, the anomalies move eastward. The line joining the positive anomalies is a wave crest, while the neighboring line connecting the negative anomalies is a wave trough. The slope of the lines determines the wave velocity. It can be seen that the velocity has varied. Specifically, the wave traveled around the Earth in 4–5 years at the beginning of the time period and in 7 years at its end. Moreover, its motion was nonuniform. For example, the crest was nearly fixed at the 150°E meridian in 1984–1985.

The US Climate Prediction Center publishes the monthly Climate Diagnostics Bulletin, which presents time–longitude sections of some oceanic and atmospheric characteristics. Slow equatorial waves can be seen in many of these sections. They exist simultaneously in the ocean and the atmosphere. Specifically, they are most pronounced in 850-hPa wind velocity anomalies in the atmosphere and in 20 °C isotherm depth anomalies in the ocean (Climate Diagnostics Bulletin).

Figure 9.8 displays a section of 850-hPa zonal wind speed anomalies from the Climate Diagnostics Bulletin. The evolution of a slow wave during the 1996 La Niña and 1997–1998 El Niño events can be seen in the figure. The maximum positive (westerly) wind anomalies were at the 100°E meridian in October 1996 and, moving eastward, reached the South American coast (80°W) in July 1998. The wave traveled an extent of 180° in 1.8 years, that is its revolution period around the Earth was roughly 3.6 years. In these years, the anomalies of atmospheric pressure, outgoing longwave radiation, depth of the 20°C isotherm for 5°N–5°S in the

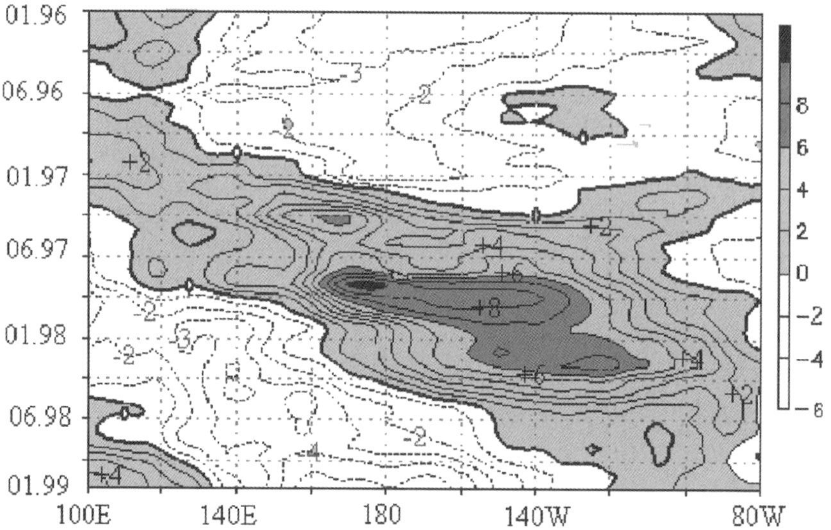

Figure 9.8 Time–longitude section of the zonal winds velocity anomalies at the 850-hPa isobaric surface in the zone of 5°N–5°S from January 1996 to December 1998. The interval between isolines is 1 m/s. Broken and solid lines are the anomalies of the east and west winds, respectively.

Pacific Ocean, and other characteristics traveled eastward at the same velocity (Climate Diagnostics Bulletin).

In the 1980s, waves moved more slowly, and the revolution period was 4.8 years. In 1990–1993 a wave was nearly motionless, but in 1994 it ran around the Earth in 1.2 years. Thus, the wave velocity can vary from 0 to more than 1 m/s. The revolution period of multiyear waves is a multiple of the Chandler period, the lunar nodal tide period, or the fundamental period of the Earth's forced nutation. This suggests that these waves are associated with long-period tides and the free and forced nutations of the Earth. That is why they were called nutation waves in (Sidorenkov, 1999, 2000b).

Nutations can be forward and reverse (see Section 5.4.3). The general solution of the wave equation also involves waves propagating in positive and negative directions. It would be natural to expect that multiyear waves can be forward and backward as well, that is eastward- and westward-propagating, respectively. Indeed, they can easily be found by close inspection of time–longitude sections. Let us analyze Figure 9.9, which, like Figures 9.7 and 9.8, shows the time–longitude section of the 925-hPa zonal wind speed anomalies in the equatorial belt constructed from the 1982–1994 NCEP/NCAR reanalysis. Joining the centers of the positive anomalies, which are wave crests, gives lines inclined rightward (heavy lines) and leftward (light lines). The slope of the heavy lines suggests that the wave crests travel eastward, while a reverse situation occurs for the thin lines (i.e. the wave crests travel westward). Similarly, inspecting the negative anomalies of u, that is wave troughs, we see that they move in the same manner as the crests: in forward and backward directions. Figures 9.7 and 9.8 present only forward SST anomaly waves, but backward ones can easily be found if desired.

Thus, forward waves moving from west to east and backward waves traveling from east to west are observed in the equatorial ocean and atmosphere. To a certain degree, they determine the long-term variability in hydrometeorological characteristics of the ocean and the atmosphere. Large positive anomalies are formed where forward and backward wave crests meet, and negative anomalies are formed in areas of meeting wave troughs. The areas and times when crests or troughs meet correspond to the longitudes and moments of the onset of El Niño or La Niña. Therefore, the ENSO phenomenon results from the interference of forward and backward multiyear waves traveling simultaneously in the atmosphere and the ocean.

The multiyear waves can be described by the expression

$$u = A_p \exp i(\sigma_p t - k_p \lambda) + A_r \exp i(\sigma_r t + k_r \lambda) \quad (9.5)$$

where A is the amplitude, σ is the circular frequency, k is the wave number, λ is the east longitude, and p and r are the indices denoting forward and backward waves, respectively. At a fixed time t, the expression $u = A\exp i(\sigma t + k\lambda)$ describes u as a function of λ. If t varies but the phase remains a constant ($\sigma t + k\lambda = \text{const.}$), then the distribution of u remains unchanged. In this case $\frac{d\lambda}{dt} = \dot{\lambda} = -\frac{\sigma}{k}$ is the wave angular velocity. When k is a scalar, $\dot{\lambda}$ is known as the phase velocity. If $\sigma < 0$, then $\dot{\lambda} > 0$; that is the wave travels in the positive direction (eastward). If $\sigma > 0$, then $\dot{\lambda} < 0$; that is the wave travels in the negative direction (westward).

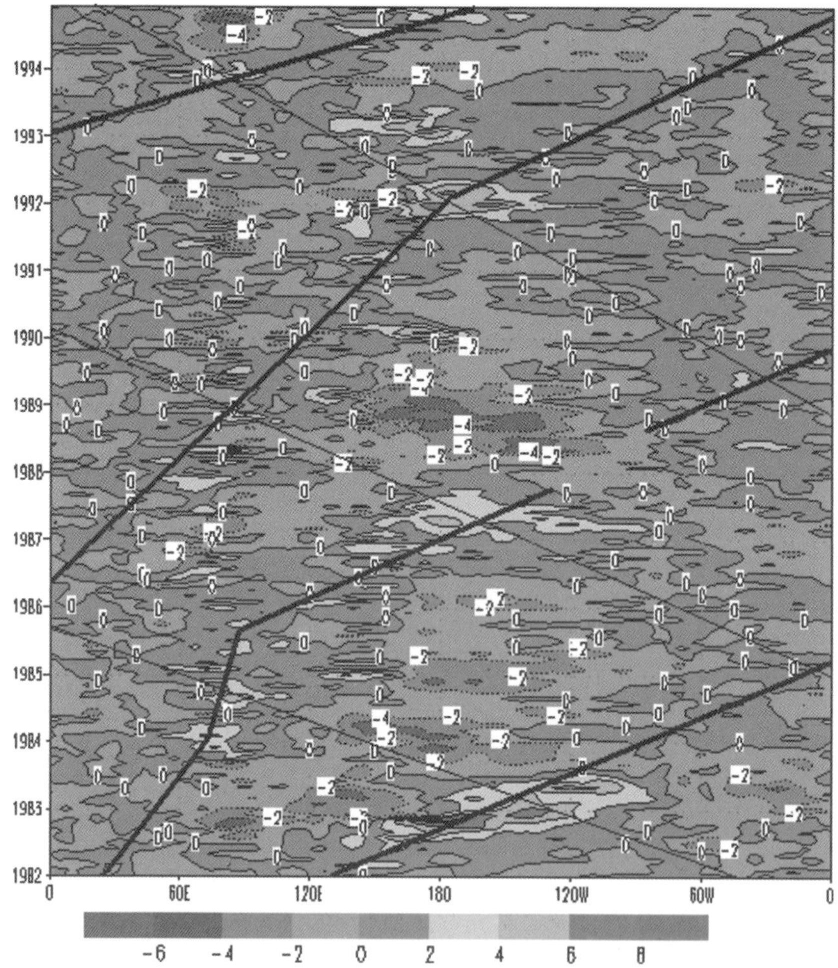

Figure 9.9 Time–longitude section of the zonal winds velocity anomalies at the 925 hPa isobaric surface in equatorial zone from January 1982 to December 1994. The interval between isolines is 1 m/s. Broken and solid lines are the anomalies of the easterly and westerly winds, respectively. Thick and thin lines are the trajectories of the direct and reverse waves, respectively.

The first and second terms on the right-hand side of (9.5) describe a forward and a backward wave, respectively. If their parameters are identical, they yield a standing wave: $u = 2A\cos k\lambda\ \exp(i\sigma t)$. If the amplitudes of waves (9.5) are different, then the standing wave is superimposed by a wave propagating in the same direction as the larger amplitude wave. Its amplitude is equal to the difference between the amplitudes of the original waves.

When the parameters of the forward and backward waves are different, sum (9.5) gives a traveling wave. The anomaly field produced by two waves with the same frequency but with different amplitudes and phases is

$$u = A_1 \exp i(\sigma t + k_1\lambda) + A_2 \exp i(\sigma t + k_2\lambda) = (A_1 e^{ik_1\lambda} + A_2 e^{ik_2\lambda})e^{i\sigma t} \\ = A_R e^{i\varphi_R} e^{i\sigma t} \tag{9.6}$$

It is well known that the square of the resulting amplitude A_R is

$$A_R^2 = (A_1 e^{ik_1\lambda} + A_2 e^{ik_2\lambda})(A_1 e^{-ik_1\lambda} + A_2 e^{-ik_2\lambda}) \\ = A_1^2 + A_2^2 + A_1 A_2 (e^{i(k_1-k_2)\lambda} + e^{-i(k_1-k_2)\lambda}) \\ = A_1^2 + A_2^2 + 2A_1 A_2 \cos(k_1 - k_2)\lambda \tag{9.7}$$

Thus, the total energy is the sum of the energies of the original waves $A_1^2 + A_2^2$ plus the additional interference term $2A_1A_2\cos(k_1 - k_2)\lambda$, which can be positive or negative. The interference effect is determined by the difference between the phases of two waves arriving at the point under consideration.

An analysis of the sections presented suggests that the wave numbers of the fundamental multiyear waves are equal to unity, the phase velocities vary from $2\pi/9$ to $2\pi/1$ rad/year, and the linear velocities range from 14 to 127 cm/s.

The detection of multiyear waves in time–longitude sections is difficult for two major reasons. First, a large number of multiyear waves travel simultaneously in different directions in the ocean and the atmosphere. They interfere to create a complex interference pattern. Second, when moving through a medium, multiyear waves, like any other geophysical waves, give rise to various internal processes (circulation cells, vertical convection, mass and heat transfer, tidal rises and falls near obstacles, etc.), whose intensity depends on local conditions. This is why the wave pattern is always highly diverse. For example, multiyear waves are easily seen in the Pacific, less pronounced in the Atlantic and Indian oceans, and hardly visible over the continents (*Climate Diagnostics Bulletin*), (Sidorenkov, 1991c).

To the best of my knowledge, multiyear waves have not been previously mentioned in publications, although patterns and descriptions of their regional responses in the atmosphere and the ocean can be found in many studies (White and Peterson, 1996; Jacobs and Mitchell, 1996; Trenberths, 1991; Rasmusson, 1991; Petersen and White, 1998). An analysis of 1980–1996 observations in (White and Peterson, 1996; Jacobs and Mitchell, 1996) revealed the Antarctic circumpolar wave (ACW) in the fields of SST, surface pressure, wind velocity, and sea surface height. It travels eastward around Antarctica at a velocity of about 0.1 m/s and completes a revolution in 8–10 years. Since the ACW has a wave number of 2 (its length is half the length of the Antarctic Circle), it leads to oscillations of about 4–5 years in the atmospheric and oceanic characteristics in this region, which are similar to ENSO cycles.

Beginning with (Bjerknes, 1969), processes leading to variability in the ENSO range are usually called teleconnections. For example, in (Petersen and White, 1998) sequences of interannual anomaly fields for SST, SLP, and precipitable water over the Southern Hemisphere in 1982–1994 were presented, in which the evolution of multiyear waves could be observed. However, they were interpreted as slow teleconnections between the tropical ENSO and ACWs through anomalous vertical convection and a regional overturning cell in the troposphere.

When the crests of multiyear waves reach a barrier such as the coast of America or Africa, they create a tide of warm surface water. When a wave trough arrives, a tidal fall and upwelling are observed. The tidal rises and falls enhance anomalies in the hydrometeorological characteristics. The longer the distance a wave travels over the ocean, the higher the tidal rise and the larger the anomalies. This is why El Niño events are weak in the Atlantic off the African coast and are very strong in the Pacific off the American coast. For atmospheric multiyear waves, the continents are not insurmountable barriers, so the coastal tidal effects of multiyear waves are less pronounced.

Thus, the interference of forward and backward multiyear waves and their interaction with the obstacles lead to El Niño and La Niña in the oceans, to variations in the SO phases in the atmosphere and, ultimately, to cycles in the SOI and SST spectra that are multiples of the Chandler period (1.2 years) and the forced lunisolar nutation period.

The spectral-temporal diagrams of SOI constructed in (Schneider and Schonwiese, 1989; Sidorenkov 1998, 2000e, 2001) and the wavelet and waveform analyses in (Astaf'eva and Sonechkin, 1995; Gu and Philander, 1995; Wang and Wang, 1996) reveal the instability of SOI cycles and, hence, the periods of multiyear waves. Generated at certain frequencies, cycles may be damped out after some time and may be excited again at different frequencies. This is why 6-year ENSO cycles prevailed until the mid-twentieth century (Figure 9.2), while 4.9-year ENSO cycles were dominant later (Figure 9.3). This instability is also typical of multiyear waves.

9.4 Modern ENSO Models

Beginning with Bjerknes' pioneering works (Bjerknes, 1966, 1969), the ENSO is viewed as a self-sustained oscillation in which the SST anomalies in the equatorial Pacific influence the strength of the trade winds. The latter drive oceanic flows, which generate SST anomalies. This concept underlies the modern studies.

The basic technique for studying the physics of the ENSO is numerical simulation. The models used at present include intermediate coupled atmosphere–ocean models consisting of a shallow-water ocean model and a simple atmospheric model (Zebiak and Cane, 1987; Schopf and Suarez, 1988), hybrid coupled models consisting of an ocean general circulation model and a simple atmospheric model (Neelin, 1990; Latif and Villwock, 1990), and coupled general circulation models in which the ocean and atmospheric models are based on the primitive equations and involve fairly complete physical parametrizations (Philander *et al.*, 1992; Mechoso *et al.*, 1993).

The behavior of models is usually analyzed depending on the seasonal cycle amplitude, the model parameters characterizing the coupling of the ocean to the atmosphere, the degree of the interaction between thermocline variations and SST anomalies, and so on. Most models reproduce interannual variability in the frequency range of the ENSO.

It was shown in (Neelin, 1990) that variations in the model temporal coupling parameters and seasonal forcing lead to simultaneous oscillations with a period of 3–4 years arising as Hopf bifurcations at the equator. Instability is facilitated by a decrease in upwelling, an increase in the wind stress response to SST anomalies in the atmospheric model, and by an increase of the vertical temperature gradient in the ocean surface layer. For a stronger coupling, a second bifurcation arises that gives rise to additional 5- to 6-month oscillations, which complicate the temporal evolution of the model.

A theoretical ENSO model, consisting of prognostic equations for SST and thermocline variations, was derived in (Wang and Fang, 1996). The model describes the nonlinear interaction between mixed-layer thermodynamics and upper-ocean dynamics through wind stress and upwelling. The thermocline variations memorize the effect of SST on wind and mediate the atmospheric feedback for SST in a nonlinear form via vertical temperature advection. For an annual mean basic state, the model exhibits a limit cycle with a period of about 34 months. The asymmetry in its evolution leads to secondary superharmonic oscillations. When the basic state varies annually, the limit cycle becomes a strange attractor. Due to the presence of chaos, the major power peaks expand, spread, and shift toward low frequencies. The annually varying basic state not only generates irregularities in the oscillation period but also tends to lock the ENSO period to 3-, 4-, and 5 years.

Thus, all the modern models treat ENSO as self-sustained oscillations of the coupled ocean–atmosphere system without paying attention to the fact that the actual ENSO spectrum contains components that are multiples of the Chandler period (1.2 years) and the lunisolar nutation period (18.6 years) rather than of the annual period. The 18.6-year period is the principal one of many zonal tidal and lunisolar nutation periods. The Chandler period is the free nutation period or the polar motion period. It is determined not by the driving-force frequency but rather by the compression and elastic properties of the Earth. The presence of cycles that are multiples of 18.6 and 1.2 years in the atmospheric and oceanic variability suggests that the atmosphere and the ocean exhibit free and forced nutation motion together with the Earth (Sidorenkov, 2000b, 2000c, 2002).

9.5 The Model of Nonlinear Excitation

During an ENSO, air and water masses are redistributed between the Eastern and Western Hemispheres. The exchange occurs most intensively between the southern subtropical part of the Pacific Ocean and the eastern part of the Indian Ocean. The swing in the oscillations of sea level in the eastern and western parts of the tropical zone of the Pacific Ocean is $\approx$50 cm. Oscillations of atmospheric pressure at the antipodal centers of the action of the Southern oscillation reach 3 hPa at the ENSO frequency. These pressure anomalies are distributed over the globe (Figure 7.1) in the way that is required to excite the polar wobble: the sign of the anomaly over the southwestern part of the Pacific and central Asia is opposite to the sign over the

Indian Ocean and North America (these anomalies are described by the tesseral spherical harmonic $P_2^1(\theta,\lambda)$).With this pressure distribution, the axis of the largest moment of inertia of the Earth (i.e. the axis of the Earth's figure) should deviate toward Asia in a La Nino events and toward North America in an El Nino events. Deviations of the axis of the Earth's figure from the rotation axis will necessarily give rise to free nutation of the Earth.

Let us now turn from a qualitative to a quantitative treatment of the problem.

The excitation of the polar wobbles can be estimated using the equation (Munk and MacDonald, 1960)

$$\frac{\mathrm{i}}{\underline{\sigma}}\frac{\mathrm{d}\underline{m}}{\mathrm{d}t}+\underline{m}=\underline{\Psi} \tag{9.8}$$

Here, $\underline{\sigma}=\sigma+\mathrm{i}\beta$, $\sigma=\frac{2\pi}{T}$ is the frequency of the free motion of the poles; $T=1.2$ years is the Chandler period; β is the damping decrement; i is the square root of -1; $\underline{m}=m_1+\mathrm{i}m_2$ (m_1 and m_1 are the direction cosines of the Earth's instantaneous angular rotation vector); and $\underline{\Psi}=\chi_1+\mathrm{i}\chi_2$ (χ_1 and χ_2 are the components of the effective angular momentum functions) (Barnes *et al.*, 1983; Sidorenkov, 2002).

To explain the observed wobble of the Earth's pole with a period of one year, it is sufficient for the atmosphere pressure oscillation to have an amplitude of 1.6 hPa; that is $P(\theta,\lambda)=1.6\sin 2\theta\sin(\lambda-\lambda_0)$ hPa (Munk and MacDonald, 1960). As noted above the amplitude of pressure oscillations during an ENSO reaches 3 hPa, so that excitation of the free nutation of the Earth is quite feasible.

The Chandler wobble of the poles induces a polar tide in the atmosphere and ocean. For example, the static polar tide in the ocean has the form (Munk and MacDonald, 1960)

$$\zeta=-\frac{1+k-h}{g}\frac{\Omega^2R^2}{2}\sin 2\theta(m_1\cos\lambda+m_2\sin\lambda) \tag{9.9}$$

where k and h are Love numbers, Ω is the magnitude of the Earth's angular rotation vector, g is acceleration of Earth's gravity, θ is the colatitude, and λ is the longitude. In the atmosphere, the polar tide is described by the same expression, since

$$\Delta P=\int_0^{\zeta_a}\rho g\mathrm{d}z \tag{9.10}$$

but with a different amplitude. We should stress that the amplitude is a function of the polar deviation $\underline{m}$.

The typical amplitude of the static polar tide in the ocean is 0.5 cm. However, analysis of tide records indicates a resonant spectral density peak at period T (Munk and MacDonald, 1960). A 14-month periodicity in the Atlantic subtropical circulation has repeatedly been noted (Lappo, Gulev, and Rozhdestvenskiy, 1990; Shuleykin, 1965).

By analogy with Equations 9.9 and 9.10, we might expect that, in the case of linear interactions of nutation waves, the amplitude A of the exciting function $\underline{\Psi}$ for the atmosphere and the ocean must be a function of $\underline{m}$. However, the presence of superharmonics of the Chandler period (≈2.4, 3.6, 4.8, and 6.0 years) and subharmonics of the forced nutation period (≈2.3, 3.7, 4.7, and 6.2 years) in the ENSO and atmospheric angular momentum spectra indicate the nonlinear interaction of the forced and free nutation motions of the Earth, ocean, and atmosphere. Therefore, we assume that the amplitude A_k of each harmonic with number k of the exciting function $\underline{\Psi}$ has a power-law dependence on the value $\underline{m}$: $A_k = \mu_k \underline{m}^{\alpha_k}$, where α_k is a power-law exponent different from unity and μ_k is a proportionality coefficient (Sidorenkov, 2002a, 2002b).

Using this assumption, we can approximate the function $\underline{\Psi}$ by the sum of N harmonics:

$$\underline{\Psi} = \sum_{k=1}^{N} \mu_k \underline{m}^{\alpha_k} \exp[\omega_k i(t - t_{0k})] \approx \sum_{k=1}^{N} \mu_k \underline{m}^{\alpha_k} \exp(\omega_k i t) \tag{9.11}$$

Here, k is the number of the harmonic, ω_k is its frequency, and $\omega_k t_{0k}$ is the initial phase, which is set equal to zero. For superharmonics of the Chandler frequency, $\omega_k = \sigma / n_k$, where $n_k = 2, 3, 4, \ldots$ are superharmonics numbers.

We were not able to find a solution of Equation 9.8 with the function $\underline{\Psi}$ in the form (9.11). Therefore, we simplified expression (9.11):

$$\underline{\Psi} = \underline{m}^{\alpha} \sum_{k=1}^{N} \mu_k \exp(\omega_k i t) \tag{9.12}$$

Then, the solution of Equation 9.8 with the function $\underline{\Psi}$ given by (9.12) at $m(0) = m_0$ takes the form

$$\begin{aligned}\underline{m}\,(t) &= \Bigg\{ \underline{m}_0^{1-\alpha} \exp[(1-\alpha)\underline{\sigma} i t] - i\underline{\sigma}(1-\alpha)\exp[(1-\alpha)\underline{\sigma} i t] \\ &\quad \times \int_0^t \exp[(\alpha-1)\underline{\sigma} i \tau] \left(\sum_{k=1}^{N} \mu_k \exp(\omega_k i \tau) \right) d\tau \Bigg\}^{\frac{1}{1-\alpha}} \\ &= \Bigg\{ \underline{m}_0^{1-\alpha} \exp[(1-\alpha)\underline{\sigma} i t] + \sum_{k=1}^{N} \frac{\mu_k (1-\alpha)\underline{\sigma}}{[(1-\alpha)\underline{\sigma} - \omega_k]} [\exp(i\omega_k t) \\ &\quad - \exp[(1-\alpha) i \underline{\sigma} t]] \Bigg\}^{\frac{1}{1-\alpha}} \end{aligned} \tag{9.13}$$

The first term in braces of (9.13) describes the free motion of the pole due to its initial deviation $\underline{m}_0$. The second term (first sum) describes the overall forced motion of the pole with the excitation frequencies ω_k. The third term (second sum) describes the polar motion caused by the function $\underline{\Psi}$ with the fundamental frequency $\underline{\sigma}$.

Expression (9.13) describes the result in the most general form. To make this equation easier to understand, we make the following simplifications: (a) we neglect the initial deviation of the pole; that is we set $\underline{m}_0 = 0$, and (b) we neglect damping; that is we set $\underline{\sigma} = \sigma$. In this case, expression (9.13) takes the form

$$\underline{m}\,(t) = \left\{ \sum_{k=1}^{N} \frac{\mu_k(1-\alpha)\sigma}{[(1-\alpha)\sigma-\omega_k]} [\exp(\mathrm{i}\omega_k t) - \exp[(1-\alpha)\mathrm{i}\,\sigma\, t]] \right\}^{\frac{1}{1-\alpha}}$$
$$= \left\{ \sum_{k=1}^{N} \frac{\mu_k(1-\alpha)\sigma}{\mathrm{i}[\omega_k-(1-\alpha)\sigma]} 2\sin\frac{[\omega_k-(1-\alpha)\,\sigma]t}{2} \exp\frac{[\omega_k+(1-\alpha)\sigma]}{2}\mathrm{i}t \right\}^{\frac{1}{1-\alpha}} \tag{9.14}$$

For $(1-\alpha)\sigma = \omega_k$, the term with number k in (9.14) is an indeterminate form of the type 0/0. For superharmonics of the Chandler frequency, the exponent $\frac{1}{1-\alpha} = n_k$ is equal to a positive integer number ($n_k = 2, 3, 4, \ldots$). Let us suppose, for example, that $\omega_1 \to (1-\alpha)\sigma$. Then, the sine of the small angle in the first term of (9.14) can be replaced by the angle itself and will cancel from the numerator and denominator. All other terms (with numbers $k = 2, 3, \ldots, N$) remain unchanged, and we obtain

$$\underline{m}(t) = \left\{ -\mu_1(1-\alpha)i\sigma\, t\exp(i\omega_1 t) + \sum_{k=2}^{N} \frac{\mu_k(1-\alpha)\sigma}{[(1-\alpha)\sigma-\omega_k]} [\exp(i\omega_k t) \right.$$
$$\left. -\exp[(1-\alpha)i\sigma\, t]] \right\}^{\frac{1}{1-\alpha}} \tag{9.15}$$

The same result will be obtained if we formally evaluate the indeterminate form in (9.14) using L'Hopital's rule. The coefficient in the first exponential term increases with time t. All other terms have constant coefficients. Thus, at sufficiently large t, we can neglect all terms in braces except for the first.

In another time interval, a periodicity with frequency ω_k may dominate. In this case, analysis of the behavior of $\underline{m}$ gives a result similar to (9.15). Summing the resonant contributions from all terms, we obtain

$$\underline{m}(t) \approx \left\{ \sum_{k=1}^{N} [-\mu_k(1-\alpha)\mathrm{i}\sigma\, t\exp(\mathrm{i}\omega_k t)] \right\}^{\frac{1}{1-\alpha}}$$
$$= \sum_{k=1}^{N} (\mu_k\omega_k)^{n_k} t^{n_k} \exp\left[\mathrm{i}\left(\sigma t + \frac{3n_k\pi}{2}\right)\right] \tag{9.16}$$

Here, we took into account the relations $(1-\alpha)\sigma = \omega_k$, and $\frac{1}{1-\alpha} = n_k$.

Thus, for $\sigma = n_k\omega_k$, that is when the frequency of free oscillations of the pole is an integer multiple of the excitation frequency, a combinational resonance arises. The perturbing forces act synchronously with the proper oscillations of the Earth, giving rise to an intense swinging of the Earth about its rotation axis. The amplitude of the polar oscillations increases with time according to a power law (i.e. proportional to t^{n_k}), the more rapidly, the lower the frequency of the superharmonics excitation component. Even low-power oscillations of the exciting function $\underline{\Psi}$ can lead to significant oscillations of the instantaneous angular rotation vector $\underline{m}$ of the Earth. As the amplitude of the polar oscillations increases, resistance leading to damping of the oscillation grows, so that an infinite increase of the amplitude becomes impossible.

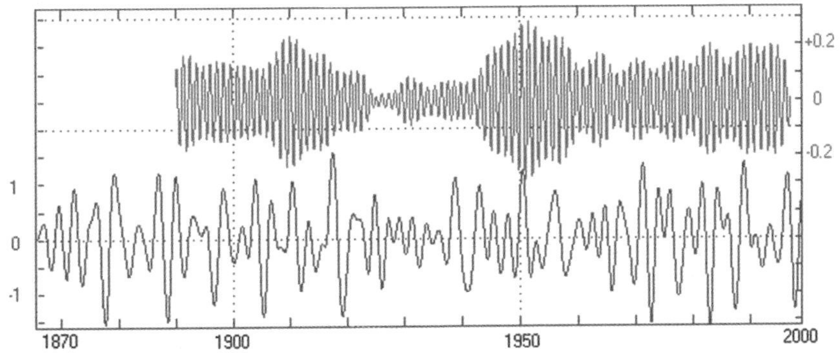

Figure 9.10 Time series of SOI (bottom) and the Chandler term of the Earth's pole coordinate y (top) in 1866–2000.

It is well known that the intensities of QBO, ENSO, and polar motion vary with time. D.M. Sonechkin (Vlasova, 1987) pointed out that extrema of QBO easterlies alternate with a period of 14 years. It was noted in (Gu and Philander, 1995; Wang and Wang, 1996) that 4-year ENSO periods prevailed in 1872–1890, while 3-year ENSO periods dominated in 1890–1910. After 1910, the dominant period switched to 7 years. The dominant period was 5–6 years from 1940 to 1960, 2 years in the 1960s, and 4–5 years in 1971–1995. The ENSO amplitudes were large in 1870–1915, small in 1915–1950, and again large after 1960. A similar pattern was observed in the CW amplitudes.

To demonstrate the instability of CW and ENSO, we calculated (with the help of a filter) the time series of the Chandler component of the North Pole's coordinate y over the entire 110-year observed series and compared it to the SOI time series (Figure 9.10). An analysis of the plot shows that the CW amplitudes in 1890–1915 and 1947–1960 were three and five times larger than in 1925–1943. It can be clearly seen that CW is amplitude modulated. The period between the amplitude maxima (beat period) is 40 years. This suggests that CW is the sum of two oscillations with very close periods. They can be calculated using the formula

$$2\cos\left(\frac{1}{T_1}-\frac{1}{T_2}\right)\pi t\cos\left(\frac{1}{T_1}+\frac{1}{T_2}\right)\pi t=\cos 2\pi\frac{1}{T_1}t+\cos 2\pi\frac{1}{T_2}t \tag{9.17}$$

where T_1 and T_2 are the periods of the first and second oscillations, $\frac{1}{T_1}-\frac{1}{T_2}$ is the beat frequency, and $\frac{1}{2}\left(\frac{1}{T_1}+\frac{1}{T_2}\right)$ is the carrier frequency. Since $T_1 = 1.2$ years and the beat period is 40 years ($\frac{T_1T_2}{T_2-T_1} = 40$), it is easy to see that $T_2 = 1.24$ years. This side period is associated with a peak in the power spectrum of the pole motion (Figure 9.6).

In 1921–1938, the SOI variations decreased noticeably. Simultaneously, the polar motion was damped out as well. Specifically, the Chandler wobbles (CW) amplitude decreased by several times, the CW period lengthened, and the CW phase changed. These findings demonstrate that ENSO is consistent with the polar motion. Likely, time variations in the ENSO intensity lead to instability in CW excitation. The instability of ENSO is likely caused by the influence of the seasonal cycle phase on the nutation motion of the atmosphere and the ocean.

The QBO period in stratospheric winds was first about 26 months (until 1964), then increased up to 30 months, and next (starting in 1990) again decreased to 27 months (Sidorenkov, 1998). Possibly, there is an inversely proportional coupling between the QBO period and the CW amplitude: the QBO period decreases for large CW amplitudes and increases for small CW amplitudes. In 1921–1940, the CW amplitudes were extremely small, and we can expect a considerable increase in the QBO periods. Unfortunately, this conclusion cannot be checked because of the lack of observations in that period.

Polar motion is a major but not a unique manifestation of the Chandler period. It is well known that the lunar nodes precess westward around the ecliptic, completing a revolution in 18.61 years. Lunar perigee moves eastward, completing a revolution in 8.85 years. Because of these opposite motions, a node meets perigee in exactly 6 years:

$$\frac{1}{18.61} + \frac{1}{8.85} = \frac{1}{6.0}$$

The Earth, moving eastward around the Sun, overtakes lunar perigee every 412 days, which is close to the Chandler period. Subtracting the node and lunar perigee frequencies from the Earth's annual frequency gives the exact Chandler frequency:

$$\frac{1}{1.0} - \left(\frac{1}{18.61} + \frac{1}{8.85}\right) = \frac{1}{1.2} \tag{9.18}$$

All these coincidences suggest that the Earth's daily rotation rate and even processes occurring on the Earth have become locked to the cycles of the Earth–Moon–Sun system over the billions years of the Solar System's evolution. On the basis of Equation 9.18 Avsyuk (1996) supposes that the Chandler wobble arises because of the gravitational effect of the Moon on the inner solid core of the Earth.

10
Mechanical Action of the Atmosphere on the Earth's Rotation

10.1
Friction and Pressure Torques

The atmosphere is in unceasing motion relative to the Earth's surface. At movement of air near a terrestrial surface the frictional stress τ develops that is proportional to the square of the wind speed and coincides in direction with the wind direction. When westerly winds blow (the air outruns the Earth) the frictional stress produces a positive force moment $\tau\ R \sin\theta\ dS$ accelerating the Earth's rotation. Here R is the radius of the Earth, θ is the 90° colatitude, $dS = R^2 \sin\theta\ d\theta\ d\lambda$ is an elementary area, and λ is the longitude. With easterly winds, that is, when the air lags behind the motion of the Earth, the moment of the frictional force retards its rotation. As is known, easterly winds predominate at low latitudes while westerly winds predominate at temperate latitudes, so that in the equatorial zone the atmosphere slows the Earth's rotation while in the zone of temperate latitudes it accelerates it. The resultant moment of the frictional forces, while small in comparison with the magnitude of each of the components individually, is sufficient, as will be clear from what follows, to produce the observed changes in the Earth's rotation rate.

The frictional stress τ, which is uniquely determined from the height distribution of the wind in the planetary boundary layer of the atmosphere, includes both the turbulent stress and the aerodynamic pressure of the air on the irregularities of the underlying surface. If the irregularities were distributed uniformly enough over the Earth (so that they could be characterized statistically by one roughness parameter) then no complications would arise with the calculation of the total force of the action of the atmosphere on the Earth. The irregularities are far from uniformly distributed over the Earth, however. Mountain ranges of the Cordillera and Andes type stretch out for tens of thousands of kilometers in the meridional direction, whereas they extend for only a few hundred kilometers in the latitudinal direction. Such individual ranges are a kind of "sail" standing in the path of winds. They cannot yet be characterized statistically by some roughness parameter. Therefore, in order to determine the total force of mechanical action one must, in addition to the frictional stress, also take into account the force of atmospheric pressure on each range separately. An insuperable difficulty arises here – how to separate this pressure force

The Interaction Between Earth's Rotation and Geophysical Processes. Nikolay S. Sidorenkov

ISBN: 978-3-527-40875-7

from that part of it that is already involved in the stress of turbulent friction as the force of aerodynamic pressure on the irregularities that are characterized statistically by a roughness parameter?

Let us compute the moment of the pressure forces acting on a mountain range. Let the vertical elementary area $d\sigma = Rd\theta\, dR$ of the Earth's surface be oriented along meridian, and let a pressure drop $P_W - P_E$ exists between its west and east sides, where P_W and P_E are the pressures on the west and east sides of the area, respectively. The moment of the pressure forces applied to the area is obviously equal to $(P_W - P_E)$ $R \sin\theta\, d\sigma$. The force moment acting on the entire range is $\iint_\sigma (P_W - P_E) R \sin\theta\, d\sigma$ where σ is the area of the projection of the range onto the vertical plane directed along the meridian. When $P_W > P_E$ the moment of the pressure forces accelerates the Earth's rotation, while when $P_W < P_E$ it retards it.

Integrating the moments of frictional forces over the Earth's entire surface and summing the moments of the pressure forces over all separately standing mountain ranges. We get the acceleration of the rotation rate of the absolute rigid Earth (see Section 4.4)

$$\Omega \frac{d\nu_3}{dt} \approx \frac{d\omega}{dt} = \frac{1}{C}\left[\iint_S \tau R \sin\theta\, ds + \sum_{i=o}^{N} \iint_{\sigma_i} (P_W - P_E)_i R \sin\theta\, d\sigma\right] \tag{10.1}$$

where $\nu_3 \approx \frac{\omega_3 - \Omega}{\Omega}$ is the dimensionless deviation of the angular velocity of the Earth's rotation; $\omega_3 \approx \omega$ is the projection of instant angular velocity to the polar axis OX_3; Ω is the mean value of the angular velocity, t is the time, C is the polar moment of inertia of the Earth. N is the number of mountain ranges, S is the area of the Earth's surface, and σ_i is the surface area of the ith mountain range. Later, our problem will consist in calculating the moments of the frictional stress forces, and ultimately in determining the nonuniformity of the Earth's rotation due to the mechanical action of the atmosphere on the Earth.

10.2
Mechanical Interaction of the Atmosphere with the Underlying Surface

The Earth's surface greatly influences the relative movement of air. The bottom layer of the airflow "sticks" to the Earth's surface and partly loses its momentum, that is decelerated due to the aerodynamic drag on surface irregularities. Because of the turbulent viscosity, the deceleration involves a rather thick layer of the atmosphere. According to observational data, the dynamic influence of the underlying surface extends as high as 1.5–2 km above the Earth's surface, depending on the surface roughness, atmospheric stratification, wind velocity, and some other factors. The atmosphere's bottom layer, in which the forces of turbulent friction play an important role, alongside the pressure gradient and the Coriolis force, is the planetary boundary layer. Let the thickness of the layer be denoted by H.

Steady zonal horizontally homogeneous air motion in the planetary boundary layer is described by the equations (Gandin *et al.*, 1955):

$$\frac{d}{dz}k\frac{d\bar{v}}{dz} + 2\omega\cos\theta\bar{u} = \frac{1}{R\rho}\frac{\partial P}{\partial\theta}$$
$$\frac{d}{dz}k\frac{d\bar{u}}{dz} - 2\omega\cos\theta\bar{v} = \frac{1}{\rho R\sin\theta}\frac{\partial P}{\partial\lambda} \tag{10.2}$$

where $k = k_z$ is the coefficient of turbulent viscosity; z is the height; $\bar{u}$ and $\bar{v}$ are the zonal and meridional components of the mean wind directed from west to east and from north to south, respectively; P is the atmospheric pressure. Hereafter, signs of averaging are omitted for short. Note that, in low latitudes, the inertial forces neglected in Equations 10.2 are comparable with the Coriolis force. Therefore, Equations 10.2 do not describe the real air motion, at least in a latitudinal zone of 5° North to 5° South.

Estimations show that the pressure gradient and Coriolis force in the atmosphere's lowermost layer as thick as 0.1 H are negligibly small in comparison with the turbulent friction. This layer is called the surface layer. Motion in the surface layer occurs because its lower layers are entrained into the upper layers moving under the action of the pressure gradient force. This is why the surface layer is sometimes referred to as the nonpressure layer.

Thereby, the steady zonal horizontally homogeneous air motion in the surface layer can be described by the following equations:

$$\frac{d}{dz}k\frac{du}{dz} = \frac{1}{\rho}\frac{d\tau_\lambda}{dz} = 0$$
$$\frac{d}{dz}k\frac{dv}{dz} = \frac{1}{\rho}\frac{d\tau_\theta}{dz} = 0 \tag{10.3}$$

where τ_λ and τ_θ are the zonal and meridional components of friction stress in the surface layer, respectively. It is convenient to introduce the complex velocity $w = u + iv$ and the friction stress $\tau = \tau_\lambda + i\tau_\theta$. Then systems (10.2) and (10.3) are reduced to the following equations:

$$\frac{d}{dz}k\frac{dw}{dz} + 2i\omega\cos\theta w = \frac{i}{R\rho}\frac{\partial P}{\partial\theta} + \frac{1}{\rho R\sin\theta}\frac{\partial P}{\partial\lambda} \tag{10.4}$$

$$\frac{d}{dz}k\frac{dw}{dz} = \frac{1}{\rho}\frac{d\tau}{dz} = 0 \tag{10.5}$$

As follows from Equation 10.5, friction stress in the surface layer is constant at a given time point (independent of the height).

Observations show that, in the surface layer, the coefficient of turbulent viscosity increases almost linearly with height. Above the surface layer, the growth slows down. Some researchers believe that at some height the growth of k is replaced by its decrease. But, in general, the coefficient of turbulent viscosity changes insignificantly above the surface layer. The above-stated can be illustrated by a profile of the turbulent viscosity coefficient, based on observations in Leipzig (Matveev, 1965).

z, m	50	100	200	300	400	500	700	900
k, m^2/s	10.4	12.8	14.1	14.1	13.1	11.8	9.3	5.5

As the details of the wind field do not interest us, we suppose that the turbulence factor in the surface layer increases linearly with height, remaining constant in other parts of the planetary boundary layer, that is:

$$k = \begin{cases} k_1 z' & z' \leq h \\ K & z' \geq h \end{cases} \tag{10.6}$$

where k_1 is the turbulence factor at a height of 1 m; h is the thickness of the surface layer; $z' = z - d$ is the height counted off from some reference level (height) of displacement $z = d$. We also assume that the pressure gradient and density of air in the planetary boundary layer do not depend on the height.

Double integrating of Equation 10.5 with respect to height between the limits z_0 and z' gives the following formula for the wind velocity at the height $z' \leq h$:

$$w(z') = \frac{w_*^2}{k_1} \ln \frac{z' + z_0}{z_0} \tag{10.7}$$

where $w_* = \sqrt{\tau/\rho}$ is the dynamic velocity or the friction speed; z_0 is the parameter of roughness or the height that is counted off from the upper boundary of the displacement layer and at which the mean wind velocity would become zero, if the logarithmic law is applicable up to this height.

The coefficient of turbulent viscosity k at a given height depends on many factors, the thermal stratification and the dynamic velocity of wind being principal. Thereby, the coefficient changes considerably with time. For example, in summer, the value of k in afternoons can be 10–12 times greater than that at nights. In the case of the indifferent stratification, for a given height of the surface layer and at a given time point, the coefficient of turbulent viscosity can be supposed to depend only on the dynamic velocity. As is generally known, that dependence is:

$$k = \kappa w_* z' \tag{10.8}$$

where κ is the Karman constant. Based on numerous calculations, $\kappa = 0.4$. In view of (10.8), expression (10.7) becomes:

$$w(z') = \frac{k_1}{\kappa^2} \ln \frac{z' + z_0}{z_0} = \frac{w_*}{\kappa} \ln \frac{z' + z_0}{z_0} \quad z' \leq h \tag{10.9}$$

If $w(z')$ is known, expression (10.9) makes it possible to easily determine the friction stress:

$$\tau = \frac{\kappa^2}{\left(\ln \frac{z' + z_0}{z_0}\right)^2} \rho w^2(z') \tag{10.10}$$

Within the limits of the surface layer, the friction stress is independent of the height. Equating the real parts of the left and right parts of expression (10.10), we find the zonal component of friction stress:

$$\tau_\lambda = \frac{\kappa^2}{\left(\ln \frac{z' + z_0}{z_0}\right)^2} \rho u^2(z') \tag{10.11}$$

Numerous experiments on direct measurement of friction stress at the surface give the following:

$$\tau_\lambda = c_z \rho u^2 \tag{10.12}$$

where c_z is the coefficient of friction attributed to the height at which the wind velocity z is measured. Comparison of relations (10.12) and (10.11) gives the following expression for the coefficient of friction:

$$c_{z'} = \frac{\kappa^2}{\left(\ln \frac{z'+z_0}{z_0}\right)^2} \tag{10.13}$$

Expression (10.13) shows that $c_{z'}$ is independent of the wind velocity, depending only on the height z' at which it is measured. For a fixed height of wind velocity measurement, the coefficient of friction depends on the underlying surface's roughness that is statistically characterized by the roughness parameter z_0.

Table 10.1 shows the values of the roughness parameter z_0 (Sutton, 1953; Matveev, 1965) and the values of the coefficient of friction for natural surfaces for the indifferent stratification (the wind velocity is measured at the height z'). The coefficient of friction is calculated by formula (10.13).

The coefficient of friction between the air and the water surface is determined by its roughness parameter. Obviously, the roughness parameter on the water surface

Table 10.1 Roughness z_0 and coefficient of friction $c_{z'}$ characteristic of natural surfaces (indifferent stratification; the velocity of wind is measured at a height of z').

Surface	z_0, cm	Coefficient of friction $c_{z'}$		
		$z' = 2$ m	$z' = 50$ m	$z' = 100$ m
Ice	0.001	0.001	0.0007	
Desert	0.03	0.002	0.0011	
Snow	0.08	0.002	0.0013	
Meadow with grass up to 1 cm in height	0.1	0.003	0.0014	
Plain with scanty grass up to 10 cm in height	0.7	0.005	0.0020	
Dense grass up to 10 cm in height	2.3	0.008	0.0027	
Airfield	2.5	0.008	0.0028	
Wheat field	5	0.012	0.0034	
Grass up to 60–70 cm in height, at the wind velocity:				
$u_2 = 1.5$ km/s	9.0	0.016	0.0040	
$u_2 = 3.5$ km/s	6.1	0.013	0.0036	
$u_2 = 6.2$ km/s	3.7	0.010	0.0031	
Bush	10		0.004	0.003
Woody plain	100		0.010	0.005
Large city	400		0.024	0.015
Hilly, irregular terrain	500		0.028	0.017
Highland	1000		0.050	0.029

depends on waves caused mainly by wind. Consequently, the coefficient of friction is bound to depend on the wind velocity. At present, there are conflicting conclusions concerning that dependence. It is only established that the dependence is weak (Deacon and Webb, 1962; Garratt, 1977). In the light of this, the coefficient of friction for the sea surface is hereafter believed to be independent of the wind velocity.

To calculate the frictional stress τ_λ, we need to know not only the coefficient of friction but also the velocity of the zonal wind at some height $z' \leq h$. However, scanty data on the wind field in the surface layer of the global atmosphere is insufficient for the task of calculating the total moment of the friction forces that the whole atmosphere imparts to the Earth. Therefore, in order to calculate the total moment, it is necessary to derive wind characteristics from the atmospheric pressure field, which can be determined more reliably than the wind field because of its weak variability.

Following Guldberg and Mohn, in the first, very rough approximation, for solids, the force of friction can be taken instead of the force of viscosity. In that case, Equation 10.4 is of the form:

$$-\mu w + 2\mathrm{i}\omega \cos\theta w = \frac{\mathrm{i}}{\rho R}\frac{\mathrm{d}P}{\mathrm{d}\theta} + \frac{1}{\rho R \sin\theta}\frac{\mathrm{d}P}{\mathrm{d}\lambda} \quad (10.14)$$

where μ is the Guldberg–Mohn coefficient of friction that is determined by the following relation:

$$\mu = 2\omega \cos\theta \operatorname{ctg}\gamma \quad (10.15)$$

where γ is the angle between the wind direction and the atmospheric-pressure gradient direction. Solving Equation 10.4 with respect to w in view of relation (10.15) and separating the real and imaginary parts gives the zonal and meridional components of the wind velocity:

$$u_\lambda = \frac{\sin^2\gamma}{2\rho\omega R \cos\theta}\frac{\mathrm{d}P}{\mathrm{d}\theta} - \frac{\sin 2\gamma}{2\rho\omega R \sin 2\theta}\frac{\mathrm{d}P}{\mathrm{d}\lambda} \quad (10.16)$$

$$v_\theta = -\frac{\sin^2\gamma}{\rho\omega R \sin 2\theta}\frac{\mathrm{d}P}{\mathrm{d}\lambda} - \frac{\sin 2\gamma}{4\rho\omega R \cos\theta}\frac{\mathrm{d}P}{\mathrm{d}\theta} \quad (10.17)$$

In this way, the zonal wind velocity is determined in some works (Pariiski and Berlyand, 1953; Sidorenkov, 1963).

The second approximation is the well-known Ekman–Åkerblom model. Equation 10.4 is the input equation for this model. Let us introduce an equation for the geostrophic wind $\underline{G}$:

$$\underline{G} = G_\lambda + \mathrm{i}G_\theta = \frac{1}{2\mathrm{i}\omega\cos\theta}\left(\frac{\mathrm{d}P}{\rho R \sin\theta \mathrm{d}\lambda} + \frac{\mathrm{i}\mathrm{d}P}{\rho R \mathrm{d}\theta}\right) \quad (10.18)$$

Let us assume that $\underline{G}$, ρ and K do not depend on the height. Then Equation 10.4 can be re-arranged as follows:

$$\frac{\mathrm{d}^2\Phi}{\mathrm{d}z^2} + 2\mathrm{i}a^2\Phi = 0 \quad (10.19)$$

where

$$\Phi = \underline{w} - \underline{G};\ a = \sqrt{\frac{\omega \cos \theta}{K}};\ 2\mathrm{i} = (1+\mathrm{i})^2$$

Solving Equation 10.19, which is a linear homogeneous second-order equation with constant coefficients, gives:

$$w = C_1 \mathrm{e}^{(1-\mathrm{i})az'} + C_2 \mathrm{e}^{-(1-\mathrm{i})az'};\quad z' \geq h \tag{10.20}$$

where C_1 and C_2 are the complex constants of integration. As at $z' \to \infty$ $\underline{w} \to \underline{G} \neq \infty$, the constant $C_1 = 0$. The constant of integration C_2 can be determined either by equalizing the wind velocities calculated by (10.9) and (10.20), respectively, at a level of $z' = h$, or, as Ekman and Åkerblom did, assuming that air "sticks" to the surface (at $z = 0$, $\underline{w} = 0$) if the turbulent viscosity coefficient constancy condition is applied to the whole planetary boundary layer. In the former case, we have for C_2:

$$C_2 = \left(\frac{k_1}{\kappa^2}\ln\frac{h+z_0}{z_0} - \underline{G}\right)\mathrm{e}^{a(1-\mathrm{i})h}$$

and in the latter:

$$C_2 = -\underline{G}$$

Substituting the values of the constants C_1 and C_2 into expression (10.20) gives for the first case:

$$\underline{w}\,(z') = \underline{G}(1-\mathrm{e}^{-(1-\mathrm{i})a(z'-h)}) + \frac{k_1}{\kappa^2}\ln\frac{h+z_0}{z_0}\mathrm{e}^{-(1-\mathrm{i})a(z'-h)};\ z' \geq h \tag{10.21}$$

and for the second case:

$$\underline{w}(z) = \underline{G}(1-\mathrm{e}^{-a(1-\mathrm{i})z}) \tag{10.22}$$

Equating the real parts on the left and right sides of relations (10.21) and (10.22), we obtain the zonal wind velocity for the first case:

$$u(z') = G_\lambda\{1-\mathrm{e}^{-a(z'-h)}\cos a(z'-h)\} + \frac{k_1}{\kappa^2}\ln\frac{h+z_0}{z_0}\mathrm{e}^{-a(z'-h)}$$
$$\times \cos a(z'-h) + G_\theta \mathrm{e}^{-a(z'-h)}\sin a(z'-h) \tag{10.23}$$

and for the second case:

$$u(z) = G_\lambda(1-\mathrm{e}^{-az}\cos az) + G_\theta \mathrm{e}^{-az}\sin az \tag{10.24}$$

Similarly, for the first case, the meridional wind velocity is:

$$v_\theta(z') = G_\theta(1-\mathrm{e}^{-a(z'-h)}\cos a(z'-h)) + \frac{k_1}{\kappa^2}\ln\frac{h+z_0}{z_0}\mathrm{e}^{-a(z'-h)}$$
$$\times \sin a(z'-h) - G_\lambda \mathrm{e}^{-a(z'-h)}\sin a(z'-h) \tag{10.25}$$

and for the second case:

$$v(z) = G_\theta(1-\mathrm{e}^{-az}\cos az) - G_\lambda \mathrm{e}^{-az}\sin az \tag{10.26}$$

Here:

$$G_\lambda = \frac{1}{2\omega \cos \theta \rho R} \frac{dP}{d\theta} \text{ and } G_\theta = -\frac{1}{2\omega \cos \theta \rho R \sin \theta} \frac{dp}{d\lambda} \quad (10.27)$$

The wind profile described by relation (10.22) is the Ekman–Åkerblom profile. It is a particular case of the wind profile described by relation (10.21). It is known that the Ekman–Åkerblom wind profile has little relation to the actual situation in the surface layer. Thus, near the Earth's surface, the wind deviation angle is much less, and the wind velocity is much more, than the theoretical value. However, the farther from the Earth's surface (higher than the surface layer), the more the profile agrees with the theoretical values.

Note that there are more rigorous theories of the planetary boundary layer (Monin and Yaglom, 1965; Zilitinkevich, 1970; Haltiner and Martin, 1957). For example, Zilitinkevich showed (Zilitinkevich, 1970) that the dynamic velocity w_* should only depend on the Rossby number and the stratification parameter. However, despite its elegance, this theory is difficult to use because of the uncertainty of its parameters. Therefore, we have chosen the above-explained semiempirical theory that is less strict but more suitable for calculations.

10.3 Implementation of Calculations

In (Sidorenkov, 1979) a 20-year series of fluctuations in the Earth's rotation rate was calculated using the method of the torque approach. Maps of the monthly average atmospheric pressure at sea level, which have been compiled since 1956 at the Branch of World Weather Analysis of the Hydrometeorological Center of the USSR, served as the initial data. The archive of data on the monthly average atmospheric pressure at sea level for the 242 months from November 1956 to December 1976 at the nodes of a coordinate grid formed by the intersection of meridians every 20° and parallels every 5° was compiled. The angular accelerations of the Earth's rotation were determined by the numerical integration of Equation 10.1.

The zonal components of the stress of turbulent friction were calculated by the formula

$$\tau = \frac{\kappa^2}{\left(\ln \frac{h+z_0}{z_0}\right)^2} \rho u^2(h) = c\rho u^2(h) \quad (10.28)$$

where z_0 is the roughness parameter, ρ is the air density, and c is the coefficient of surface friction, which is referred to the height $z' = h$ of the wind speed measurement. The zonal wind speed u at the upper boundary of the surface layer ($z' = h$) was determined from the Ekman–Akerblom equation

$$u(h) = \frac{1 - \exp\left(-\sqrt{\frac{\omega \cos \theta}{K}} h\right) \cos \sqrt{\frac{\omega \cos \theta}{K}} h}{2\rho\omega \cos \theta} \frac{dP}{Rd\theta} \quad (10.29)$$

Here, K is the coefficient of turbulent viscosity; h is the thickness of the surface layer of the atmosphere; P is the atmospheric pressure.

The calculations of characteristics of the Earth's rotation rate by meteorological data were roughly simplified. Therefore, the calculated series of theoretical deviations ν^{T} are inexact. They are no more than appropriate indices of the atmosphere's mechanical effect on the Earth. Figure 10.1 presents the comparison of the results of calculations with the data of astronomical observations. It shows that they are similar in large detail. Probable errors may primary be associated with a rough assumption that the coefficient of turbulent viscosity K is constant (Sidorenkov, 1979). The multiyear variations in the theoretical deviations ν^{T} agree well with the observational deviations ν^{A}. In particular, such features as the acceleration of the Earth's rotation in 1958–1961 and 1972–1975 and the retardation in 1962–1971 and in 1976, which were observed by astronomical methods, are also well reproduced by the above calculations.

Consequently, the multiyear variations in the Earth's rotation rate are due to the mechanical action of the atmosphere on the Earth. The atmosphere creates the moments of frictional and pressure forces that are applied to the Earth's surface and change the Earth's rotation rate. The theory allows one to calculate with satisfactory accuracy the multiyear variations in the Earth's rotation rate, using the available global data on the pressure and wind fields. In other words, the decadal fluctuations

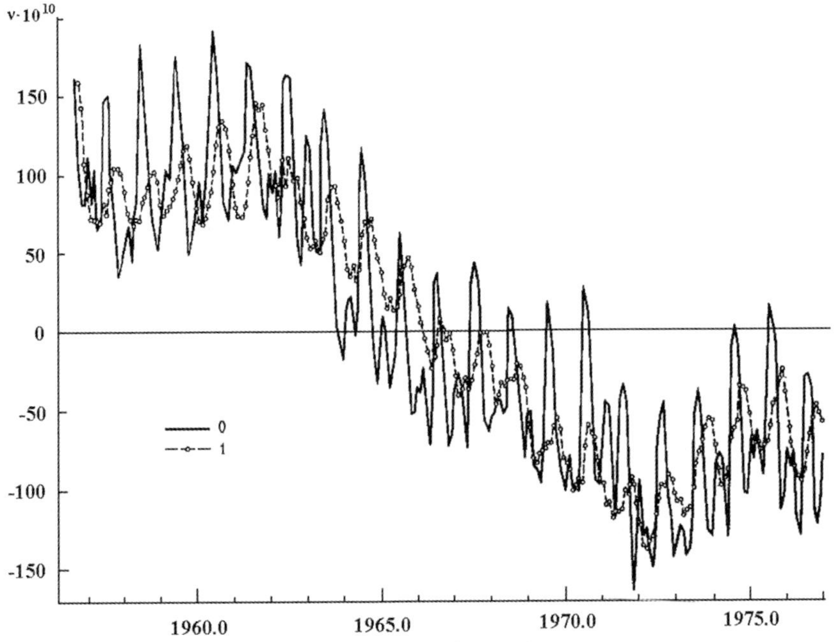

Figure 10.1 Variations in the mean monthly values of the Earth's rotation rate for the last 20 years; (0) based on astronomical data; (1) theoretical values.

in the Earth's rotation rate are likely to be due to the flow of small portions (sometimes positive and sometimes negative) of the angular momentum, which are transferred through the surface layer of the atmosphere to the Earth.

10.4 Mechanism of the Continental Drift on Decadal-Long Time Scale

10.4.1 Hypothesis

It is well known that seasonal variations in the Earth's rotation are determined by the redistribution of the angular momentum between the atmosphere and the Earth. When the moment of the atmosphere is increasing then the moment of the Earth is decreasing and vice versa (Lambeck, 1980; Sidorenkov, 2002a, 2002b, 2005). This regularity is well seen on Figure 7.10 where the time series of the angular momentum of the atmosphere is compared with the time series of the angular momentum of the Earth taken with the opposite sign.

Thus, the nontidal irregularities of the Earth rotation are mainly due to the exchange between the angular momentum of the solid lithosphere and its fluid environment – the atmosphere and the hydrosphere. This exchange occurs due to the moments of the frictional forces and pressure forces pushing on mountain ranges. Special Bureau for the Atmosphere carries out the monitoring of the exchange of the angular momentum by both the momentum approach (that is, by the evaluation of the effective functions of the atmospheric and oceanic angular momentum), and the torque approach (that is, the evaluations of the torque resulting from the wind and current stresses and pressures).

Calculations of the friction and pressure momentum forces are performed for the entire Earth surface as a whole. However, the lithosphere is cracked on a set of the lithosphere plates. The atmosphere and ocean are acting on the lithosphere plates, and only then is this action transmitted to the Earth. What is the result of the atmospheric action on the lithosphere plates? Let us recall that under the lithosphere, there is a layer of the lower viscosity – the asthenosphere in which the lithosphere plates are capable to float. Continents are frozen into the oceanic plates, and they may passively move with them (Trubitsyn and Rykov, 1998; Trubitsyn, 2000). The lithosphere plates float in the asthenospherical substratum. On the decade time scale, the lithosphere plates can move in the horizontal direction under the effect the friction and pressure (acting on mountain ranges) forces. The plates are in motion under the action of the friction stresses and pressure, which the atmosphere and ocean produce on the exterior surface of the plate. The viscous cohesive force with the asthenosphere on the soles and faces of the plates decelerates their movement, but the exterior forces overcome this resistance. Therefore, when calculating the torque, it is necessary to carry out the integration not only for the entire Earth surface but also separately for every lithosphere plate. The moment of forces affecting on an individual plate determines the vector of the movement of the plate (Sidorenkov, 2002a, 2002b, 2004).

10.4.2 Evidence

A good example of this is the situation in the Drake Passage (Figure 10.2). Strong westerly winds dominate in the 40° S – 50° S. They generate the powerful Antarctic Circumpolar Current (ACC) in the Southern Ocean. The South America, the Antarctic Peninsula, and the underwater lithosphere present a barrier for ACC. Westerly atmospheric winds and oceanic currents have replaced this barrier downstream and have shifted this lithosphere bridge to the east by 1500 km. This process resulted in the formation of the Scotia Sea (the South-Antilles hollow). It is bordered along the perimeter by the remains of the lithosphere bridge in the form of the South-Antilles ridge and numerous islands, the arc of the South Sandwich Islands being the principal of them. This ridge, at the drifting in the eastward stream, has crumpled the oceanic lithosphere and has formed the deep South-Sandwich trench (Figure 10.2).

Let us present one more piece of evidence for the benefit of our hypothesis. The atmospheric circulation has a remarkable feature: at the latitudes of 35° N and 35° S, the wind direction alters to the opposite one. Easterly winds predominate in the tropical belt between these latitudes, and westerly winds in the moderate and high latitudes. According to this, the stresses of friction on the surface of the lithosphere is directed to the opposite sides. Therefore, the maximum stress in the lithosphere should concentrate near the latitudes of 35° N and 35° S. These bands should exhibit an increased seismic and tectonic activity. Really, in the Northern Hemisphere, in this band, continuous mountain ranges are extending through the Mediterranean Sea, Middle East, Iran, Pamir, Tibet, Japan and USA. Here, Earthquakes and eruptions of volcanoes occur most frequently. In the Southern Hemisphere, the band of the sign change in wind direction is located over the World Ocean. Therefore, the seismic and tectonic processes do not manifest themselves.

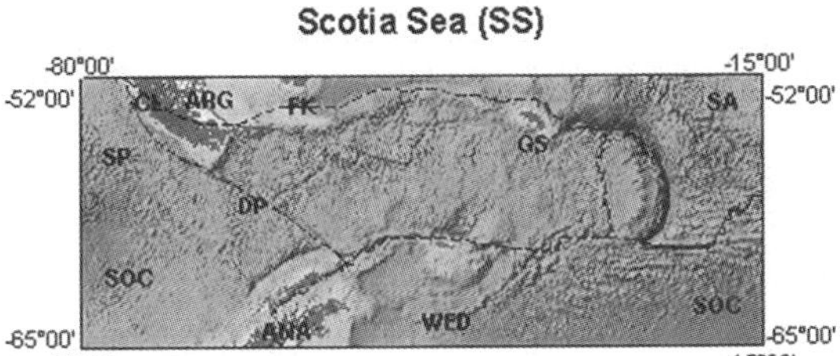

Figure 10.2 Westerly atmospheric winds and oceanic currents have broken down the South America –Antarctic Peninsula lithosphere bridge and have shifted it eastward by 1500 km. The color map can be looked on the site: http://www.walrus.wr.usgs.gov/infobank/gazette/html/regions/ss.html.

10.4.3 Estimations

Now let us estimate the order of magnitudes of the atmospheric and oceanic forces effecting on a separate plate and of the stresses of the interaction between plates. At the common wind velocity ($u = 10\,\text{m/s}$) the friction stress τ on the surface of the plate is $\tau = \rho u^2 = 0.004 \times 1.27\,\text{kg/m}^3\ (10\,\text{m/s})^2 = 0.5\,\text{N/m}^2$, the area of the plate is $\approx 2 \times 10^{13}\,\text{m}^2$; therefore, the total atmospheric force effecting on a separate plate, is $\approx 10^{13}$ N. Under the effect of this force the plate interacts with the circumjacent plates through the frontal contacts. The interaction takes place only at the sites of adhesion of plates, and the area of contacts may be small. The total atmospheric force concentrates on this small area. Therefore, the stresses may reach such high values (10^6–$10^7\,\text{N/m}^2$), at which the discontinuity and displacement of plates from each other occur. The discontinuity triggers the seismic waves. Thus, the mechanical action of the atmosphere and ocean on the lithosphere plates controls the relative movements of the lithosphere plates and can cause the Earthquakes and volcanic activity.

There is a substantial body of publications in which strong correlations between the seismicity and the variations in the atmospheric indices, as well as between the seismicity and the fluctuations in the Earth's rotation (Zharov, Konov, and Smirnov, 1991; Gorkavyi *et al.*, 1994a; Gorkaviy, Trapeznikov, and Fridman, 1994b; Barsukov, 2002) are found. Our hypothesis explains these correlations. The atmospheric and oceanic circulation is the initial cause of both the whole class of Earthquakes and the variations in the Earth rotation. Note that the variations in the Earth rotation are very small ($\delta\omega/\omega \approx 10^{-8}$) and do not affect the geophysical processes (Sidorenkov, 1961, 2002a).

It was found by Sidorenkov (2002a, 2002b, 2004) that the variations observed in the Antarctic ice sheet mass agreed with the mass's variations required for the explanation of the decade (5–100 year) fluctuations of the Earth's rotation rate and the secular polar motion. However, this agreement proved to be only the qualitative one. As to the quantitative agreement, the variations observed in the ice masses proved to be 28 times less than the required variations. Sidorenkov (1980a, 1991a, 2002a, 2002b) has proposed that the lithosphere drifts over the asthenosphere. The Earth's layers that are deeper than the asthenosphere don't take part in the formation of the observed decade fluctuations. The lithosphere's moments of inertia are 28 times less than the moment of inertia of the whole Earth and therefore the variations in the Antarctic ice mass exactly correspond to the mass's variations required for the explanation of the decade fluctuations in the lithosphere's angular rotation rate.

The state of the ice sheets in the Antarctic and Greenland depends on the climatic variations. Therefore, the decadal fluctuations in the Earth's rotation may also correlate with the fluctuations in the climatic characteristics and indices. This relationship has been found in (Lambeck, 1980; Sidorenkov, 2002a, 2002b). There is a close correlation between the Earth's rotation fluctuations and the frequencies of the atmospheric circulation forms, the anomalies of the hemisphere-averaged air temperature, and many another climate characteristics (Sidorenkov, 2002a, 2002b;

Sidorenkov and Shveikina, 1994, 1996; Sidorenkov and Orlov, 2008). These relationships are explained given the assumption that the lithosphere drifts along the asthenosphere.

The differential rotation of the lithosphere and mantle is considered in many papers devoted to the tectonics of plates (Ricard, Dogliony, and Sabadini, 1991). Chuikova and Maksimova (2005) have found the uncompensated masses and stresses at the crust and upper mantle. They supposed the crust's movements caused by the crust pressing on the mantle for both the isostatic and gravity nonequilibrium.

Thus, the research results and observations confirm the hypothesis about the movement of the lithosphere plates under the impact of the atmospheric and oceanic circulation on the decade time scale. The total effect of the movement of all lithosphere plates is interpreted by geophysics as the decadal fluctuations of the Earth's rotation.

10.4.4 Model

Our hypothesis can be mathematically described similarly to the Trubitsyn model of the mantle convection with floating continents (Trubitsyn, 2000).

It is known that the motion of a rigid body is defined by the motion of its center of mass and by the rotation with respect to the center of mass. To deduce the differential equations of the plate motion, we use the theorem of the movement of the center of mass of the plate:

$$m\frac{\mathrm{d}\boldsymbol{V}_{0i}}{\mathrm{d}t} = \boldsymbol{F}_i \qquad (10.30)$$

and the theorem of the angular momentum:

$$\frac{\mathrm{d}\boldsymbol{H}_i}{\mathrm{d}t} = \boldsymbol{L}_i \qquad (10.31)$$

where m is the mass of the plate; $\boldsymbol{V}_{0i}$ is the instantaneous velocity vector of the center of mass; $\boldsymbol{H}_i$ is the angular momentum of the plate; $\boldsymbol{F}_i$ is the external force; and $\boldsymbol{L}_i$ is the total force moment (torque), which is the sum of the moments q_i of forces σ_j applied to separate elements of the plate surface

$$q_k = \varepsilon_{ijk}(x_i - x_{i0})\sigma_j \qquad (10.32)$$

The angular momentum of the plate may be determined as

$$H_i = I_{ij}\omega_j \qquad (10.33)$$

Here, I_{ij} is the moment of inertia tensor of the plate,

$$I_{ik} = \int \rho[(x_l - x_{l0})^2\delta_{ik} - (x_i - x_{0i})(x_k - x_{0k})]\mathrm{d}W \qquad (10.34)$$

Let us consider only the horizontal movement of the center of mass of the plate and its rotation around of the vertical axis. In this case, Equations 10.30–10.31 are reduced to the system of three equations:

$$m\frac{du_1}{dt} = \iint(-p\delta_{1j} + \tau_{1j} + f_{1j})n_j ds \tag{10.35}$$

$$m\frac{du_2}{dt} = \iint(-p\delta_{2j} + \tau_{2j} + f_{2j})n_j ds \tag{10.36}$$

$$I_{33}\frac{d\omega_3}{dt} = \iint\varepsilon_{ij3}(x_i - x_{0i})(-p\delta_{jk} + \tau_{jk} + f_{jk})n_k ds \tag{10.37}$$

where, x_i are the coordinates of an arbitrary point of the continent; x_{0i} are the coordinates of the instantaneous center of mass of the plate; δ_{ij} is the Kronecker symbol (equal to 1 at $i = j$ and 0 at $i \neq j$); ε_{ijk} is the Levy–Civita symbol that is equal to 0 (if any two indexes coincide) or 1 (at an even transposition of indexes with respect to (1,2,3)) and -1 (if this transposition is uneven); p is the pressure; τ_{jk} are the friction stresses of the atmosphere and ocean on the exterior surface of the plate; f_{jk} are the viscous stresses of the asthenosphere on the submerged surface of the plate; n_j is the unit vector of the outward normal to the surface of the plate; ds is the absolute value of an elementary surface area of the solid continent.

Taking into account that:

$$\frac{dx_1}{dt} = u_1, \quad \frac{dx_2}{dt} = u_2, \quad \frac{d\varphi}{dt} = \omega_3 \tag{10.38}$$

we can, using the given coordinates of the center of mass of the plate $x_1(t)$, $x_2(t)$, $\varphi(t)$ and values $p(t)$, $\tau_{jk}(t)$ and $f_{jk}(t)$ calculate the linear velocities $u_1(t)$, $u_2(t)$ of the translational motion and the angular velocity $\omega_3(t)$ of the rotation of the plate. Equations of motion (10.36)–(10.38) are necessary to write out for each plate.

The model allows us to calculate the linear velocities $u_1(t_1)$ and $u_2(t_1)$ of the translational horizontal motion of the plate's center of mass and the angular velocity $\omega_3(t_1)$ of the rotation of the plate around the vertical axis. The calculations are performed for moment t_1, using the values of the frictional stress $\tau_{jk}(t_1)$ and the pressure $p(t_1)$ forces of the atmosphere and the ocean and the force $f_{jk}(t_1)$ of the interaction between the plate and the viscous asthenosphere. The force $f_{jk}(t_1)$ is applied to the submerged surface of the plate. Knowing these velocities and the initial coordinates of the plate $x_1(t_1)$, $x_2(t_1)$, $\varphi(t_1)$, it is possible to find its position in the subsequent instant $t_2 = t_1 + \Delta t$: $x_1(t_2) = x_1(t_1) + u_1(t_1)\Delta t$, $x_2(t_2) = x_2(t_1) + u_2(t_1)\Delta t$, $\varphi(t_2) = \varphi(t_1) + \omega_3(t_1)\Delta t$. Then, using new values $\tau_{jk}(t_2)$, $p(t_2)$ and $f_{jk}(t_2)$, we calculate $u_1(t_2)$, $u_2(t_2)$ and $\omega_3(t_2)$ and determine the position of the plate for the following instant t_3. The calculations are performed up to the final moment of time. The time step depends on the discretization of calculations of the friction and pressure forces of the atmosphere and the ocean.

11
Decadal Fluctuations in Geophysical Processes

11.1
Discussion on Conceivable Hypotheses

The conclusion about the decadal fluctuations in the Earth's rotation rate being due to the mechanical effect of the Earth's atmosphere (see Section 10.3) allows one to considerably reduce the number of geophysical processes that could be responsible for the above fluctuations. There are evidently three hypotheses for explaining the decadal fluctuations in the Earth's rotation:

(a) the redistribution of the angular momentum in the Earth–atmosphere system;
(b) the inflow of the angular momentum to the atmosphere from above (the magnetosphere or cosmic space);
(c) the redistribution of water between low and high latitudes.

If we suppose that in the case of the decadal fluctuations, as well as in the case of seasonal variations, the angular momentum redistributes between the Earth and the atmosphere, then drastic variations should take place in the angular momentum of the zonal winds.

For example, in 1870–1903, the Earth's rotation slowed down by 820×10^{-10} (in the dimensionless units of ν), which is equivalent of the Earth's angular momentum loss by $48 \times 10^{25}\,\mathrm{kg\,m^2/s}$. If the same redistribution occurred only in the Earth–atmosphere system (as it occurs in the case of seasonal variations), then the relative angular momentum of the atmosphere should have grown by $48 \times 10^{25}\,\mathrm{kg\,m^2/s}$. This means that the velocities of westerly winds should gradually have strengthened and the easterly winds weakened by about 20 m/s on average. However, significant decadal variations in the intensity of zonal circulation have not been detected yet. It is unlikely that they have simply remained unnoticed.

Having the data on the atmospheric pressure for 1956–1976, we have attempted to determine how the atmospheric pressure field changes with the changes in the mode of the Earth's rotation; in particular, we have studied the difference in the pressure fields during the periods of the Earth's rotation acceleration (1958–1961 and 1972–1975) and retardation (1962–1971). Using the same pressure data at the nodes

The Interaction Between Earth's Rotation and Geophysical Processes. Nikolay S. Sidorenkov

ISBN: 978-3-527-40875-7

of the coordinate grid, which are used in Section 10.3, we have calculated the arithmetic means of the atmospheric pressure $P(\theta,\lambda)$ for the eight years of acceleration $P_1(\theta,\lambda)$ and the ten years of retardation in the Earth's rotation rate $P_2(\theta,\lambda)$ and have found the difference $\Delta P(\theta,\lambda) = P_1(\theta,\lambda) - P_2(\theta,\lambda)$. It has turned out that the field of these differences has a clear sectorial structure with two maxima (at meridians 140° and 0°E longitude) and two minima (at meridians of 80° and 260°E longitude), the differences being small (they vary within the limits of plus or minus several tenths of hectoPascal.) This fact, being of interest in itself, is uninformative for our case. The changes in the Earth's rotation rate are mainly determined from the zonal pressure anomalies averaged over the latitudes. In this connection, the pressure differences used in (Sidorenkov, 1979) are averaged over the latitudinal circles, and the zonal differences $\Delta P(\theta)$ have been calculated. It has been found that the differences $\Delta P(\theta)$ are positive in the 15°–40°N and 0°–10°S latitudinal zones. Their values do not exceed +0.2 hPa. In the zones of high and temperate latitudes, $\Delta P(\theta)$ are negative and reach − 0.7 hPa.

Thus, the acceleration periods differ from the retardation ones by a small positive pressure anomaly at the low latitudes. This corresponds to a slight weakening of the easterly surface winds, which retards the rotation, and a slight strengthening of the westerly winds, which accelerates the Earth's rotation. The small values of differences $\Delta P(\theta)$ and of the increments of velocity are indicative of the absence of significant changes in the atmospheric circulation that are required for satisfying the balance in the angular momentum. At the same time, they satisfactorily reflect the changes in the flow of the angular momentum through the surface layer of the atmosphere, which cause the variations in the mode of the Earth's rotation.

Thus, the facts are indicative of the existence of the angular momentum transfer through the surface layer of the atmosphere, which leads to the multiyear variations in the Earth's rotation rate. But the corresponding changes in the angular momentum of the atmosphere that are required to satisfy the balance are not observed. Therefore, the hypothesis (a) can safely be rejected.

The hypotheses (b) and (c) are based on the supposition that the atmosphere receives the portions of either the positive or the negative angular momentum, either from above (from the circumterrestrial space) or from the Earth (in the process of water redistribution). Let us first dwell on hypothesis (b).

The angular momentum can be transported to the Earth at the anisotropic flow of the solar wind, as well as at the expense of the electromagnetic interaction between the geomagnetic and the interplanetary magnetic fields. The assessment of the order of magnitude of these effects presents no special problem.

It is known that the Earth is constantly blown off by the solar wind – the flow of charged particles emitted by the Sun. At steady winds, there are 1–10 particles in 1 cm^3 at the Earth's orbit, the velocity of their movement being 300–400 km/s. At gusty winds, the concentration of particles increases up to several tens of particles in 1 cm^3 and their velocity – up to 800 km/s. The intensity of the interplanetary magnetic field near the Earth's orbit does not usually exceed 10γ.

In order to assess the upper limit of the torque produced by the solar wind with respect to the Earth's rotation axis, let us assume that the flow of the solar wind particles is anisotropic to the extent that it falls only on one hemisphere, for example – on the Eastern Hemisphere. Let the concentration of particles in the flow be $\delta = 100$ particles/cm^3 and the velocity $v = 1000$ km/s. Then

$$N = \frac{1}{2}\pi R_0^2 \delta v < 6 \times 10^{27} \text{ particles/s} \tag{11.1}$$

falls onto this hemisphere. Here, $R_0 = 6.37 \times 10^6$ m is the Earth's radius. Taking into account that the mean mass of one particle is approximately equal to 2×10^{-27} kg, it is easy to find the value of the inflowing momentum: $J = N2 \times 10^{-27} \cdot v < 1.2 \times 10^7$ kg m s^{-2}. The torque M produced by this flow of momentum will be not greater than

$$M_3 = JR_0 < 8 \times 10^{13} \text{ kg m}^2 \text{ s}^{-2} \tag{11.2}$$

As follows from the third equation of system (4.25), this torque is capable to accelerate the Earth's rotation by the value of

$$\frac{d\omega_3}{dt} = \Omega \frac{d\nu_3}{dt} = \frac{M_3}{C} = 10^{-24} \text{ s}^{-2} \tag{11.3}$$

where $C = 8.11 \times 10^{44}$ kg m^2 is the moment of the Earth's inertia.

As is mentioned in Section 2.5, the accelerations observed have the order of magnitude of 10^{-19}–10^{-20} s^{-2}. Thus, under the most favorable conditions the solar wind can cause the angular accelerations that are by 4–5 orders of magnitude smaller than the observed ones. More precise estimations of the effect of the solar wind, which are obtained with account for its interaction with the Earth's magnetosphere (Coleman, 1971; Hirshberg, 1972), have led to similar conclusions. Hence, the decadal fluctuations in the Earth's rotation rate cannot be explained by the solar-wind effect.

In principle, the electromagnetic interaction between the geomagnetic and interplanetary magnetic fields can also produce the torque with respect to the axis of rotation (Livshits, Sidorenkov and Starkova, 1979). Indeed, at the external field $\boldsymbol{F}$, the Earth will be subjected to the torque

$$\boldsymbol{M} = [\boldsymbol{L} \times \boldsymbol{F}] \tag{11.4}$$

Here, $\boldsymbol{L} = 8.1 \times 10^{25}$ g$^{1/2}$ cm$^{3/2}$ is the magnetic moment of the Earth. Under the most favorable conditions, when $\boldsymbol{F} = 10\gamma$ ($1\,\gamma = 10^{-5}$ g$^{1/2}$ cm$^{1/2}$ s^{-2}) and $\boldsymbol{F} \perp \boldsymbol{L}$, the torque affecting the Earth will not exceed 8×10^{14} kg m^2 s^{-2}, or will be smaller by 3–4 orders of magnitude than the torques required for creating the fluctuations observed in the Earth's rotation.

Thus, neither the solar wind nor the interplanetary magnetic field can be responsible for the angular momentum flow through the surface layer of the atmosphere and, as a consequence, for the decadal fluctuations in the Earth's rotation rate. Hence, the hypothesis (b) is doubtful.

Now, let us consider hypothesis (c) for the effect of the redistribution of water between the low and high latitudes. In this case, the mechanism of the angular momentum exchange between the atmosphere and the Earth may be represented as follows. The angular momentum $m_e R^2 \cos^2\varphi_e \omega$ enters the atmosphere through evaporation, the angular momentum $m_p R^2 \cos^2\varphi_p \omega$ abandons it through precipitation. Here m_e and m_p are the masses of evaporated and precipitated water, respectively; while φ_e and φ_p are their mean latitudes. Water is not stored in the atmosphere; therefore, $m_e = m_p = m$. The water cycle ultimately leads to the fact that the angular momentum Δh_3 equal to the difference

$$\Delta h_3 = m_e R^2 \cos^2\varphi_e\, \omega - m_p R^2 \cos^2\varphi_p\, \omega = m R^2 \omega (\cos^2\varphi_e - \cos^2\varphi_p) \qquad (11.5)$$

remains in the atmosphere. When $\varphi_e < \varphi_p$, the atmosphere receives a positive portion of the angular momentum, whereas at $\varphi_e > \varphi_p$ it receives a negative portion of the angular momentum.

Let us assess the order of magnitude of the Earth's rotation acceleration due to the redistribution of water. The global flow of water vapor from the Earth's surface into the atmosphere at the expense of evaporation is about 1.2×10^{15} kg/day (Kulikov and Sidorenkov, 1977), or $\frac{dm}{dt} = 1.4 \times 10^{10}$ kg s^{-1}. Let us assume that water evaporates at the equator and precipitates at the poles, that is, $\cos^2\varphi_e - \cos^2\varphi_p = 1$. Accounting that $R = 6.37 \times 10^6$ m and $\omega = 7.29 \times 10^{-5}$ s^{-1}, we obtain from (11.5):

$$\frac{dh_3}{dt} = \frac{dm}{dt} R^2 \omega \approx 4 \times 10^{19}\ \text{kg m}^2\ \text{s}^{-2} \qquad (11.6)$$

As follows from the third equation of system (4.25), this flow of the angular momentum accelerates the Earth's rotation by the value of:

$$\frac{d\omega_3}{dt} = \Omega \frac{d\nu_3}{dt} = -\frac{1}{C}\frac{dh_3}{dt} = -5 \times 10^{-19}\ \text{s}^{-2} \qquad (11.7)$$

Hence, the redistribution of water may be responsible for the flows of the angular momentum through the surface layer of the atmosphere and thus, for the decadal fluctuations in the Earth's rotation rate.

The angular momentum cannot be stored in the atmosphere because the atmosphere is capable of holding only that amount L of the angular momentum that corresponds to the power M of the atmospheric thermal engines ($L \sim M = \eta W$), where the efficiency η and the solar radiation power W are almost constant from year to year. Consequently, the angular momentum received by the atmosphere in one way or another must flow down through the surface layer of the atmosphere to the Earth. The Earth accumulates this portion of the angular momentum, which leads to changes in the angular velocity of the Earth's rotation. When the positive angular momentum flows to the Earth, its rotation accelerates; when the negative angular momentum flows to the Earth, its rotation decelerates. When calculating the moments of the friction and pressure forces, we make allowance for the above flows of the angular momentum and detect the effects of these flows.

We would underline that in the process of water redistribution over the Earth, the angular momentum of the Earth–atmosphere system is conserved, and the redistribution of water provides only a gradual change in the Earth's angular momentum and a multiyear variation in the Earth's rotation rate. At the exchange of the angular momentum with the outer space, the balance of the angular momentum in the Earth–atmosphere system is not conserved because this system is unclosed.

11.2
Theory of Estimations of the Global Water Exchange Effect on the Earth's Rotation

The effect of water redistribution has been discussed in the literature for more than a century. The variations in the World Ocean level and the accumulation of ice in Antarctica and Greenland were among the first phenomena used to explain the abrupt fluctuations in the Earth's rotation rate, the ideas of which arose in the end of the nineteenth century (Thomson, 1882; Munk and MacDonald, 1960). The effects of the above two factors were most appropriately studied in (Sidorenkov, 1980a, 1982b; Sidorenkov and Svirenko, 1988; Sidorenkov *et al.*, 2005).

The redistribution of water over the Earth's surface causes changes in the components of the inertia tensor of the lithosphere and, as a consequence, variations in the vector's components of instantaneous angular velocity. This relationship is described by the following system of equations (Munk and MacDonald, 1960)

$$-\frac{1}{\sigma}\frac{\mathrm{d}\nu_2}{\mathrm{d}t}+\nu_1=\frac{n_{13}}{C-A}=-\frac{R^2}{2(C-A)}\iint_S \zeta(\theta,\lambda,t)\sin 2\theta\cos\lambda\,\mathrm{d}s \quad (11.8)$$

$$\frac{1}{\sigma}\frac{\mathrm{d}\nu_1}{\mathrm{d}t}+\nu_2=\frac{n_{23}}{C-A}=-\frac{R^2}{2(C-A)}\iint_S \zeta(\theta,\lambda,t)\sin 2\theta\sin\lambda\,\mathrm{d}s \quad (11.9)$$

$$\delta\nu_3=-(1+k')\frac{\delta n_{33}}{C}=-(1+k')\frac{R^2}{C}\iint_S \zeta(\theta,\lambda,t)\sin^2\theta\,\mathrm{d}s \quad (11.10)$$

Here, we use the Cartesian coordinate system Ox_i rigidly fixed to the Earth. Its centre O lies at the centre of the Earth's mass, and the axes have the following direction: x_1 is directed along the prime meridian, x_2 is aligned with the 90°E meridian, and x_3 is directed along the Earth's spin axis. Dimensionless values $\nu_1=\omega_1/\Omega$, $\nu_2=\omega_2/\Omega$ and $1+\nu_3=\omega_3/\Omega$ are the direction cosines of the instantaneous Earth's rotation axis; ω_1, ω_2 and ω_3 are the components of the vector of instantaneous angular velocity; Ω is the mean angular velocity of the Earth equal to 7.29×10^{-5} radians per sidereal second; $\sigma=2\pi/1.18$ is the Chandler frequency; t is the time; C and A are the polar and equatorial planetary moments of inertia of the Earth which are referred to the principal axes; n_{13}, n_{23} and n_{33} are the variable parts of the components of the Earth's inertia tensor; $k'=-0.3$ is the load deformation coefficient; $\zeta(\theta,\lambda,t)$ is the deviation of the specific (that is, per unit area) amount of water or ice at point $\{\theta,\lambda\}$ at moment t ($\zeta(\theta,\lambda,t)=0$ when $\nu_1=\nu_2=\nu_3=0$); θ is the colatitude; λ is the east longitude; R is the mean radius of the Earth; $\mathrm{d}s=R^2\sin\theta\,\mathrm{d}\theta\,\mathrm{d}\lambda$; S is the area of the entire Earth; δ is the sign of the increment.

It is convenient to combine Equations 11.8 and 11.9 into one

$$\frac{\mathrm{d}}{\mathrm{d}t}(\nu_1+\mathrm{i}\nu_2)-\mathrm{i}\sigma(\nu_1+\mathrm{i}\nu_2)=\frac{-\mathrm{i}\sigma}{C-A}(n_{13}+\mathrm{i}n_{23})=\underline{\Psi} \quad (11.11)$$

Let the mass of the Antarctic and Greenland ice vary by the linear law $\underline{\Psi}=(z_1+\mathrm{i}z_2)t$ and let $\nu_1+\mathrm{i}\nu_2=0$ at $t=0$. Then, according to (Sidorenkov,

2002a, p. 334), the solution of Equation 11.11 will be

$$\nu_1 + i\nu_2 = (z_1 + iz_2)t + \frac{z_1 + iz_2}{i\sigma}(1 - e^{i\sigma t}) \tag{11.12}$$

This means that the pole makes the circular motion with the Chandler frequency and a linear motion toward the meridian $\lambda = \mathrm{arctg}(z_2/z_1)$ with the velocity $\sqrt{z_1^2 + z_2^2}$.

When expression (11.12) is averaged over the time interval multiple of the Chandler period (1.2 years), the periodic term disappears and only linear terms of the pole drift remain.

Since we are interested in the time scales larger than several years, we will use only average coordinates of the so-called "secular" polar motion that no longer contain periodic terms. Averaging the Equations 11.8–11.10 substantially weakens the restrictions imposed on the choice of a model for the Earth.

Let us dwell upon the computation of the integrals on the right side of the Equations 11.8–11.10. We divide the integration domain S (the entire Earth's surface) into four natural parts S_N: the World Ocean – S_O, Antarctica – S_A, Greenland – S_G and the rest of land – S_C: (that is $S = S_O + S_A + S_G + S_C = \sum_{N=1}^{4} S_N$, where S_N are any of the areas mentioned above).

The experience of estimating the atmospheric effects on the instability of the Earth's rotation has shown that the World Ocean levels off the surface load as an inverted barometer (Munk and MacDonald, 1960, p.100). Similarly, the mass of water running down the ice sheets of Antarctica and Greenland is evenly distributed on the entire surface of the World Ocean over the characteristic time $2\pi R(gH)^{-\frac{1}{2}} \approx 56$ h, where R is radius of the Earth; g is the acceleration due to gravity; $H \approx 4000$ m is the mean depth of the ocean; $(gH)^{\frac{1}{2}} \approx 200$ m s^{-1} is the velocity of propagation of Kelvin's wave. Note that the tsunami waves also have the same velocity.

Thus, the World Ocean levels off the spatial inhomogeneity of the increment of the specific water mass over large time intervals (>3 day). Therefore, C, over the ocean at a given time t may be assumed constant everywhere ($\zeta(\theta,\lambda) = \zeta_O = \text{constant}$) and can be put outside the integral. The dimensions of Antarctica S_A and Greenland S_G are small compared with the total area S. Thus, in a first approximation, the dependence of the specific ice mass $\zeta(\theta,\lambda)$ on the latitude θ and longitude λ can be neglected and the mean value for Antarctica and Greenland can be used

$$\zeta_N(t) = \frac{1}{S_N}\iint\limits_{S_N} \zeta(\theta, \lambda, t)\, ds$$

On the rest of land S_C, the water is dominantly in liquid phase and rapidly flows down to the World Ocean ($T < 1$ year). The mass of glaciers on the rest of land is a mere 2% of ice on the Earth. Consequently, a similar assumption can be formally accepted for the rest of land S_C.

Thus, any integral on the right side of equation system (11.8–11.10) can be represented as the sum of four integrals

$$\iint\limits_{S} \zeta(\theta, \lambda, t)\Phi(\theta, \lambda)\, ds = \sum_{N=1}^{4} \zeta_N(t) \iint\limits_{S_N} \Phi(\theta, \lambda)\, ds \tag{11.13}$$

The integrals of the form $\int\int_{S_N} \Phi(\theta, \lambda)\, ds$ are taken with respect to the given boundaries (the World Ocean, Antarctica, Greenland, and rest of land). They are time independent and can be computed by numerical integration (Sidorenkov, 1982, 2002).

In the case of a long-term irregularity of the Earth's rotation and the secular motion of the pole, it is possible to use Equations 11.8–11.10 averaged over the time interval larger than the Chandler period. In this case, we may ignore the terms $d\nu_i/dt$, because $\frac{1}{\sigma}\frac{d\nu}{dt} \ll \nu$. Finally, the system of averaged Equations 11.8–11.10 with the account for (11.13) becomes reduced to the system of three algebraic equations:

$$\begin{aligned} \nu_1 \times 10^{11} &= 24149\zeta_O + 1337\zeta_A - 2102\zeta_G - 23387\zeta_C \\ \nu_2 \times 10^{11} &= 37714\zeta_O + 3820\zeta_A + 1909\zeta_G - 43443\zeta_C \\ \nu_3 \times 10^{12} &= -26746\zeta_O - 87\zeta_A - 28\zeta_G - 9268\zeta_C \\ 0 &= 71436\zeta_O + 2820\zeta_A + 414\zeta_G + 25330\zeta_C \end{aligned} \qquad (11.14)$$

Here, ζ_O, ζ_A, ζ_G, and ζ_C are the area-averaged (in accordance with (11.13) increments of the specific water masses (g cm^{-2}) in the World Ocean, Antarctica, Greenland, and rest of land, respectively. The last equation in (11.14) is the equation of the global water balance. Its coefficients are equal to the areas of the Earth's surface occupied by Oceans, Antarctica, Greenland, and rest of land. The entire surface area of the Earth is taken to be 100 000 conventional units.

The system of four algebraic Equations (10.14) provides the opportunity to solve both (a) the direct problem of determining the secular polar motion and variations in the Earth's rotation from the given time series of ζ_O, ζ_A, ζ_G, ζ_C and (b) the inverse problem of finding the unknown parameters of the global water exchange ζ_O, ζ_A, ζ_G and ζ_C from the given pole coordinates $\nu_1(t)$, $\nu_2(t)$, and the Earth's rotation velocity $\nu_3(t)$. The solutions of the combined problems are also possible: by the well-known parameters ζ_N of the global water exchange and the $\nu_3(t)$ value to improve the less-known parameters of the global water exchange and the coordinates (ν_1 and ν_2) of the pole's secular motion.

11.3 Secular and Decadal Variations

11.3.1 Assessments and Computations

The data on the irregularity of the Earth's rotation have been available since the eighteenth century and those on the secular motion (by which the progressive and long-term variations are meant) of the North pole since the late nineteenth century (Fedorov *et al.*, 1972; Vondrak, 1999). Unfortunately, the measurements of the secular motion of the pole are unreliable (Yatskiv *et al.*, 1976). The observational data prior to

1980 are very contradictory, which casts doubt on the trajectory of the pole's secular motion and even on the possibility of this motion. We do not share this doubt (Sidorenkov, 1983).

First, if the secular motion did not exist (that is, $\nu_1 = \nu_2 = 0$), then the value ν_3 would not feature the respective variations (in fact, the ν_1, ν_2 and ν_3 values characterize variations in the same vector ω. The existence of the secular variations in ν_3 is beyond doubt; hence, we may logically expect the existence of the pole's secular motion. This is also evidenced by changes in the acceleration of the Earth's rotation in 1903, 1927, 1931, 1935, 1950, 1958, and 1962, which were followed by changes in the direction of the observed secular motion of the North pole. The absence of a closer correlation between the ν_1, ν_2 and ν_3 values can be explained by both the unreliable determination of the pole's secular motion and the specific factors that cause this motion.

Second, there is correlation between the variations in the Earth's rotation and in the masses of ice sheets in Antarctica and Greenland (Sidorenkov, 1980a, 1982b, 1987). If the rate of ice accumulation or melting in Antarctica and Greenland is constant, the components of the excitation function should change by the linear law, so that $\psi_1 + i\psi_2 = (A_1 + iA_2)t$, where A_1 and A_2 are some constants and t is the time. Substituting this excitation function in Equation 11.11 and assuming that $\nu_1 + i\nu_2 = 0$ at $t = 0$, we find the same solution as (11.12), that is, in the case of the monotonous ice accumulation or melting, the pole will execute, apart from the circular motion with Chandler's period, the linear drift in the direction of meridian $\lambda = \mathrm{arctg}\frac{A_2}{A_1}$ with velocity $\sqrt{A_1^2 + A_2^2}$.

We have assumed that the available values of $\nu_1(t)$, $\nu_2(t)$ and $\nu_3(t)$ are exact ones and have calculated the time series of $\zeta_O(t)$, $\zeta_A(t)$, $\zeta_G(t)$, and $\zeta_C(t)$ from 1891 to 2005 (Sidorenkov, 1982b; Sidorenkov, 1987; Sidorenkov and Svirenko, 1988; Sidorenkov *et al.*, 2005). The ν_i and ζ_i summary series are given in Appendix D (Table D.1).

The values of the angular velocity components ν_1, ν_2, and ν_3 for the year under consideration were substituted into the system of Equation 11.18, after which the unknown parameters ζ_O, ζ_A, ζ_G, and ζ_C for the given year were calculated. Since the values of ν_i are averaged over six years, the calculated parameters ζ_N are also averaged. The above calculations for each year within the 1891–2005 period resulted in the "theoretical" series of deviations of the specific water masses (in oceans ζ_O, Antarctica ζ_A, Greenland ζ_G, and on rest of land ζ_C) from their values at a certain initial moment (Appendix D). This moment was assumed to be that when values ν_1, ν_2, and ν_3 were equal to zero. The beginning of records of ν_1 and ν_2 in astronomy is 1902 and that of ν_3 is approximately 1890; thus, the date of the initial moment is not exact.

We should notice that the available astronomical observations do not clearly define the secular motion of the pole. In addition, one of the forcing factors of the long-term variations in the Earth's rotation rate is the tidal deceleration that is irrelevant to the secular variations in the global water exchange. This term should be eliminated from the ν_3 values, but the exact rate of the tidal deceleration of the Earth's rotation is unknown. Hence, the theoretical data do not reliably describe the secular trend of variations in the parameters of the global water exchange.

11.3.2
Comparison of the Theoretical and Empirical Values

Let us compare the theoretical values with the empirical data. As is well known, the increment Z of the ice mass of any ice sheet is determined as the sum of the resultant water balances B over the previous years:

$$Z = \sum_i B_i$$

where i is the number of the year. The balance is $B = P - L$, that is, it is calculated as the difference between the precipitation P falling on the entire surface of the ice sheet and the total discharge L of its water resources. The discharge L is composed of the iceberg discharge, snowmelt discharge, discharge due to liquid precipitation, losses through evaporation, bottom melting, wind-blown snow, and other factors. It is clear that it is virtually impossible to determine the magnitude of L and its long-term changes.

The long-term variations in precipitation over Antarctica as a whole can be investigated either from precipitation measurements or the annual layers of snow accumulation determined from ice cores. The field material on the input part of the budget, the accumulation of snow on the ice sheet of Antarctica, was collected by Petrov (1975), who analyzed the annual values of snow accumulation, using the ice cores for nine stations. Three of them (Amundsen Scott, Little America, and Wilkes) have series since 1880; other stations have shorter series. Petrov calculated the values of P with annual discreteness – a characteristic of snow accumulation on the Antarctic ice sheet – from 1885 to 1957. No estimates of changes in L were made.

Calculations using the system of Equation 11.14 resulted in a time series of the specific mass $\zeta_A(t)$. Examination of the ice cores or annual precipitation gives the accumulation rate of P. To obtain the values equivalent to the $\zeta_A(t)$ series, the cumulative sums $\sum_{j=1}^{k} P_j$ have to be calculated. Because $P > 0$, a linear trend appears that hampers the comparison of calculated $\zeta_A(t)$ and observed $\zeta'_A(t)$ values. In reality, the inflow of the ice mass is substantially compensated by its outflow and the trend is small. It is impossible to accurately calculate the mass balance that forms this trend. Therefore, it was assumed that the calculated and observed trends were equal. Under this assumption, it was found that the discharge of ice L exceeds the mean value of snow accumulation $\bar{P}$ (averaged over the entire period of observations, 1885–2000) by 3%; hence, $B = P - L = P - 1.03 \cdot \bar{P}$, where $\bar{P} = 15\ \mathrm{g/(cm^2\ year)}$. Then, having determined the annual values of B, it is easy to calculate the integral (cumulative) curve of the increment of the specific ice mass $\zeta'_A(t)$ in Antarctica:

$$\zeta'_A(n) = \frac{Z_A}{S_A} = \frac{1}{S_A} \sum_{j=1}^{n} B_j \qquad (11.15)$$

where n is the year number.

In (Sidorenkov, 2002), the curve of the specific ice mass calculated by Petrov (1975) was used, and a good qualitative agreement between the theoretical ζ_A and empirical ζ'_A was obtained for 1891–1957. However, at the end of the 1940s and the beginning of

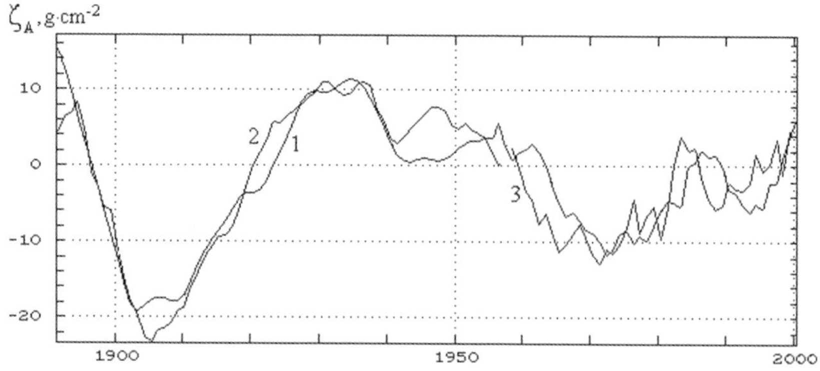

Figure 11.1 Temporal variations in the theoretical specific ice masses $\zeta_A(t)$ for Antarctica (1); the integral curves $\zeta'_A(t)$ obtained from Petrov's data for the stations of Amundsen Scott, Little America, and Wilkes (2); the average sums of precipitation from Bryazgin's data (3).

the 1950s, the ice mass decrease calculated by empirical data was much more rapid than that obtained theoretically. In this study, we calculated the empirical ζ'_A curve for 1891–1957 from the data of three stations: Amundsen Scott, Little America, and Wilkes. The time series for the other six stations from (Petrov, 1975) were too short and only deteriorated the results. The theoretical ζ_A and empirical ζ'_A curves for 1891–1957 agree well (Figure 11.1).

Unfortunately, we have not found the data on snow accumulation in Antarctica for the last few decades (from 1954 until present time). Bryazgin (1990) calculated the time series of annual precipitation P_B from the data for 11 Antarctic meteorological stations from 1958 to the present. It is known that the amount of precipitation at coastal stations is hundreds of times larger than that in the interior parts of the continent. Therefore, to reduce the data P_B to Petrov's data series, we calculated the normalized effective anomalies of the accumulation of the annual precipitation and multiplied this new series by dispersion D_P of the precipitation series for Amundsen Scott, Little America, and Wilkes stations from (Petrov, 1975):

$$Z' = \frac{P_B - 1.03\bar{P}_B}{D_B} D_P$$

where $\bar{P}_B$ and D_B are the norm and dispersion of the Bryazgin time series. Thus, we have calculated the series of the mean specific snow mass over the area of Antarctica $\zeta'_A(t)$ for the period from 1958 to 2000. This empirical $\zeta'_A(t)$ curve is shown in Figure 11.1 (curve 3). It also agrees well with the theoretical $\zeta_A(t)$ curve. The coefficient of correlation between the $\zeta_A(t)$ and $\zeta'_A(t)$ series is 0.91 ± 0.08.

It should be noted that in this study the series of the mean specific masses $\zeta'_A(t)$ for 1958–2000 is calculated differently from in (Sidorenkov, 2002a, 2002b) and the quantitative rather than qualitative agreement with the theoretical series $\zeta_A(t)$ is obtained.

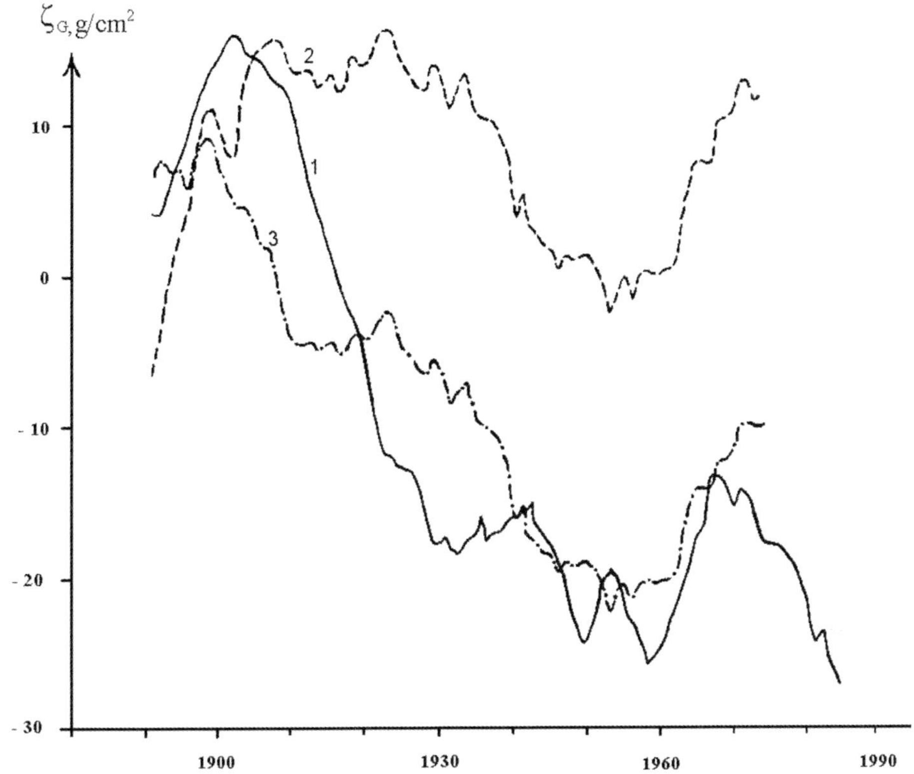

Figure 11.2 Temporal variations in the specific mass of ice ζ_G in Greenland. (1) theoretical values; (2) and (3) empirical data (Klige, 1980, 1985).

The empirical data on the variations in the ice mass in Greenland are contradictory. Figure 11.2 demonstrates two curves of ζ'_G obtained in different variants of calculations made by Klige (1980, 1985). The degree of disagreement between the empirical and theoretical curves is of the same order. However, the main features of the theoretical curve are found in both empirical curves.

It follows from Table D.1 in Appendix D that on rest of land, the deviations of the ice specific mass ζ_C from its average value for the period under study varied only little: from the minimum ($-49\,\mathrm{g/cm^2}$) in 1903 to the maximum ($+3\,\mathrm{g/cm^2}$) in 1935. This is explained by a rapid flow of water from the rest of land into the World Ocean.

Figure 11.3 presents the temporal courses of ζ_O obtained by theoretical and observational data. According to the theoretical data, the amount of water in the World Ocean rapidly increased till 1903, and then it markedly decreased till 1935, increased till 1972, and since 1973 up to the present time it continues to rapidly decrease. This theoretical temporal course of deviations ζ_O does not agree with the observed variations in the level of the World Ocean H (Klige, 1985). This disagreement may probably be due to the fact that the variations in the level of the World

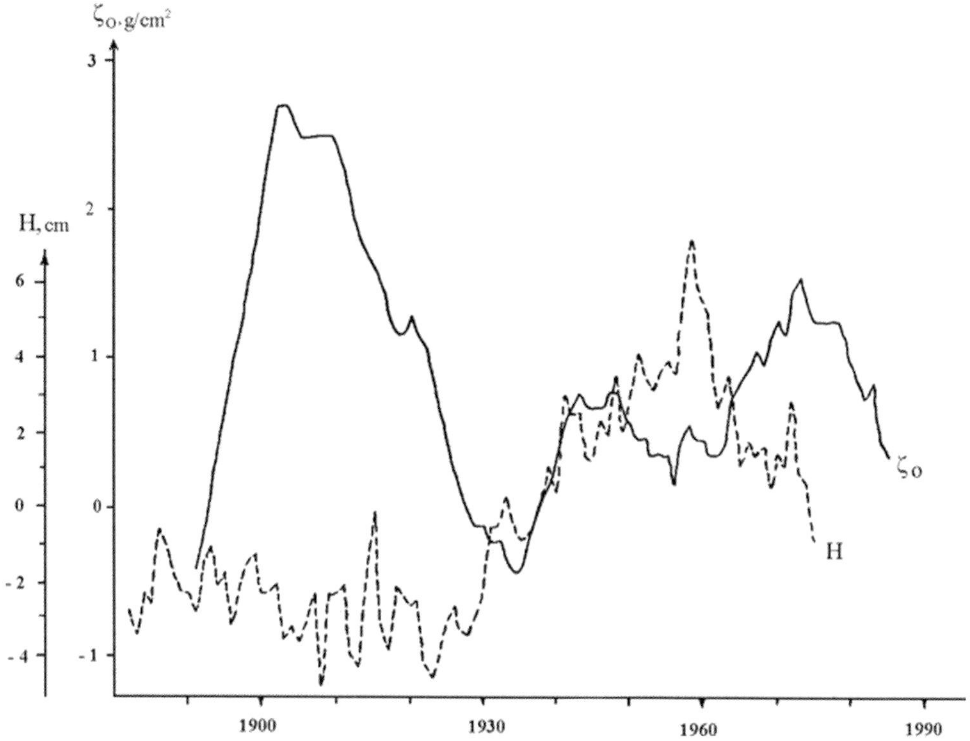

Figure 11.3 Deviations of the World Ocean specific mass of water ζ_O (theoretical values) and level *H* (Klige, 1985).

Ocean reflect not only the variations in the water mass but also the variations in the water temperature (the effect of volumetric expansion), the vertical movements of the oceanic floor and coastline, and so on.

11.3.3 Discussion of Results

The respective theoretical and empirical data for Antarctica show a good qualitative agreement, for Greenland – a satisfactory agreement, and a worse one – for the World Ocean. The latter is probably explained by an inadequate assessment of the World Ocean water mass. However, in all cases the calculated values of ζ exceed the observed ones by approximately 28 times.

These contradictory results indicate that the observed decades-long fluctuations in the Earth's rotation rate are not due to the rotation and polar motion of the whole Earth but rather to changes in the speed of drift of the lithosphere over the asthenosphere. Indeed, the moments of the like-sign forces arising in the process of fluctuations in the global water exchange operate for decades. It is possible that, with such long-term impacts, the matter of the asthenosphere underlying the lithosphere does not behave like a solid body but rather flows like a viscous fluid. Then, the decades-long global water exchange can result in the lithosphere's sliding over the asthenosphere without having a noticeable effect on the Earth's deeper layers. In astronomical observations, changes in the lithosphere's drift rate are recorded as "the irregularities in the Earth's rotation" and "polar motion." However, such apparent "irregularities" and "motions" require the redistribution of water masses that are 28 times lower than in the case of rotation of the whole Earth.

The sliding of the lithosphere over the asthenosphere is possible in the case when the action duration T is many times longer than the characteristic relaxation time τ within the asthenosphere. It is known that the relaxation time τ is determined from the relationship $\tau = \mu/\eta$, where μ is the viscous coefficient and η is the rigidity. For the asthenosphere, $\mu \approx 10^{18} - 10^{23}$ Poise and $\eta \approx 10^{12}\,\mathrm{dyn\,cm^{-2}}$. As a result, $\tau = \mu/\eta = 10^{6} - 10^{11}$ s or 0.03–3000 years. Clearly, the above-mentioned hypothesis could be accepted if we take a lower limit to the permissible values of μ. At the upper limit of viscosity, the drift of the lithosphere is hardly probable.

The hypothesis on the drift of the lithosphere over the asthenosphere is based not only on the analysis of the effect of redistribution of water between the ocean and the ice sheets in Antarctica and Greenland but also on a review of the mechanism of the angular momentum interchange between the atmosphere and the Earth (Section 9.4). The frictional forces and the pressures of the atmosphere and oceans on the lithosphere plates cause their drift over the asthenosphere. This hypothesis also agrees with the fact that there is a significant correlation between the seismic activity and the irregularities of the Earth's rotation.

We understand that the model of the Earth assumed in the system of Equations 11.8–11.11 is not adequate to the real Earth. Therefore, the system of Equation 11.14 does not allow us to obtain the absolute values of the water exchange parameters (ζ_A, ζ_G, and ζ_O); they allow us to study only temporal variations of these parameters. The configuration of the temporal variations of ζ_i does not depend on the model of the Earth (that is, on the parameters C, A, and k'). It is completely determined by the configuration of temporal changes in the Earth's rotation parameters ν_1, ν_2, and ν_3. The model parameters C, A, and k' are independent of time and do not influence the configuration of changes in the water exchange parameters ζ_A, ζ_G, ζ_O. They serve as the scale factors and influence the amplitude of variations in ζ_A, ζ_G, and ζ_O. The larger the C and A values and the less the k' value, the larger the amplitude of variations in ζ_A, ζ_G, and ζ_O. To verify this statement, calculations were carried out with three models: the absolutely solid Earth ($k' = 0$), the elastic Earth ($k' = -0.3$), and the Earth's model that consisted only of the lithosphere (Sidorenkov *et al.*, 2005). In all cases, the configuration was the same for the temporal variations in ζ_A, ζ_G, and ζ_O (for ζ_A, the minima always fell on 1903 and 1972 and the maxima – on 1891, 1934, and 2000). The amplitude of variations in ζ_A, ζ_G, and ζ_O depends significantly on the choice of a model. For the elastic Earth model, the amplitude of the calculated values of ζ_A is 28 times as large as the amplitude of observed glaciological variations; for the absolutely solid Earth, it is 20 times larger (Sidorenkov, 2002a, 2002b). The best agreement between the calculated and observed amplitudes of ζ_A was obtained in using the inertia moments of the lithosphere

(Sidorenkov *et al.*, 2005). Note that the elastic Earth model has been used in this study (Equations 11.8–11.11). Therefore, when comparing the curves in Figures 11.1–11.3, the amplitudes of variations in the calculated values $\zeta(t)$ were decreased by a factor of 28.

Some researchers believe that the fluctuations in the Earth's rotation affect the terrestrial processes, in particular, the World Ocean level, being responsible for its variations. However, the quantitative estimations show that this reverse effect is negligibly small (Sidorenkov, 1961).

11.4
Effect of Ice Sheets

As is seen from the system of Equation 11.18, the decadal fluctuations in the Earth's rotational rate depend mainly on the state of the Antarctic ice sheet (the ζ_A value), whereas the secular motion of the North pole depends on the increment of the ice mass in Greenland (the ζ_G value). Let us assess the order of magnitudes of the probable secular motion of the pole and the irregularity of the Earth's rotation. Recall that the atmospheric precipitation falling on the ice sheets of Greenland and Antarctica can remain there for centuries. The decrease in the ice masses occurs mostly due to the ice discharge into the surrounding seas. The velocity of the ice discharge is very low; therefore, its variability is also insignificant. At the same time, the annual sums of precipitation can deviate from its long-term value by twofold and even more. The annual sums of precipitation falling on the ice sheets of Antarctica and Greenland are 16 g/cm^2 and 30 g/cm^2, respectively.

If the specific mass of ice in Greenland ζ_G increases by 30 g/cm^2 per year, whereas $\zeta_A = 0$ and $\zeta_C = 0$, then it is easily calculated that the North pole will displace along meridian 141°E with the velocity equal to 18 cm yr^{-1}; the velocity of the Earth's rotation will increase by 1.26×10^{-10} per year. At the rate of ice accumulation in Antarctica equal to 16 g/cm^2 per year, the North pole will move in the direction of meridian 81°E with the velocity of 8 cm yr^{-1}, and the velocity of the Earth's rotation will increase by 5.12×10^{-10} per year. These estimates show that the variations in the annual sums of precipitation can cause the pole's displacements equal to 10–20 cm per year.

The position of the meridians, along which the mean North pole of the Earth's rotation must move, is presented in Figure 11.4, in which the arrows show the directions of the pole's displacement at the ice accumulation in Antarctica (A), Greenland (G), Antarctica and Greenland simultaneously (A + G), and on the rest of land (C). At the ice melting and decrease, the pole moves in the opposite direction. The polygonal curve depicts the observed interannual displacement of the mean pole according to the data in Table 8 from (Fedorov *et al.*, 1972). In the literature on astronomy, this pole's motion is called the secular motion.

The coordinates of the mean pole after 1968 are taken from the reports of the Central Bureau of the International Polar Motion Service and from (Vondrak, 1999).

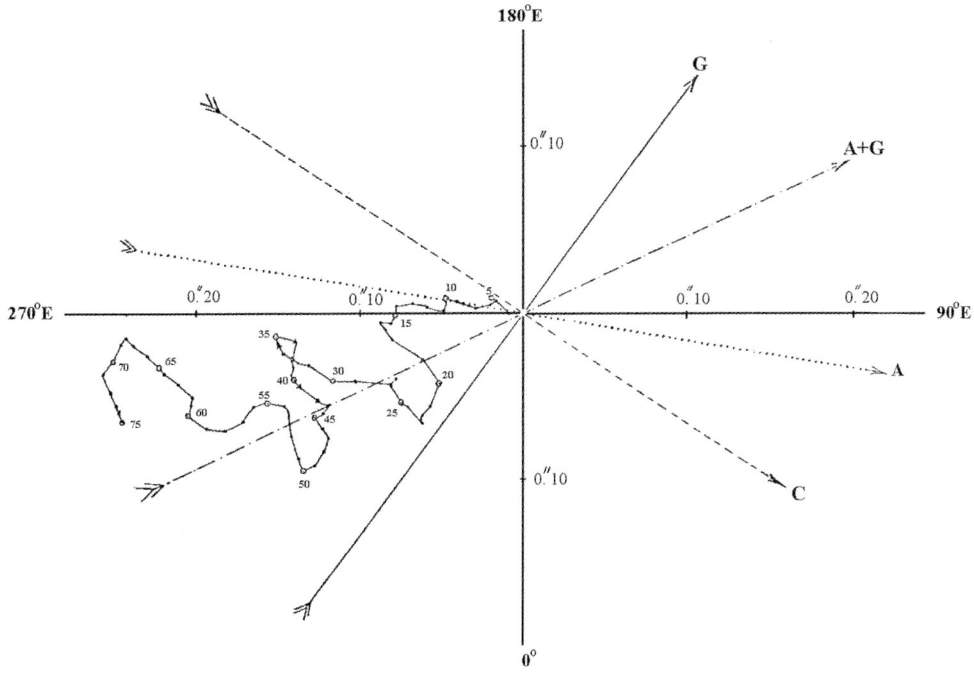

Figure 11.4 Directions of displacement of the mean North pole of the Earth's rotation due to the ice accumulation in Antarctica (A), Greenland (G), Antarctica and Greenland (A + G), and on rest of land (C).

According to the International Latitude Service (ILS), the pole moves along meridian 285° E, which is close to the trajectory of the pole in the case of simultaneous ice melting in Antarctica and Greenland (Figure 11.4). This fact deserves special attention, because many glaciologists believe that the ice sheets of Greenland and Antarctica have indeed decreased during the recent decades (Lliboutry, 1965; Kotlyakov, 1968).

The polygonal form of the curve can be explained by the fact that the secular motion of the pole is the result of summation of the variable contributions of various land areas, these contributions being a priori random.

As is known (see Section 3.1.3), since 1890 the pole is shifting in the direction of meridian 290° E with a velocity of about 10 cm yr^{-1}. Such shift, if it is real, testifies to a decrease in the ice mass. It could be caused, for example, by melting the ice in Greenland during this period, the rate of melting being near 16 g cm^{-2} yr^{-1}. Are such changes in the ice mass real?

At present, the total ice mass is about 28.4×10^{18} kg, out of which 25.7×10^{18} kg (90%) falls on the ice sheet of Antarctica, 2.5×10^{18} kg (9%) – on the ice sheet of Greenland, and only 0.22×10^{18} kg (less than 1%) is the ice mass of all other (mountain) glaciers. The areas of ice sheets in Antarctica and Greenland are 13.9×10^{12} and 1.8×10^{12} m^2, respectively. On each square centimeter of Antarctica there falls 185 kg of ice on average, in Greenland 139 kg. Using the values of ζ_A and ζ_G, it is easy to calculate the limiting shifts of the pole. If the ice sheet of Greenland melts completely, the North pole will displace by 830 m along meridian 321°E; the velocity of the Earth's rotation will decrease by 0.58×10^{-6}, (the length of a day will increase by 50 ms. In the case of disappearance of the ice sheet of Antarctica, the pole will shift by 920 m along meridian 261°E, and the velocity of the Earth's rotation will decrease by 0.59×10^{-6} (the length of day will increase by 512 ms). What are the probable characteristic periods of such shifts?

According to the paleogeographic data, glaciers can exist over tens thousands of years. For example, 25 thousand years ago, vast ice sheets still covered almost the entire East European Plain and considerable areas of West Europe and North America; and 12 thousand years ago they no longer existed. After the end of the last glacial period, there was the period of accelerating climate warming, which in the 50–20 centuries BC peaked at the period of the "Climate Optimum". Later, the climate grew worse, and during the "sub-Atlantic period" (the tenth century BC – the third century A.D.) the climate was relatively cold. The climate warming observed later resulted in the "Medieval Climate Optimum" or the "Little Climatic Optimum," at the beginning of the Middle Ages (roughly between 750 and 1200 AD). During that period, the climate in Europe was so warm that Vikings floated without hindrance to Greenland, rendered it habitable, and were engaged in the pasture sheep raising. From this follows the name of the island – Greenland, that is, the green land.

During the Medieval Climate Optimum, the mass of the ice sheet in Greenland was much smaller than it is at present. If we suppose that the ice mass in Greenland was smaller by 20% at that time than at present, then during the last thousand years the pole had to shift along meridian 141°E with a velocity of about 17 cm yr^{-1}.

After the Medieval Climate Optimum, the climate experienced the following changes: cooling in the thirteenth and fourteenth centuries, slight warming in the fifteenth and sixteenth centuries, and a new cooling in the seventeenth to nineteenth centuries. The latter cooling was given the name of the "Little Ice Age", because it was accompanied by advancing the glaciers in Europe and America. Since the second half of the nineteenth century, the climate warming began and ever since, it has been lasting off and on in the twentieth century and at present. It is likely that since that time the ice mass in Greenland is decreasing, due to which the pole is shifting in the opposite direction.

All the estimates presented above are performed using Equation 11.14, that is, for the elastic Earth. If the lithosphere drifts over the astenosphere, then the results of estimations should be greater by approximately 28 times.

11.5
Effect of Climate Changes

The state of the ice sheets in Antarctica and Greenland depends on the climate changes. Therefore, the decadal fluctuations in the Earth's rotation rate can correlate with the variations in the climatic characteristics and indices. And such correlation has been found (Lambeck, 1980; Sidorenkov and Svirenko, 1983; Sidorenkov and Orlov, 2008). Let us dwell on the relationship between the velocity of the Earth's rotation and the frequency of the types of atmospheric circulation proposed by Vangengeim (Catalogue, 1964).

The entire diversity of the forms of atmospheric circulation over the region extending from Greenland to the Yenisei River and to the north of 30°N was divided by Vangengeim (1935) into three forms: western W, eastern E, and meridional C. Form W is characterized by a slightly disturbed western–eastern air transfer. Form E is characterized by the ridge of high pressure located over the European part of the former USSR and the trough of low pressure over West Europe and West Siberia. At form C the pattern of the pressure field is opposite: the ridges are located over West Europe and West Siberia and the trough over the European part of the former USSR.

The data on the frequency of the above forms of atmospheric circulation are available from 1891 (Girs, 1971) to the present time. The method of the integral curves of the anomalies of frequencies of the circulation forms is convenient for revealing the long-term variations in the atmospheric circulation. These integral curves are constructed by way of calculating the anomalies of frequencies of the above circulation forms and the cumulative sums of the anomalies; the graphs of these sums are the integral curves. The ascending branch of the integral curve corresponds to the periods of prevalence of positive anomalies, and the descending branch to the periods of prevalence of negative anomalies. These periods are called epochs.

Sidorenkov and Svirenko (1983) have found that at a low frequency of form C, the Earth's rotation accelerates (Figure 11.5), and vice versa; the coefficient of correlation between the integral anomalies of form $\sum C'$ and the deviations of the length of day δP is 0.80. This close correlation counts in favor of the effect of the global water exchange on the interannual irregularity of the Earth's rotation and the polar motion, and finally in favor of the hypothesis of the drift of the lithosphere over the astenosphere.

There is close correlation between the decadal fluctuations in the Earth's rotation rate and those in the global air temperature (Jean-Pierre and Jolanta, 1992; Sidorenkov and Shveikina, 1994), the regional characteristics of precipitation and cloudiness (Sidorenkov and Shveikina, 1996), and even in the catches of food fish in the Pacific Ocean (Klyashtorin and Sidorenkov, 1996). Figure 11.5 demonstrates the temporal variations in the length of the terrestrial day and the air temperature in the Northern Hemisphere for 1891–1998. The comparison of two curves shows their strong correlation.

The relationship between the velocity of the Earth's rotation and the climate was noticed by Newton (1972), who has found that the prolonged fluctuation in the rate of the secular deceleration of the Earth's rotation coincided with the Medieval Climate Optimum. Note also that the most drastic disturbances in the Earth's rotation over the recent 300 years, which were observed near 1870 and 1935 (see Figure 3.9), coincided with the end of the Little Ice Age and the Arctic warming, respectively.

The climatic variations have characteristic periods that cover decades or much longer time intervals. The irregular changes in the length of day are also likely to have characteristic time periods of an order of decades and millenniums. Therefore, the

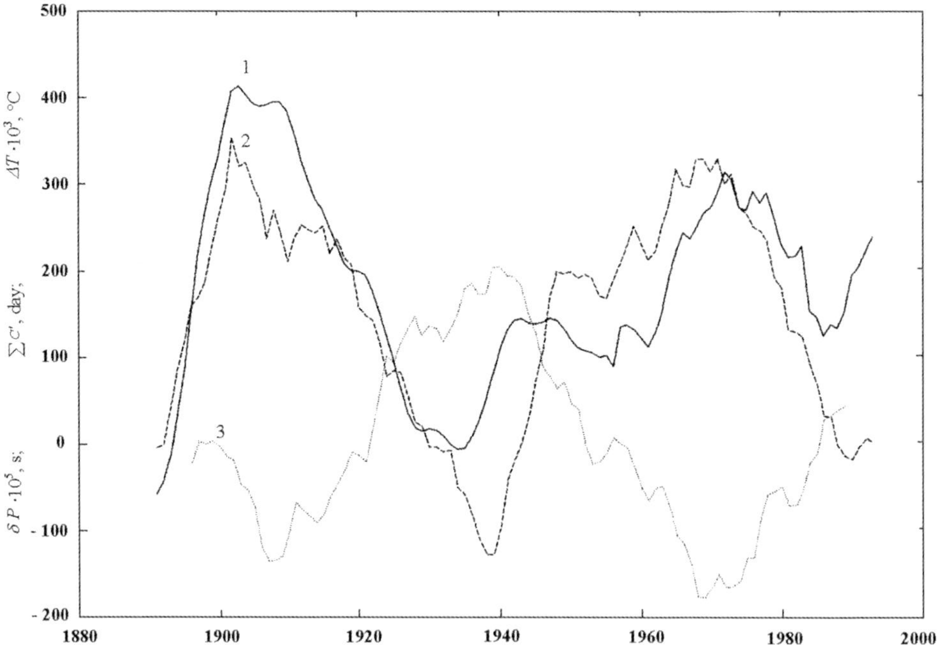

Figure 11.5 Synchronous changes in the length of day, δP (curve 1), the cumulative sums of anomalies of the circulation form C (curve 2), and of the ten-year running anomalies of the Northern Hemisphere's air temperature ΔT (after elimination of a trend and a 1000-fold magnification) (curve 3).

estimations of the rate of the secular deceleration of the Earth's rotation, which are usually performed for the time intervals less than 2000 years, are unlikely to be associated only with the tidal deceleration. For example, a tendency for decelerating the Earth's rotation, which is observed during two recent centuries, can be explained by the process of melting the ice sheets in Greenland and Antarctica. The mean direction of the secular motion of the pole corresponds to the above process.

The redistribution of water between the World Ocean and the ice sheets in Antarctica and Greenland causes changes in the Earth's gravitational field. If the irregularities in the Earth's rotation are also due to variations in the global water exchange, then the changes in the length of day and the pole's shifts should correlate with the changes in the Earth's gravitational field. The drift of the lithosphere over the astenosphere, if it exists, should be followed by the regular vertical movements of the Earth's crust and the seismic phenomena. It is necessary to have purposeful geodesic observations that could substantiate or demolish these suppositions. Further progress in studying the nature of the decadal fluctuations in the Earth's rotation depends on the improvement (increase in volume and accuracy) of hydrometeorological and geophysical observations.

The long-term variations in δP can now be determined very accurately. Though their nature is not quite clear, our many-year experience shows that the long-term variations in δP present a unique nature-born integral index of the global climate changes.

The periods of the Earth's rotation acceleration (of the decrease in the length of day δP) coincide with the epochs of the negative anomalies of form C of the atmospheric circulation and of the positive anomalies of (W + E). During these periods, the ice mass in Antarctica increases, the intensity of the zonal circulation weakens, the temperature growth in the Northern Hemisphere accelerates, the positive anomalies of the global cloudiness prevail, and the catches of food fish in the Pacific Ocean augment. During the periods of deceleration of the Earth's rotation (of the increase in δP), form C is observed more frequently and (W + E) more seldom than on average, the ice mass in Antarctica decreases, the air temperature in the Northern Hemisphere increases slower, the observed negative anomalies of the global cloudiness, the catches of food fish in the Pacific Ocean are decreasing from year to year.

Thus, as was mentioned above, there are close relationships between the long-term fluctuations in the Earth's rotation, on the one hand, and the variations in the ice sheet in Antarctica, epochs of atmospheric circulation, global air temperature, regional precipitation and cloudiness, and even in the catches of food fish in the Pacific Ocean, on the other hand.

This close correlation can be used for the extrapolation of $\sum C'$, $\sum(W+E)'$, and ΔT into the past or future, that is, for diagnosis or prognosis of the anomalies of the atmospheric circulation forms or epochs and of the variations in the air temperature anomalies. The point is that the irregular long-period oscillations inherent in the Earth's rotational rate have a representative time period of about 70 years. The acceleration that had started in 1973 came to an end in 2003, and the Earth's rotational rate began to decelerate (the length of day began to increase). According to the experience of the past decades, this deceleration has to continue until 2039 (± 3 years). As the periods of deceleration of the Earth's rotational rate coincide with the ascending branch of the integral curve $\sum C'$ and the descending one of the curve $\sum(W+E)'$, the frequency of the form C during the period of 2004–2039 (± 3 years) is to be above the norm, and that of the form (W + E) – below the norm.

11.6 Effect of the Earth's Core

The Earth's mantle and liquid core can exchange the angular momentum. The moment of inertia of the core C_C is equal to 9×10^{36} kg m^2, and the absolute angular momentum $\tilde{N}_C \Omega = 66 \times \cdot 10^{31}$ kg m^2 s^{-1}. The decadal fluctuations of the Earth's rotation rate have the order of magnitude of 10^{-8}. If they are due to the redistribution of the angular momentum between the mantle and the core, then it holds:

$$C_M \omega_M + C_C \omega_C = \text{const} \tag{11.16}$$

where C_M and C_C are the axial moments of inertia for the mantle and the core, respectively; ω_M and ω_C are the angular velocities of the mantle and core rotations, respectively. Differentiating (11.16) and passing to increments, we find:

$$\Delta\omega_C = -\frac{C_M}{C_C}\Delta\omega_M \approx -8\Delta\omega_M \tag{11.17}$$

This means that the fluctuations in the angular velocity of the core must have the opposite sign and exceed by 8 times the observed amplitudes of the decadal fluctuations in the Earth's rotation rate. The decadal fluctuations in the Earth's rotation rate have the order of magnitude of 1×10^{-8}. Hence, the fluctuations in the core must be 8×10^{-8}, that is over the year the core can be behind or ahead of the mantle by 0.07°. Indirect evidence of the circulation of substance in the liquid core is the westward drift of the nondipole part of the Earth's magnetic field. This drift shows that the external parts of the liquid core rotate slower than the Earth's mantle. Even in 1950 (Bullard *et al.*, 1950) the drift rate was found to be ~0.18° per year. If we accept this value, then the fluctuations in the rate of the westward drift would be less than 0.07/0.18, or about 38%. The observations confirm the existence of fluctuations of this order of magnitude.

In (Vestine and Kahle, 1968; Cain, Schmitz and Kluth, 1985) a close correlation was obtained between the fluctuations in the angular velocities of the Earth's rotation and of the rate of westward drift of the eccentric geomagnetic dipole, with the characteristic time period of about 60 years (Figure 11.6). With this, the onset of extrema of the eccentric dipole drift rate lags by ~7 years. It is likely that during this period the magnetic signal passes through the mantle and reaches the Earth's surface.

Braginsky (1970, 1972) has shown that in the Earth's core, the magnetohydrodynamic oscillations of the torsional type occur; they cause the fluctuations in the rates of westward drift of the eccentric geomagnetic dipole and of the Earth's rotation. The frequency of these fluctuations is described by the formula for the Alfven waves,

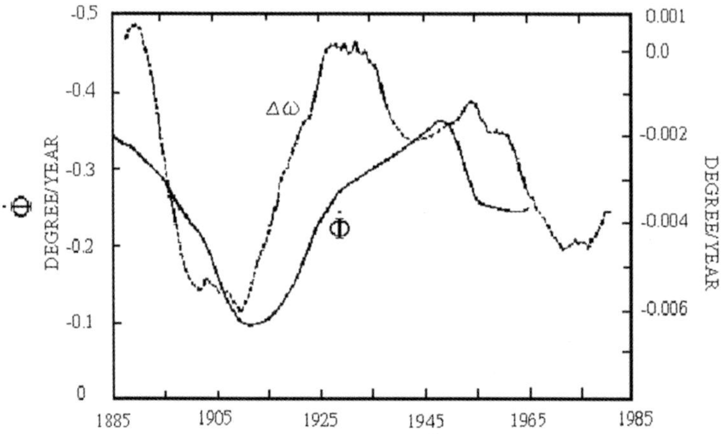

Figure 11.6 Eccentric geomagnetic dipole motion $\dot{\Phi}$ (solid lines) and change in the length of day expressed as the angular deviations of rotation $\Delta\omega$ (deg/year) (dashed line) (Cain, Schmitz and Kluth, 1985).

which is known in the magnetic hydrodynamics:

$$\omega \approx \frac{B_S}{L}\sqrt{4\pi\rho}$$

where B_S is the radial component of the magnetic field in the cylindrical coordinate, L is the characteristic scale, and ρ is the density. Assuming in the core $B_S \approx 3$ Gauss, $L \approx 10^6$ m, $\rho \approx 10^4$ kg/m^3, we obtain $T \approx 60$ years.

The correlation is found between the decadal fluctuations in the Earth's rotation rate ν_3 and the derivative of the geomagnetic moment $\dot{M}$ (Pushkov and Chernova, 1972; Jin and Thomas, 1977). The fluctuations in ν_3 are ahead of the fluctuations in $\dot{M}$ by 10 years.

Thus, on the one hand, the decadal fluctuations in the Earth's rotation rate could be produced by the exchange of the angular momentum between the mantle and the liquid core of the Earth. On the other hand, there is a close relationship between the decadal fluctuations of the Earth's rotation and the changes in the climatic and glaciological characteristics. Yet, processes in the Earth's core cannot produce changes in the epochs of atmospheric circulation, air-temperature fluctuations, atmospheric precipitation, the state of glaciers, and other climatic processes and characteristics.

These contradictions could be eliminated by assuming that there is a third reason that simultaneously affects the processes in both the Earth's core and in the climatic system. This reason is the gravitational interaction of the Earth with the Moon, the Sun and other planets.

The Sun revolves around the barycenter (the centre of mass) of the Solar System along the compound curves of the fourth order (conchoids of a circle), the so-called "Pascal's limacons" (Figure 11.7). The curvature of the Sun's trajectory constantly changes, and the Sun moves with a varying acceleration. Being a satellite of the Sun, the Earth revolves around it and also moves with the Sun around the Solar System's barycenter. Like the Sun, the Earth undergoes all varying accelerations. Similar to the lunisolar tides, the accelerations disturb the Earth's orbital revolution and processes in the Earth's shells, producing the decadal fluctuations in the latter.

Landscheidt (2003) and Wilson *et al.* (2008) present much evidence to show that the changes in the Sun's equatorial rotation rate are synchronized with the changes in the Sun's orbital motion around the barycenter of the Solar System. Ian Wilson (Sidorenkov and Wilson, 2009) showed that the recent maximum asymmetries in the solar motion around the barycenter have occurred in the years 1865, 1900, 1934, 1970 and 2007. These years of maximum asymmetry in the solar motion closely correspond to the points of inflection in the Earth's length of the day (Figure 3.9). Figure 11.8 constructed by Ian Wilson, shows that, from 1700 to 2000 AD, on every occasion where the Sun has experienced a maximum in the asymmetry of its motion around the centre-of-mass of the Solar System, the Earth has also experienced a significant deviation in its rotation rate (that is, the length of day) from that expected from the long-term trends. This fact indicates that the changes in the Earth's rotation rate are synchronized with a phenomenon that is linked to the changes in the solar motion around the barycenter of the Solar System.

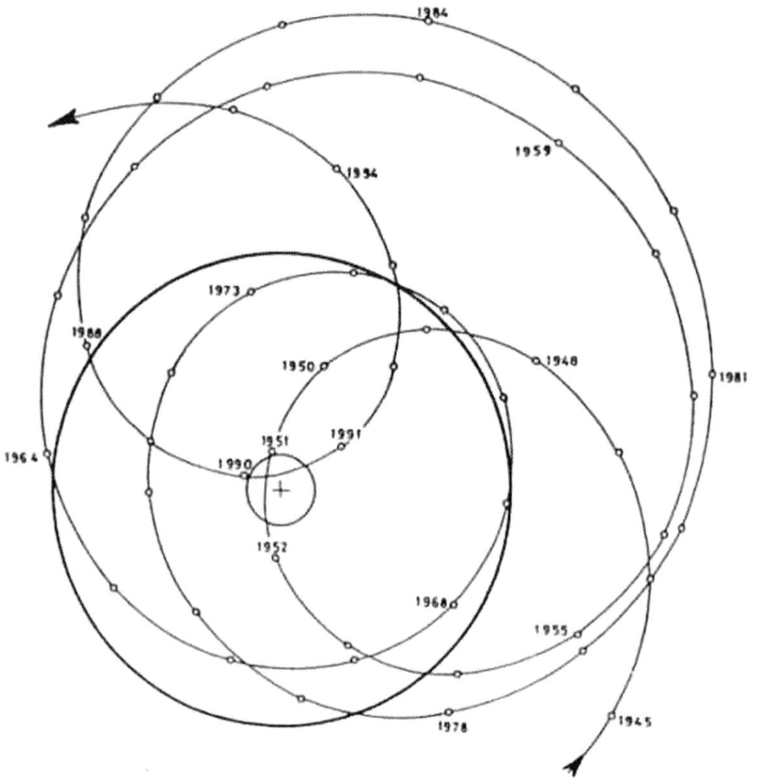

Figure 11.7 Motion of the barycenter of the Solar System in the heliocentric coordinate system for 1945–1995. The Sun's limb is marked by a thick circle. The position of the barycenter relative to the Sun's centre (a cross) is indicated by small circles (Landscheidt, 2003).

From the empirical data, Ian Wilson proposes (Sidorenkov and Wilson, 2009) that there is compelling evidence to support the contention that there is a spin-orbit coupling between the Earth's rotation rate and its motion (that it shares with the Sun) around the barycenter of the Solar System.

Furthermore, the gravity of the nonspherical nonuniform shells of the Earth (that occupy the eccentric positions) due to the Moon, the Sun and planets leads to the shifts and fluctuations in the centers of mass of the shells relative to each other, as well as to their forced transitions (Barkin, 1996, 2000).

In Barkin's and Wilson's cases, the set of phenomena that arise in the Earth's shells may be called "the generalized tides". On the one hand, the generalized tides evoke changes in the Earth's core and are related to the perennial variations in the geomagnetic field. On the other hand, they are responsible for changes in the climatic system and fluctuations in the Earth's rotation. In these cases, evidently, the decades-long variations in the rotation will correlate with all the above-mentioned geophysical and hydrometeorlogical processes.

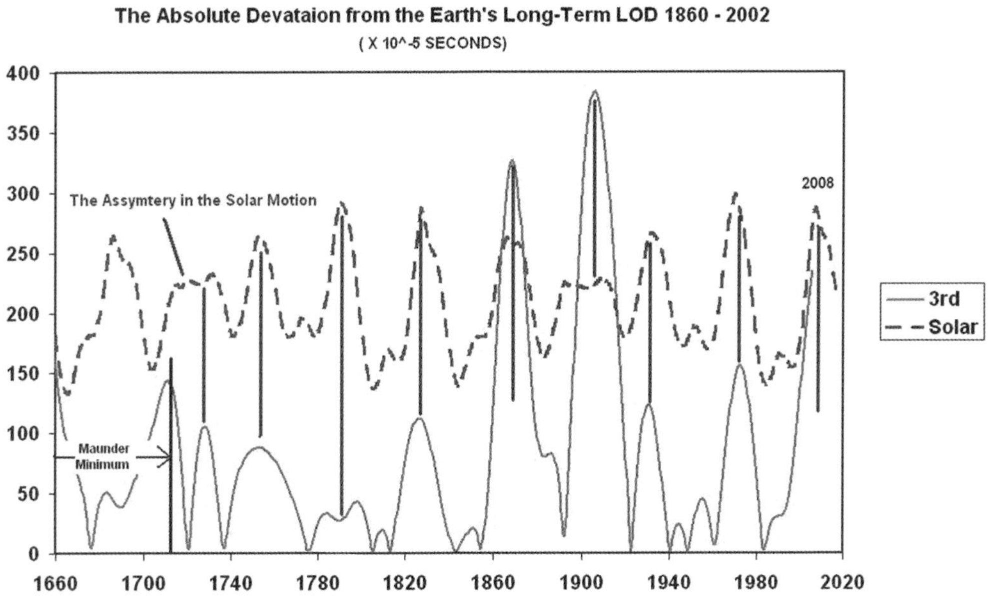

Figure 11.8 On every occasion where the Sun has experienced a maximum in the asymmetry of its motion around the centre-of-mass of the Solar System, the Earth has also experienced a significant deviation in its rotation rate (that is, the length of day) from the long-term trends.

11.7 Summary

We have shown that there are strong correlations between the decadal variations in the length of day, variations in the rate of the westward drift of the geomagnetic eccentric dipole, and variations in certain climate characteristics (the increments of the Antarctic and Greenland ice sheet masses, anomalies of the atmospheric circulation regimes, the hemisphere-averaged air temperature, the Pacific Decadal Oscillation, etc.).

From the empirical data, we argue that these correlations are due to the generalized tides that occur in Earth's shells due to the gravitational interaction of the Earth with the Moon, the Sun, and other planets.

12 Geodynamics and Weather Predictions

12.1 Predictions of Hydrometeorological Characteristics

12.1.1 Synoptic Processes in the Atmosphere

About 100 years ago Mul'tanovskii (1933) noted that synoptic processes evolve in jumps rather than continuously. Analyzing kinematic continuity charts, he observed that the locations of pressure fields persisted for several days and then transformed rapidly (during 12–36 hours). The resulting pattern again persisted for several days until the next transformation. Studying the First Natural Synoptic Region, which extends north of 30°N from Greenland to the Yenisei River, he found that a certain type of evolution persisted for several days. In 1915, Mul'tanovskii called that period the natural synoptic period (NSP). In other words, an NSP is a period of time, during which a particular synoptic process persists over a given natural synoptic region.

After upper-level pressure charts were begun to be constructed in the 1940s, it was found that an upper-air pressure and temperature pattern persists in the troposphere during an NSP. The pressure and temperature fields cause displacements of pressure systems near the Earth's surface and lead to preserved geographical locations of their centers within a natural synoptic region. At the end of an NSP, the tropospheric pressure and temperature fields are rapidly rebuilt, which leads to new locations of pressure centers and to variations in pressure system trajectories near the Earth's surface. On a continuity chart, the current NSP is transformed into another one in one or two days, that is, in a "jump" as compared with the NSP length, which ranges from 5 to about 8 days (Pagava *et al.*, 1966). The nature of NSPs remained unknown until the studies in (Sidorenkov, 2002a, 2000b, 2005).

12.1.2 Conductors of Synoptic Processes

As was mentioned above, the tidal oscillations of the Earth's angular velocity have been reliably recorded since the 1980s.

The Interaction Between Earth's Rotation and Geophysical Processes. Nikolay S. Sidorenkov

ISBN: 978-3-527-40875-7

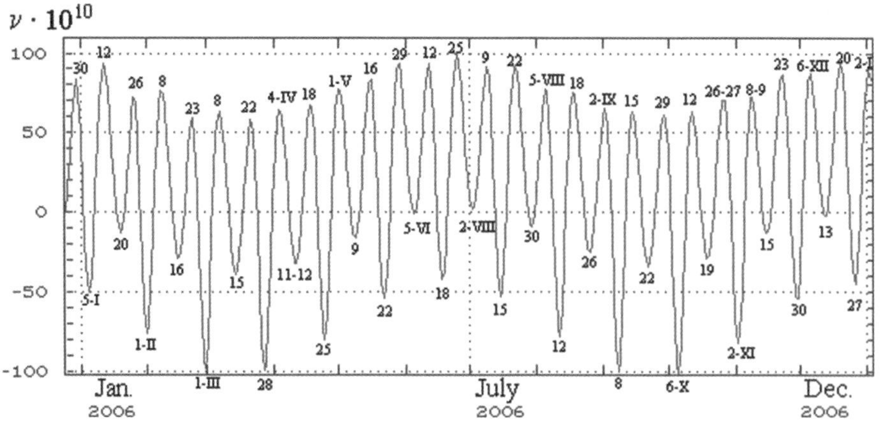

Figure 12.1 Tidal oscillations in the Earth's angular rotation velocity in 2006.

Since then the author has carried out the simultaneous monitoring of tide-generated variations in the Earth's angular velocity, the evolution of synoptic processes in the atmosphere, atmospheric circulation regimes, and time variations in hydrometeorological characteristics. It was found that most types of transitions between synoptic processes in the atmosphere occur simultaneously with variations in the Earth's angular velocity ω. It is well known that the effect of terrestrial zonal tides within a lunar month (27.3 days) leads to four modes with different period lengths in the Earth's rotation: two acceleration periods of lengths m_1 and m_3 and two deceleration periods of lengths m_2 and m_4 (Figure 12.1). The modes alternate with an average period of $m_i \approx 27.3/4 = 6.8$ days. However, since the lunar perigee and nodes move, this period varies from 5 to 8 days. In 2006, for example, deceleration was observed from September 29 to October 6; acceleration, from October 6 to 12; deceleration, from October 12 to 19; and acceleration, from October 19 to 27. Thus, the lunar month consisted of four intervals with a total duration of $7 + 6 + 7 + 8$ days (Figure 12.1). Any combinations and any real values of m_i in the range of 5 to 8 days are possible. The only unchanged characteristic is a monthly period of 27.3 days.

Based on historical data, the author examined how frequently the extrema of ω concurred with transitions between synoptic processes. The characteristics of synoptic processes of different types were taken from the Catalogue of Elementary Synoptic Processes (ESPs) by G.Ya. Vangengeim (Catalogue, 1964). Catalogues of ESP transitions and extrema (maxima and minima) of tidal oscillations in ω were prepared and analyzed over an 8-year period (from October 1, 1987, to September 30, 1995; a total of 2922 days). A statistical analysis showed that the extrema of ω coincided with ESP transitions up to ± 1 day in 76% of the cases. In 24% of the cases, the difference between an extremum of ω and the nearest ESP transition was two or more days (Sidorenkov, 2000d). Variations in the vertical component of the total tidal force (Figure 12.2) clearly reveal that dominant diurnal harmonics alternate with dominant semidiurnal harmonics at quasiweekly intervals. For diurnal tides, the amplitude is maximal when the Moon is at the largest distance from the equator

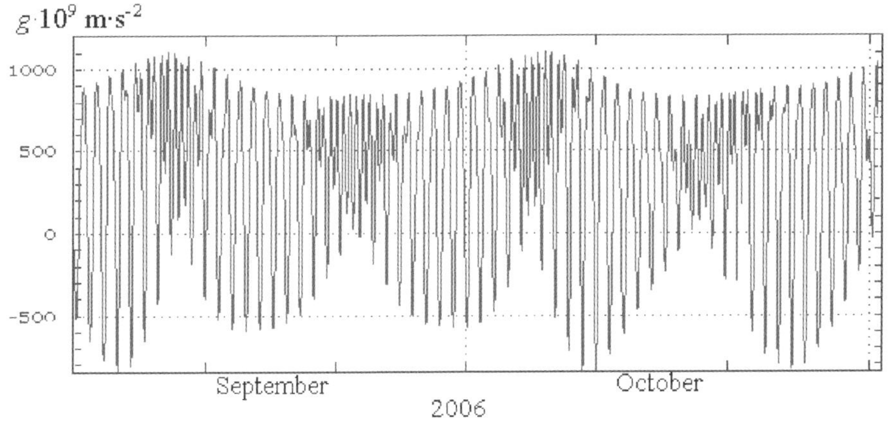

Figure 12.2 Variations in gravity due to the tides in September and October, 2006. The vertical axis represents gravity values in 0.1 microgal.

(when the Moon's declination is close to its maximum in absolute value). For semidiurnal tides, the amplitude is maximal when the Moon is at the smallest distance from the equator (when the Moon's declination is close to 0°).

Two trains of large-amplitude diurnal oscillations and two trains of small-amplitude semidiurnal oscillations are observed during a lunar month. Their duration also varies from 5 to 8 days, but the average value is always 6.8 days. The extrema of the lower envelope of diurnal and semidiurnal tides coincide with the extrema of zonal tides. This suggests that NSPs in the atmosphere form under the influence of not only long-period zonal tides but also diurnal and semidiurnal lunisolar tides.

Each quasiweekly mode of the Earth's rotation is associated with an NSP. Since the Earth's rotation modes are caused by lunisolar tides, NSPs possibly have the same cause.

This conclusion was checked by calculating the spectra of the equatorial components of the atmospheric angular momentum. The spectra revealed that lunisolar tidal harmonics are completely dominant in the variability of the atmospheric circulation (see Section 7.5). Specifically, it was found that the evolution of synoptic processes in the atmosphere is caused not only by internal dynamics but is also affected by an external "conductor" – lunisolar tides. NSPs are caused by oscillations in tidal forces, and variations in NSPs are associated with changes in the signs of tidal forces. Variations in the NSP length are caused by the frequency modulation of tidal forces due to the motion of the lunar perigee.

There is a statistically significant simultaneous correlation between the tidal oscillations of the Earth's angular velocity and transitions between synoptic processes occurring in the atmosphere. Close relations have been found between lunisolar tides and variations in meteorological characteristics. For illustrative purposes, Figure 12.3 shows the periodogram of daily air temperature anomalies in Moscow over 1960–2003.

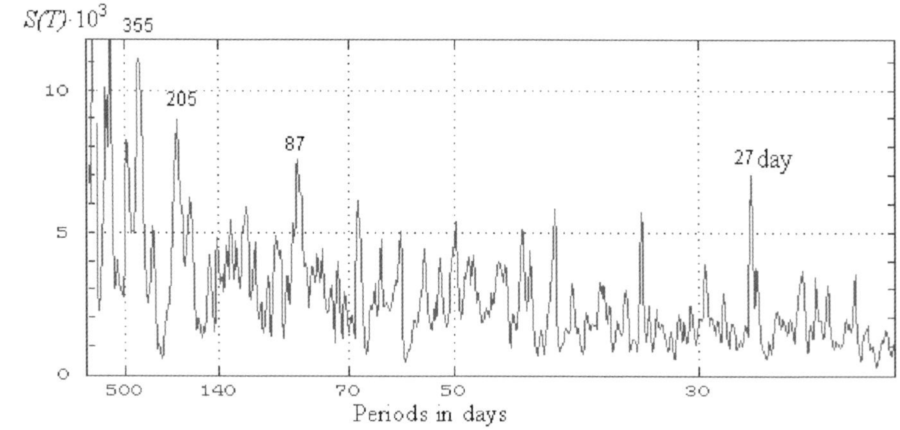

Figure 12.3 Periodogram of air-temperature anomalies in Moscow.

Inspection of the plot reveals peaks near the periods of 27, 87, 205, and 355 days, which correspond to the lunar sidereal (27.3 days) and synodic (29.5 days) periods. The 206-day cycle generates due to the revolution of the major axis of the Moon's orbit relative to the major axis of the Earth's orbit. The perigee of the lunar orbit executes one revolution within 8.85 years and the perihelion of the Earth's orbit – within one year. The perigee meets with the perihelion every 412 days. But the axes of the Moon's and the Earth's orbits became collinear every 206 days. It is just this cyclicity of the reciprocal configurations of two orbits that has effects on the terrestrial processes. A period of 355 days corresponds to 13 sidereal and 12 synodic months. This is a lunar year! Therefore, temperature anomalies are formed not only by the internal stochastic dynamics of the atmosphere but also by lunisolar tides, which recur to some extent with a lunar-year period of 355 days (Figure 12.4). This dependence is well seen in the autocorrelation function (ACF) of tidal oscillations.

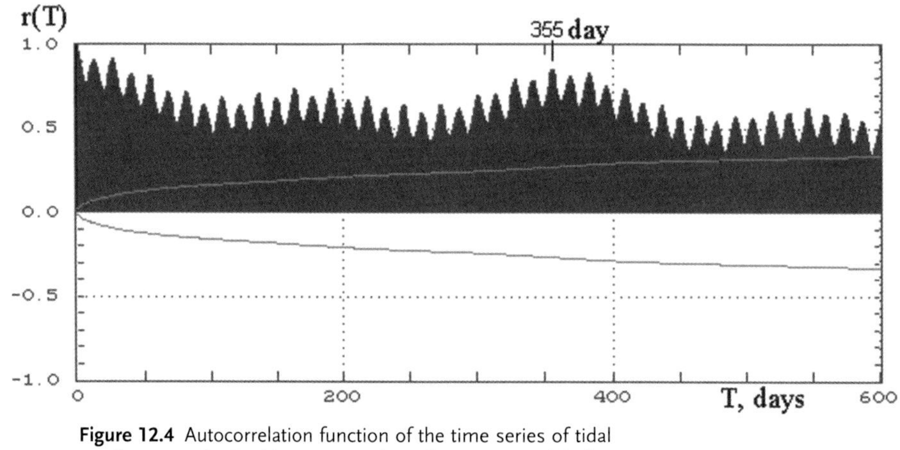

Figure 12.4 Autocorrelation function of the time series of tidal oscillations in the Earth's rotation velocity from 1962 to 2003 for a shift of 0 to 600 days.

Solar tides give a maximum correlation at a shift of 365 days, while lunar tides give a maximum correlation every sidereal month. Therefore, a maximum ACF is observed at a shift of 355 days (13 sidereal months). A large ACF maximum also occurs at a shift of 382 days (14 sidereal months). In addition to the lunar year, there are longer cycles: saros (223 synodic months), Metonic cycle (235 synodic months), inex (358 synodic months), and so forth, which also affect temperature anomalies.

Tides affect the meridional air circulation and atmospheric-pressure variations. Higher pressure values caused by lunar tides lead to positive temperature anomalies in summer and to negative temperature anomalies in winter. The temperature anomalies over a monthly cycle depend on the season. In the analysis and prediction of temperature, this dependence can be taken into account either by fixing a calendar time interval or by choosing a time interval multiple of a year.

Based on the empirical facts described above, the following method was developed for the prediction of hydrometeorological characteristics (Sidorenkov, 2002a, 2002b).

Based on the theory, oscillations in ω can be reliably calculated forward in time with any discretization for any period (Figure 12.1). An analogous period in the past with approximately the same Earth rotation mode is determined by applying correlation analysis to the values of ω predicted for the nearest year. It is assumed that the synoptic processes and the variations in temperature anomalies over the predicted time interval correlate with those over the analogous period. The temperature anomalies observed at its boundaries are taken as expected values. They are added to the corresponding normals to obtain a predicted air temperature.

This technique can produce day-to-day air-temperature predictions for a period of up to one year at any site where there are sufficiently long series of observations.

The skill score of the predicted monthly mean temperature anomalies in Moscow over 2000–2007 was found to be about 70%. The forecast errors arise mainly due to several-day long shifts in the oscillation phase. For this reason, superlong-term day-to-day temperature forecasts are not suitable for consumers interested in the exact times when temperature variations take place, but would be satisfactory for those interested in the fact of occurrence of such variations rather than in their exact time. The best skill scores were obtained for the Volga region, Central Russia, and the North Caucasus. The results based on high-latitude and sea-station data were noticeably worse.

12.2 Long-Period Variability in Tidal Oscillations and Atmospheric Processes

12.2.1 Variability in Lunar Tidal Forces

The lunar tidal force varies with time with a period of 13.65 days. It is a function of the Moon's declination and the geocentric distance, which vary with time in a complex manner. The amplitude of monthly oscillations in the Moon's declination varies from 29° to 18° with a period of 18.61 years due to the regression of the lunar nodes. The

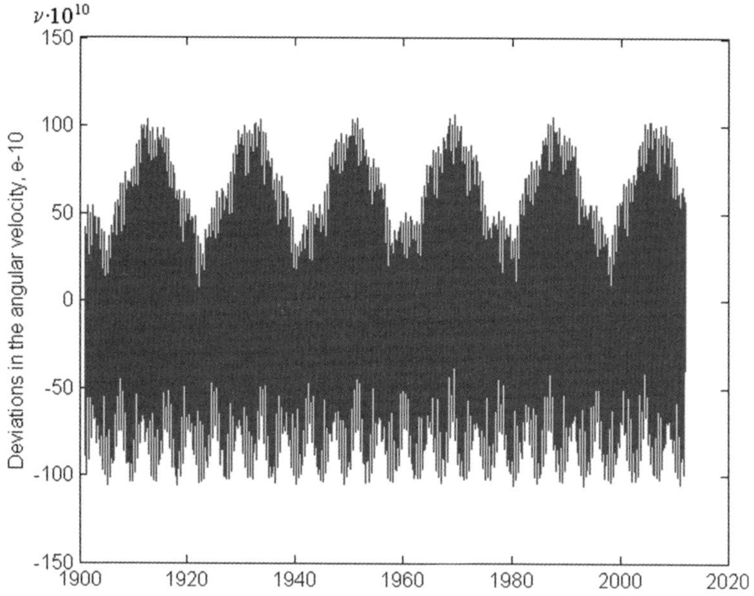

Figure 12.5 Tidal oscillations in the Earth's rotation velocity ν from 1901 to 2011. The vertical axis represents the relative deviations of ν in 10^{-10}.

lunar perigee moves with a period of 8.85 years, which leads to variations in the quasiweekly oscillation period of tidal forces in the range of 5 to 8 days. As a result, the oscillation amplitude of tidal forces varies with time with periods of 18.61 years, 8.85 years, 6.0 years, 1 year, 0.5 year, 1 month, 0.5 month, 1 week, 1 day, 0.5 day, and many other less significant periods. The variability of tidal forces is most pronounced in the Earth's rotation oscillations (Sidorenkov, 2006). Figure 12.5 shows day-to-day tidal oscillations of the Earth's rotation velocity from 1901 to 2011. It can be seen that the amplitude of fortnightly oscillations varies in a complex manner. The upper and lower envelopes describe waves with a period of 18.6 and 4.4 years, respectively. For comparison, Figure 12.6 displays the tidal oscillations in the Earth's rotation velocity calculated for the periods of minimal (1997) and maximal (2007) tidal variability. It can be seen that the amplitude of tidal oscillations in the angular velocity in 2007 is nearly twice as large as in 1997.

The temporal variability of a geophysical quantity can be conveniently characterized by its variance calculated over a sliding time window. Specifically, the variance D of a time series of tidal oscillations in the Earth's rotation velocity is calculated over a one-year sliding window as

$$D = \frac{1}{N}\sum_{i=1}^{N} (\nu_i - \bar{\nu})^2 \qquad (12.1)$$

where ν_i are the daily velocities, $\nu = \frac{1}{N}\sum_{i=1}^{N} \nu_i$ are the annual mean velocities, and $N = 365$. This formula was used to compute D over one year with a sliding step of

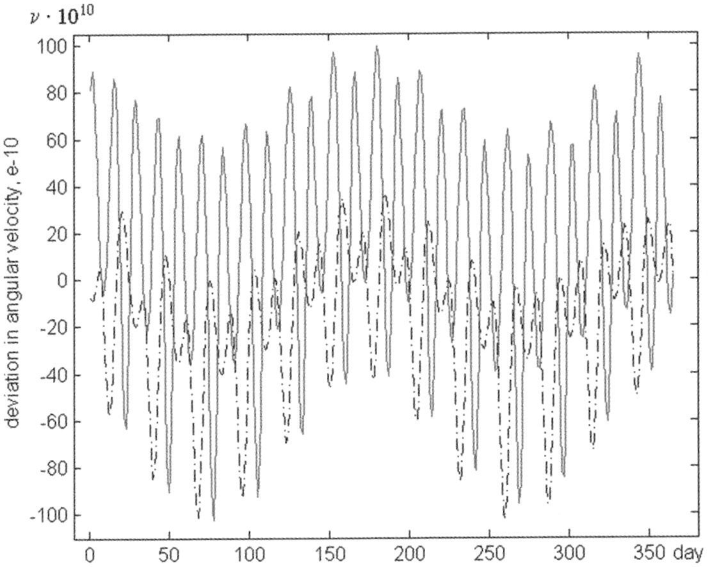

Figure 12.6 Tidal oscillations in the Earth's rotation velocity ν in 1997 (dashed) and 2007 (bold). The vertical axis represents the relative deviations of ν in 10^{-10}.

1 day. The results are shown in Figure 12.7. It can be seen that *D* undergoes threefold variations from its minimum values in 1903, 1923, 1942, 1960, 1978, and 1997 to its maximum values in 1914, 1932, 1950, 1969, 1988, and 2007. A minimum of *D* is observed when the Moon's descending node coincides with the spring equinox, and a maximum of *D* is observed when the Moon's ascending node coincides with the spring equinox (Sidorenkov, 2006).

12.2.2
Rate of Extreme Natural Processes

Due to the variability of tidal forces, the amplitude of oscillations in weather elements and marine hydrological characteristics also varies with time with the same period. However, the larger the amplitude of oscillations in hydrometeorological characteristics, the more frequent the extreme events (heat or cold waves, droughts or floods, hurricane-force winds, strong thunderstorms, hailstorms) and the greater the economic damage they cause. In other words, the rate of extreme natural processes varies according to the various oscillations of tidal forces. Industrial and social processes are also affected by varying tidal forces.

To statistically prove the existence of variability in hydrometeorological characteristics with a period of 18.6 years, one needs the time series of high-frequency hydrometeorological observations covering several dozens of 18.6-year cycles (longer than 300 years). Unfortunately, meteorologists use (monthly and annual) mean data, which do not contain any information on tidal oscillations. For example, the

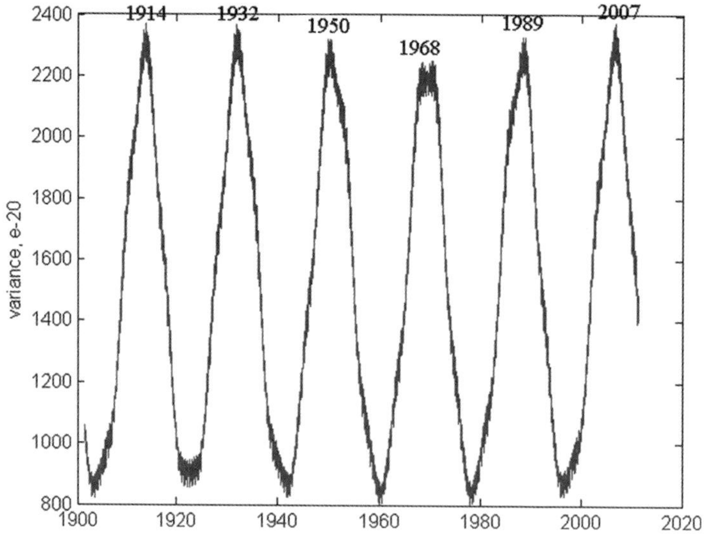

Figure 12.7 Variance of tidal oscillations in the Earth's rotation velocity calculated with a one-year sliding window for 1900–2010.

averaging of h_1 and h_2 even on a daily basis yields nearly constant values with no amplitude or frequency modulation of the daily carrier frequency. The spectra of such series contain only noise components (see Section 7.5).

Starting in 1966, weather observations are made every 3 hours. However, these series cover only two 18.6-year cycles and are not suited for a rigorous statistical analysis of the problem in question. Meteorological observations over earlier years are difficult to find. However, the author has managed to obtain the first preliminary evidence of the 18.6-year variability by analyzing time series of atmospheric angular momentum components. Specifically, the series of h_1 and h_2 from 1948 to 2006 derived by Salstein (2000) were used to calculate the variance over a three-year sliding window. Both components gave similar results. For this reason, Figure 12.8 shows only the variance of h_1. It can be seen that the maxima of the variance are consistent with those of tidal-force variability in 1951, 1969, 1987, and 2007. Only in the 1980s is the maximum of h_1 slightly ahead of the tidal-force maximum. Thus, the long-time variability of the atmospheric angular momentum is caused, to a certain degree, by the 18.6-year variations in tidal forces.

The statistics of hydrometeorological hazards (recorded by a) the Hydrometeorological Center of Russia and b) the Federal Service for Hydrometeorology and Environmental Monitoring of Russia) clearly indicate an increased rate of hazards in 1998–2007 (Figure 12.9).

Increased (decreased) tidal amplitude leads to a higher (lower) rate of extreme natural processes. A maximum of the 18.6-year variability of tidal forces was observed in 2007 (Figure 12.7). Therefore, an increased rate of extreme natural processes observed over recent years is caused not only by global warming but also by the currently observed maximum of tidal force variability. Since this variability will

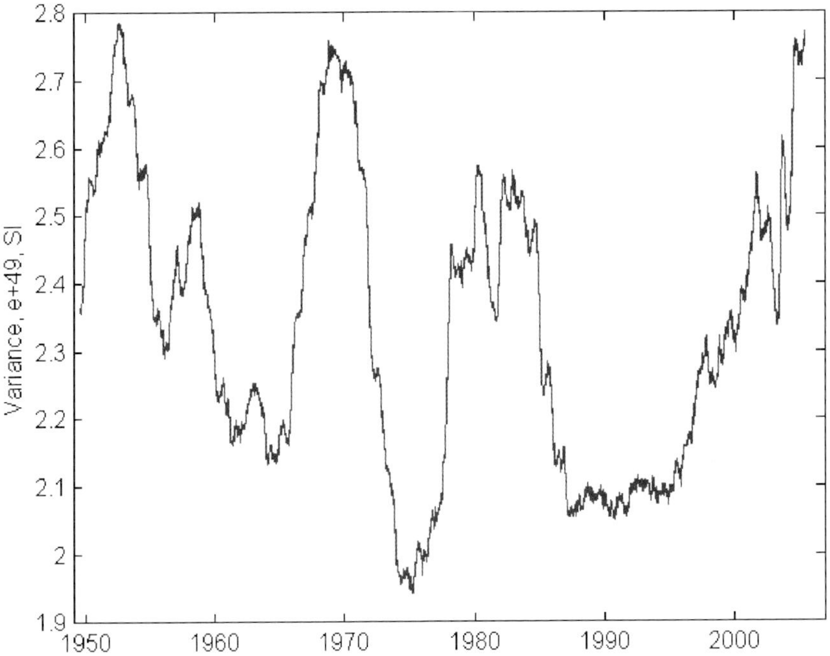

Figure 12.8 Variance of the equatorial component h_1 of the atmospheric relative angular momentum calculated with a three-year sliding window.

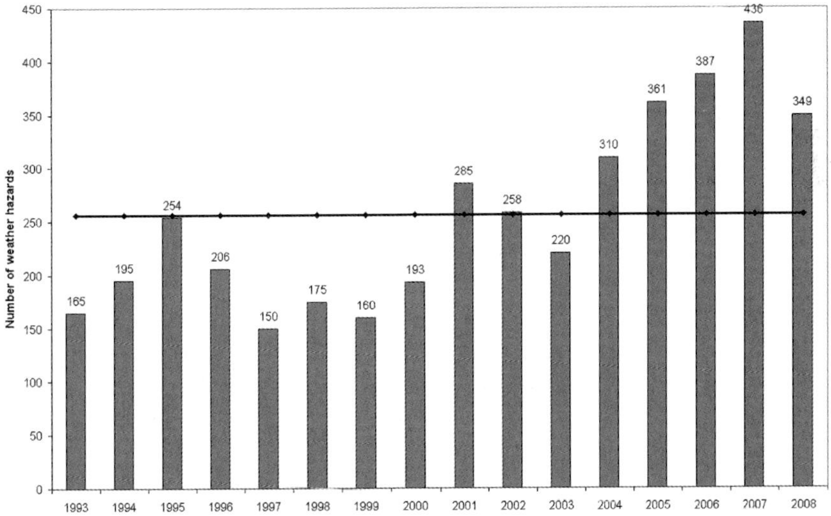

Figure 12.9 Variations in the annual number of weather hazards over the last 16 years from www.meteorf.ru:Doklad_climate_2008.pdf. The black straight line corresponds to the annual mean value over 16 years.

decrease in 2008–2016, we can expect a reduction in the rate of extreme natural processes in this period due to the influence of tidal forces.

12.3
Hydrodynamic Equations of Motion

Thus, the zonal tides significantly affect atmosphere dynamics.

This conclusion is confirmed by the author's many years' successful experience in the super-long-term forecast of temperatures, synoptic and natural processes, based on the behavior of lunisolar tides. Chapter 7 shows the spectra of the components of the atmospheric angular momentum, which also indicate that the tides play a dominating role in the variability of the global atmospheric circulation. These arguments suggest that the equations of hydrodynamics that are currently employed in global atmospheric and oceanic models are inadequate as far as the real processes are concerned because the tide-generating forces are not taken into account. Probably, that is why the predictability of atmospheric processes, based on the current global hydrodynamical models, is no more than two weeks.

As is generally known, a terrestrial reference system is used in meteorology and oceanology. The equations of relative movement are:

$$\frac{d\mathbf{V}}{dt} = -\frac{1}{\rho}\nabla p - 2\Omega \times \mathbf{V} + \mathbf{g} + \mathbf{F} \tag{12.2}$$

where **V** is the vector of the velocity of wind or current, ρ and p are the density and the pressure, respectively, Ω is the angular velocity of the Earth's rotation, **F** is the friction force, **g** is the gravity, which is a sum of the Earth's gravitational attraction and the centrifugal force. The tide-generating forces of the Moon and the Sun should be taken into account to more accurately describe atmospheric and oceanic circulation (Sidorenkov, 2002a, 2002b). These forces can be found by calculating the gradient ∇U of the tide-generating potential U. In this case, instead of (12.2) we have:

$$\frac{d\mathbf{V}}{dt} = -\nabla U - \frac{1}{\rho}\nabla p - 2\Omega \times \mathbf{V} + \mathbf{g} + \mathbf{F} \tag{12.3}$$

Let us choose a spherical coordinate system whose coordinate lines are the geocentric radiuses R, meridians φ and parallels λ. Let the unit vectors $\mathbf{e}_R$, $\mathbf{e}_\varphi$, $\mathbf{e}_\lambda$ at an observation point be directed vertically upward, northward and eastward, respectively. Then:

$$\nabla U = T_R\mathbf{e}_R + T_\varphi\mathbf{e}_\varphi + T_\lambda\mathbf{e}_\lambda = \frac{\partial U}{\partial R}\mathbf{e}_R + \frac{\partial U}{R\partial\varphi}\mathbf{e}_\varphi + \frac{\partial U}{R\cos\varphi\partial\lambda}\mathbf{e}_\lambda \tag{12.4}$$

where T are the projections of the tide-generating forces, and R is the Earth's radius. Let us now find each of the projections by differentiating the expression for U(5.14).

Bearing in mind that a change in the hour angle H is equivalent to a change in longitude λ, we have:

$$\begin{aligned} T_R &= C\left[\cos^2\varphi\cos^2\delta\cos 2H+\sin 2\varphi\sin 2\delta\cos H+3\left(\sin^2\varphi-\frac{1}{3}\right)\left(\sin^2\delta-\frac{1}{3}\right)\right] \\ T_\varphi &= C\left[-\frac{1}{2}\sin 2\varphi\cos^2\delta\cos 2H+\cos 2\varphi\sin 2\delta\cos H+\frac{3}{2}\sin 2\varphi\left(\sin^2\delta-\frac{1}{3}\right)\right] \\ T_\lambda &= C[-\cos\varphi\cos^2\delta\sin 2H-\sin\varphi\sin 2\delta\sin H] \\ &= C[\cos\varphi\cos^2\delta\cos(2H+90^\circ)+\sin\varphi\sin 2\delta\cos(H+90^\circ)] \end{aligned} \tag{12.5}$$

Here, C is the multiplier related to the geocentric distance of a perturbing body, the elastic properties and structure of the Earth. For a perfectly rigid Earth, $C = G\frac{2R}{a^2}\left(\frac{c}{d}\right)^3$, where G is the Doodson constant. It is clear from the expressions (12.4) that the projections T_R and T_φ contain the same functions of coordinates δ, H and the distance c/d of a perturbing body as the potential U. Consequently, they are related to the same tidal waves as the potential U. The projection T_λ (along parallel) has only the diurnal and semidiurnal waves whose phases are 90° greater than the phases of the potential U; and it has no long-period terms. Coefficients C of all the projections and the functions of latitude that are in the projections T_φ and T_λ are different from those of the potential U.

Also, an expansion of every projection T of the tide-generating forces can be obtained by direct calculation of the components of the gradient of the series (5.20):

$$T = \nabla\left\{\sum_{n=2}^{\infty}\sum_{m=0}^{n} P_{nm}(\sin\varphi)\sum_j C_{nmj}\cos\left[\omega_{nmj}t+\beta_{nmj}+m\lambda+(n-m)\frac{\pi}{2}\right]\right\} \tag{12.6}$$

We hope that the account for the lunisolar tidal forces in the equations of motion of the global atmosphere and oceans models will allow us to radically extend the time limits of predictability of atmospheric processes. It will be possible to do successful multimonth weather forecasts.

13
Conclusion

We are living in a time of rapid scientific progress. Undoubtedly, the problems solved in the book, as well as the developed models will soon be unable to satisfy the observation accuracy requirements that are becoming increasingly stringent and a new clarification will be required. Note that over the last twenty years new methods of space geodesy and astronomy have increased the accuracy of determination of the universal time, the coordinates of the Earth's pole and nutation corrections 100-fold. Nowadays the accuracy is $0''.0001$ of arc for the pole coordinates and nutation and 0.000005 s for corrections to the UT1 scale, which corresponds to several millimeters on the Earth surface.

Besides these fantastic successes in monitoring the Earth's rotation rate instabilities, a grandiose multiyear project of the National Center for Environmental Prediction (NCEP) and the National Center for Atmospheric Research (NCAR) is completed, and all meteorological and aerologic observation from 1948 to the present time reanalyzed. Based on NCEP/NCAR reanalyses, David Salstein calculated the values of the components of the atmospheric angular momentum and effective functions of the atmospheric angular momentum with a 6-hour step for a period from 1948 to the present time. Recently, effective functions of the angular momentum of the ocean have been calculated as well.

These innovations give an ample opportunity to study instabilities of the Earth's rotation rate, atmospheric and ocean processes, first, within the diurnal periods. Using these data, the author has already discovered diurnal nutation of the vector of the angular momentum of the atmosphere with a wide spectrum of the lunisolar tidal oscillations transformed by the atmosphere.

Comparison between the time series of the Earth's rotation rate, which were obtained using modern methods, and semiempirical models of ocean tides shows that oceans play a dominating role in excitation of diurnal and semidiurnal oscillations of the Earth's rotation rate and monthly variations in polar motion. However, large high-frequency variations in the Earth's rotation still remain unexplained.

Fundamental discoveries are potentially possible in researches into decadal fluctuations. The point is that the decadal fluctuations in the Earth's rotation are closely connected with epoch changes in atmospheric circulation, climatic

The Interaction Between Earth's Rotation and Geophysical Processes. Nikolay S. Sidorenkov

ISBN: 978-3-527-40875-7

characteristics, the condition of ice sheets in Antarctica and Greenland, and so on. However, no significant angular momentum exchange between the Earth and the atmosphere has been revealed. The angular momentum is believed to be redistributed between the mantle and the liquid core. If so, the core controls the atmospheric processes, climate changes and glaciers, which is rather improbable. These contradictions are eliminated if it is assumed that there is a third reason, "generalized tides", which simultaneously affects processes in both the Earth's core and the climatic system. Resolving these inconsistencies promises revolutionary discoveries and a breakthrough in understanding geophysical processes.

The problem of the interannual variations is waiting to be solved as well. It is still unclear why the spectrum of variations in the polar motion is similar to that of the quasibiennial oscillation in equatorial stratospheric wind; why the cycle of the El Niño–Southern Oscillation (ENSO), and that of the quasibiennial oscillation (QBO), is a multiple of Chandler's period, and how those oscillations are connected with free and forced nutation of the Earth and the lunisolar tides.

Some instabilities in the Earth's rotation are caused by the tidal forces, some, by processes in geospheres. Reflecting these processes, the instabilities can serve as their indicators. Based on variations in the Earth's rotation, we can track variations in zonal tides, changes in the angular momentum of winds, changes in global and hemisphere temperature of air, epoch change in atmospheric circulation, evolution of ice sheets and other climatic characteristics. The nature has created this integrated index of natural processes and we should take advantage of it. The reader has found some ways in this book, others he/she can obtain by studying how some particular data are correlated with the series of parameters of the Earth's rotation given in the Appendices and at http://www.hpiers.obspm.fr/eop-pc/.

Appendices

The Interaction Between Earth's Rotation and Geophysical Processes. Nikolay S. Sidorenkov

ISBN: 978-3-527-40875-7

Appendix A

A.1 Spherical Analysis

We use spherical analysis, that is, expansion into a series in terms of spherical functions, to solve many problems we consider here. It is known that the continuous function of the colatitude θ and east longitude λ can be uniquely expanded into a series by surface spherical harmonics:

$$\begin{aligned} f(\theta,\lambda) &= \sum_{n=0}^{\infty}\sum_{m=0}^{n}(a_{nm}\cos m\lambda + b_{nm}\sin m\lambda)P_{nm}(\theta) \\ &= a_{00} + a_{10}\cos\theta + (a_{11}\cos\lambda + b_{11}\sin\lambda)\sin\theta + a_{20}\left(\frac{3}{2}\cos^2\theta - \frac{1}{2}\right) \\ &+ (a_{21}\cos\lambda + b_{21}\sin\lambda)\cdot\frac{3}{2}\sin 2\theta + (a_{22}\cos 2\lambda + b_{22}\sin 2\lambda)3\sin^2\theta + \ldots \end{aligned} \tag{A.1}$$

Here, a_{nm} and b_{nm} are the coefficients of the spherical harmonic of degree n and order m, and $P_{nm}(\theta)$ are the associated Legendre functions calculated by the formula:

$$P_{nm}(\theta) = \frac{\sin^m\theta}{2^n\cdot n!}\frac{d^{n+m}(\cos^2\theta - 1)^n}{(d\cos\theta)^{n+m}} \tag{A.2}$$

Note that somewhat different definition of the Legendre function is accepted in a number of works (Jeffreys and Swirles, 1966; Munk and Macdonald, 1960):

$$\frac{(n-m)!}{n!}P_{nm} \tag{A.3}$$

The surface spherical functions are the eigenfunctions for a sphere. They look like waves (Figure A.1). The spherical functions are referred to as either the zonal functions (if they depend only on the latitude) or the sectoral functions (if they depend only on the longitude) or the tesseral functions (if they depend on both the latitude and longitude). The greater the degree of the function, the smaller the area of the sphere occupied by each wave and the finer the details described by the function.

The Interaction Between Earth's Rotation and Geophysical Processes. Nikolay S. Sidorenkov

ISBN: 978-3-527-40875-7

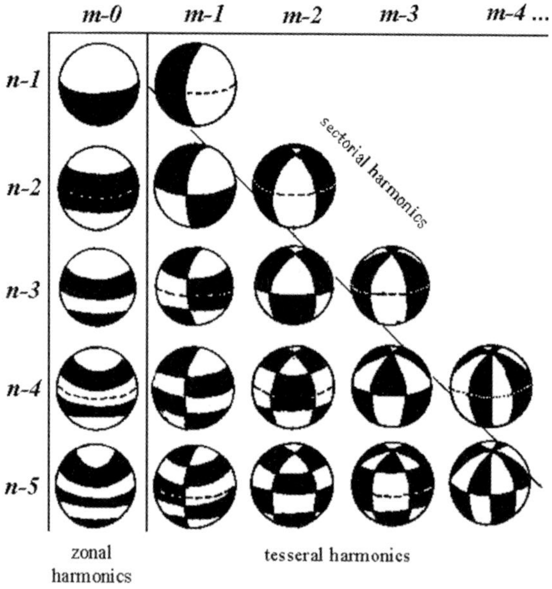

Figure A.1 Types of surface spherical functions.

The spherical functions are orthogonal, that is, the surface integral of the product of two spherical functions of different degrees over the spherical surface is equal to zero. Whereas the integral of the squared spherical functions is:

$$\frac{1}{4\pi}\int_0^{\pi}\int_0^{2\pi} P_{nm}^2 \begin{pmatrix} \cos 2\,m\lambda \\ \sin 2\,m\lambda \end{pmatrix} dS = \frac{(n+m)!}{(2n+1)(n-m)!(2-\delta_{om})} \qquad (A.4)$$

where $\delta_{om} = 1$, if $m = 0$, and $\delta_{om} = 0$, if $m \neq 0$. It is clear that the value of this integral quickly increases with the increasing index m. This results in some difficulties in calculating the coefficients a_{nm} and b_{nm}. The normalized Legendre functions $\bar{P}_{nm}$ are frequently used to avoid the difficulties:

$$\bar{P}_{nm} = N_{nm} P_{nm} \qquad (A.5)$$

where

$$N_{nm}^2 = \frac{(2n+1)(n-m)!(2-\delta_{om})}{(n+m)!} \qquad (A.6)$$

In this case, we have for any indexes n and m:

$$\frac{1}{4\pi}\int_0^{\pi}\int_0^{2\pi} \bar{P}_{nm}^2 \begin{pmatrix} \cos 2\,m\lambda \\ \sin 2\,m\lambda \end{pmatrix} \mathrm{d}S = 1 \qquad (A.7)$$

Using the orthogonality condition and Equation A.7 for the normalized Legendre functions, we obtain the expressions to determine the normalized coefficients of the expansion, $\bar{a}_{nm}$ and $\bar{b}_{nm}$:

$$
\begin{aligned}
\bar{a}_{no} &= \frac{1}{4\pi}\int_0^{\pi}\int_0^{2\pi} f(\theta,\lambda)\bar{P}_{no}\mathrm{d}S;\ b_{no}=0 \\
\begin{pmatrix}\bar{a}_{nm}\\ \bar{b}_{nm}\end{pmatrix} &= \frac{1}{4\pi}\int_0^{\pi}\int_0^{2\pi} f(\theta,\lambda)\bar{P}_{nm}\begin{pmatrix}\cos m\lambda\\ \sin m\lambda\end{pmatrix}\mathrm{d}S
\end{aligned}
\tag{A.8}
$$

Obviously, if we used the non-normalized Legendre functions like (A.2) or (A.3) we would have different values of the coefficients a_{nm} and b_{nm}. This should be borne in mind when comparing the results of the calculation. The coefficients $\bar{a}_{nm}$ and $\bar{b}_{nm}$ of the expansion of the function $f(\theta,\lambda)$ into a series in terms of the normalized spherical functions are related to the coefficients a_{nm}, b_{nm} of the expansion of the same function into a series (A.1) in terms of the general (non-normalized) spherical functions like (A.2); this relation is as follows:

$$a_{nm} = N_{nm}\bar{a}_{nm};\quad b_{nm} = N_{nm}\bar{b}_{nm} \tag{A.9}$$

As expansions into a series in terms of the spherical functions are frequently used hereafter, Table A.1 lists the values of the following factors: $(N_{nm})^2$, Legendre functions P_{nm} and the polynomials resulted from the multiplication of P_{nm} by either $r^n\cos m\lambda$ or $r^n\sin m\lambda$, for various n and m from 0 to 4.

Table A.1 Values of normalizing factors N_{nm}^2, Legendre functions $P_{nm}(\cos\theta)$ and polynomials for various indexes n and m.

n	*m*	N_{nm}^2	P_{nm}	Polynomials
0	0	1	1	1
1	0	3	$\cos\theta$	z
1	1	3	$\sin\theta$	x; y
2	0	5	$\frac{3}{2}\cos^2\theta-\frac{1}{2}$	$\frac{1}{2}(2z^2-x^2-y^2)$
2	1	$\frac{5}{3}$	$\frac{3}{2}\sin 2\theta$	$3zx$; $3zy$
2	2	$\frac{5}{12}$	$3\sin^2\theta$	$3(x^2-y^2)$; $6xy$
3	0	7	$\frac{5}{2}\cos^3\theta-\frac{3}{2}\cos\theta$	$\frac{1}{2}z(2z^2-3x^2-3y^2)$
3	1	$\frac{7}{6}$	$\frac{3}{2}\sin\theta(5\cos^2\theta-1)$	$\frac{3}{2}x(4z^2-x^2-y^2)$; $\frac{3}{2}y(4z^2-x^2-y^2)$;
3	2	$\frac{7}{60}$	$15\sin^2\theta\cos\theta$	$15z(x^2-y^2)$; $30\,xyz$
3	3	$\frac{7}{360}$	$15\sin^3\theta$	$15(x^3-3xy^2)$; $15(3x^2y-y^3)$
4	0	9	$\frac{1}{8}(35\cos^4\theta-30\cos^2\theta+3)$	$\frac{1}{8}\left[8z^4-24z^2(x^2+y^2)+3(x^2+y^2)^2\right]$
4	1	$\frac{9}{10}$	$\frac{5}{2}\sin\theta(7\cos^3\theta-3\cos\theta)$	$\frac{5}{2}x\left[4z^3-3z(x^2+y^2)\right]$; $\frac{5}{2y}\left[4z^3-3z(x^2+y^2)\right]$
4	2	$\frac{1}{20}$	$\frac{15}{2}\sin^2\theta(7\cos^2\theta-1)$	$\frac{15}{2}(x^2-y^2)(6z^2-x^2-y^2)$; $15xy(6z^2-x^2-y^2)$
4	3	$\frac{1}{280}$	$105\sin^3\theta\cos\theta$	$105z(x^3-3xy^2)$; $105z(3x^2y-y^3)$
4	4	$\frac{1}{2240}$	$105\sin^4\theta$	$105(x^4-6x^2y^2+y^4)$; $420\,xy\,(x^2-y^2)$

Expansion into a series in terms of the spherical functions allows the main features of the field under study to be compactly described. It is easy to see from (A.8), for example, that the coefficients in expression (A.1) have the following physical interpretation. a_{00} is the average value of the function f on the globe; a_{10} is the difference in the function f between the Northern and Southern Hemispheres; a_{20} is the difference in the function f between the low and high latitudes; a_{11}, b_{11} and a_{22}, b_{22} are the sectoral perturbations with wavelengths of 360° and 180°, respectively; a_{21} and b_{21} are the differences in the distribution of the function f in comparison with the rotationally symmetric distribution, and so forth.

Appendix B

B.1
The Figure of the Earth

If the Earth were a liquid rotating planet, its figure would have a gravitational equipotential surface, with geopotential U_0 consisting of gravitational potential and potential of centrifugal forces at the surface. It is clear that in that case the Earth's figure would depend on the distribution of density inside the planet and its angular rotation velocity. Therefore, though the real Earth is not strictly in hydrostatic equilibrium, its figure is determined on the basis of the condition:

$$W + \Phi = U_0 = \text{const} \tag{B.1}$$

The gravitational external potential of the Earth is determined (Zharkov *et al.*, 1996; Zharkov and Trubitsyn, 1980)

$$W = \frac{\gamma M}{r}\left[1 + \sum_{n=2}^{\infty}\sum_{m=0}^{n}\left(\frac{R_e}{r}\right)^n (a_{nm}\cos_{m\lambda} + b_{nm}\sin m\lambda)P_{nm}\right] \tag{B.2}$$

Where γ is the gravitational constant; r is distance; M and R_e are the mass and equatorial radius of the Earth; a_{nm} and b_{nm} are the Stokes coefficients.

The potential of centrifugal forces is equal to

$$\Phi = \frac{\omega^2 R^2}{3} - \frac{\omega^2 R^2}{3} P_{20}(\cos\theta) \tag{B.3}$$

This figure (Figure B.1) is called the geoid.

Let us define the geoid oblateness. If expressions (B.2) and (B.3) are taken into consideration, expression (B.1) gives the following formulae for U_0 at the equator ($R = R_e$, $\theta = 90°$) and at the poles ($R = R_P$, $\theta = 0$), respectively, with an accuracy up to the expansion terms with $n > 2$:

$$U_0 = \frac{\gamma M}{R_e} - \frac{\gamma M}{2R_e} a_{20} + \frac{\Omega^2 R_e^2}{2}, \quad \text{and} \quad U_0 = \frac{\gamma M}{R_P} + \frac{\gamma M R_e^2}{R_P^3} a_{20}$$

The Interaction Between Earth's Rotation and Geophysical Processes. Nikolay S. Sidorenkov

ISBN: 978-3-527-40875-7

Figure B.1 Photograph of the Earth and the Moon from space.

Subtracting the latter equation from the former and bearing in mind that $R_e \approx R_P$, we have the following expression for the geoid oblateness, with an accuracy up to the terms of the first order with respect to α:

$$\alpha = \frac{R_e - R_P}{R_e} = -\frac{3}{2}a_{20} + \frac{q}{2} + 0(\alpha^2) \tag{B.4}$$

Here, a_{20} is the potential expansion coefficient; $q = \frac{\Omega^2 R_e^3}{\gamma M}$ is the parameter equal to the ratio of the centrifugal acceleration to the gravitational acceleration at the equator. Bearing in mind that for the liquid planet with the density distribution similar the real Earth's

$$a_{20} = -\frac{k_f \Omega^2 R_e^3}{3\gamma M}$$

we obtain, from (B.4), the following formula for the Earth oblateness:

$$\alpha = (1 + k_f)\frac{1}{2}\frac{\Omega^2 R_e^3}{\gamma M} = h_f \frac{1}{2}\frac{\Omega^2 R_e^3}{\gamma M} \tag{B.5}$$

Here, k_f and h_f are the Love numbers of the second order for the liquid Earth. The same formula for the oblateness can be also derived if the following considerations are taken into account. Under the action of the centrifugal potential, φ the surface of a liquid planet changes with respect to its center by as much as:

$$h_f \frac{\varphi}{g} = -h_f \frac{\Omega^2 R^2}{3g} P_{20} \tag{B.6}$$

where g is the gravitational acceleration. On the other hand, it is known (Zharkov and Trubitsyn, 1980; Jeffreys, 1959) that the equation of the surface of a liquid rotating planet in hydrostatic equilibrium is as follows:

$$R = R_e\left(1 - \frac{\alpha}{3} - \frac{2}{3}\alpha P_{20}\right) + 0(\alpha_f^2) \qquad \text{(B.7)}$$

Here, $R_e\left(1 - \frac{\alpha}{3}\right)$ is the mean radius of the planet; $\frac{2}{3}\alpha R_e P_{20}$ is the radial displacement of the planet surface. Equating the radial displacements in formulae (B.6) and (B.7), we obtain the desired formula (B.5).

Let us compare the values of the parameters of the Earth's figure obtained from observation and theoretical data. Analysis of satellite observations gives the estimate of the real Earth's surface oblateness (Zharkov and Trubitsyn, 1980):

$$\alpha = \frac{1}{298.257}$$

The corresponding "secular" Love numbers are as follows:

$$h_S = \frac{2\alpha}{q} = 1.934;\ k_S = h_S - 1 = 0.934 \qquad \text{(B.8)}$$

If we knew a law of the density distribution inside the Earth, we could calculate the oblateness α_f, which the Earth would have if it were in hydrostatic equilibrium (Zharkov and Trubitsyn, 1980). To illustrate this, let us estimate the hydrostatic oblateness α_f for two extreme models of the Earth's structure: a homogeneous Earth and an Earth whose total mass is concentrated in its center. In the former case, we have $a_{20} = -\frac{2}{5}\alpha$ and substitute it into (B.4), which gives $\alpha_f = \frac{5}{4}q = \frac{1}{231}$ and, accordingly, $h_f = \frac{5}{2}$. In the latter case (the total mass is concentrated in the center of the Earth), $a_{20} = 0$, $\alpha_f = \frac{q}{2} = \frac{1}{577}$ and, accordingly, $h_f = 1$. Obviously, the true values of the hydrostatic oblateness and the Love numbers are between these limits. In accordance with the hydrostatic theory, the oblateness of a hydrostatically equilibrated planet, with a accuracy up to the terms of the first order of smallness by α, is related to the polar moment of inertia as follows (Jeffreys, 1959):

$$\alpha_f = \frac{5}{2}q\left[1 + \frac{25}{4}\left(1 - \frac{3}{2}\frac{N_{33}}{MR_e^2}\right)^2\right]^{-1} \qquad \text{(B.9)}$$

Formula (B.9) gives the approximate value of α_f that should be refined. Jeffreys (1963) calculated the necessary correction by using the numerical method based on the Earth's density distribution. As a result, he obtained the following value of hydrostatic oblateness:

$$\alpha_f = \frac{1}{299.67} \qquad \text{(B.10)}$$

The corresponding Love numbers are:

$$h_f = 1.925,\ k_f = 0.925 \qquad \text{(B.11)}$$

Comparison of the parameters of the liquid rotating planet in hydrostatic equilibrium (B.10) and (B.11) with the parameters of the real Earth $k_S = 0.936$

and (B.8) reveals some differences. In particular, the actual oblateness is 0.5% more than the hydrostatic one. The Love numbers k_S and h_S are greater than the Love numbers k_f and h_f of the liquid Earth. Some researches (Zharkov and Trubitsyn, 1980; Mac Donald, 1963; McKenzie, 1966) make much of these differences. They relate the differences to the delay in changing the oblateness of the Earth's in response to the Earth's rotation slowing down.

It is known that tidal friction regularly slows down the Earth's rotation at a rate of $\dot{\Omega} = -55 \times 10^{-23}\ \mathrm{s}^{-2}$ (the length of a day increases as much as 0.002 s per century). From Equation (B.5), it is easy to obtain an expression for the delay τ in changing the Earth's oblateness in response to changes in the angular velocity:

$$\tau = \frac{\alpha - \alpha_f}{\alpha} \cdot \frac{\Omega}{2\dot{\Omega}} \tag{B.12}$$

Substituting the values of α and α_f of (B.8) and (B.10) gives $\tau = 10^7$ years, that is, the current oblateness of the Earth is the hydrostatic oblateness of the planet with the angular velocity equal to the angular velocity of the Earth 10^7 years ago. Based on the delay τ, we can estimate viscosity η of the lower mantle because the time of relaxation τ of the nonequilibrium part of the equatorial swell is proportional to viscosity η and inversely proportional to the shear modulus $\mu \approx 10^{11}$ Pa:

$$\eta \approx \tau\mu \approx 10^{25}\ \mathrm{Pa\,s} = 10^{26}\ \mathrm{Poise} \tag{B.13}$$

If the Earth was strictly in hydrostatic equilibrium, all the coefficients a_{nm} and b_{nm} in the expansion of gravitational potential W(B.2) would be zero except for the zonal coefficients a_{n0} with even orders n, that is:

$$W_f = \frac{\gamma M}{r}\left\{1 + \frac{R_e^2}{r^2} a_{20} P_{20} + \frac{R_e^4}{r^4} a_{40} P_{40} + \frac{R_e^6}{r^6} a_{60} P_{60} + \ldots\right\} \tag{B.14}$$

And, as follows from the theory (Zharkov and Trubitsyn, 1980; Jeffreys, 1959), the values of the even zonal coefficients should be proportional to $\alpha^{n/2}$:

$$a_{20} \approx \alpha,\ a_{40} \approx \alpha^2,\ a_{60} \approx \alpha^3, \ldots,\ a_{no} \approx \alpha^{n/2}$$

Satellite observations do not confirm this law. All the zonal coefficients, beginning with a_{30}, and many of the tesseral coefficients are approximately of the same order of magnitude, being equal to something between 1 and 9 multiplied by 10^{-6}, that is, they are of the same order of magnitude as α^2. An exception is $a_{20} \approx \alpha$. Thus, the gravitational field of the Earth deviates from the gravitational field of a hydrostatically balanced planet by a value comparable with the squared oblateness. These deviations indicate that the Earth's figure deviates by the value of the same order of magnitude ($\alpha^2 R \approx 70$ m) from the equilibrium figure of a rotating liquid planet.

As is generally known, the deviations of the Earth's figure and gravitational field are caused by anomalies of mass distribution at the surface of and inside the Earth. The difference in the Earth's figure that is related to the difference between the actual oblateness and hydrostatic one makes $\delta(R_e - R_P) = (\alpha - \alpha_f)R_e = 100$ m, that is, it is of the same order of magnitude as the nonequilibria described by other terms of the expansion. It is clear that abnormal mass distribution (for example, the thawing of the ice sheets covering Antarctica and Greenland) may cause the difference as well. Therefore, there is no sufficient reason for believing that the differences $\alpha - \alpha_f$, $h - h_f$, $k - k_f$ are caused by the delay in changes in the oblateness of the Earth's as a result of the secular slowing down of the Earth's rotation.

Thus, with a relative accuracy up to the order of the squared oblateness, the real Earth's form does not differ from the equivalent model of a liquid rotating planet, and the secular Love numbers are equal to the Love numbers for the liquid Earth.

Based on the differences in the terms of the expansions of the gravitational potential between the real Earth and a hydrostatically equilibrium planet, we can estimate the gravitational energy caused by the Earth's nonequilibrium. It turns out to be equal 10^{24} J (Zharkov and Trubitsyn, 1980). Note that the gravitational energy caused by the Earth nonequilibrium to support the irregularity in the speed of the Earth's rotation would only last 100 years because the power of observed fluctuations of the angular velocity is 10^{15} W.

Appendix C

Acronyms

ν	The Earth's rotation dimensionless velocity (see Equation 2.1)
l.o.d.	The length of the terrestrial day
cpd	cycle per day
dB	decibel
G	gauss
hPa	hectopascal
hr	hour
J	joule
K	kelvin
ms	millisecond(s)
μs	microsecond(s)
N	newton
P	poise
Pa	pascal
rad	radian
s	second
St	stokes
W	watt
yr	year
AAM	Atmospheric Angular Momentum
CIO	Conventional International Origin
CPC	Climate Prediction Cente
CW	Chandler Wobble
DORIS	Doppler Orbit determination and Radiopositioning Integrated on Satellite
ECMWF	European Centre for Medium-Range Weather Forecasts
ENSO	El Niño-Southern Oscillation
EOP	Earth Orientation Parameters
ESP	Elementary Synoptic Process
FCN	Free Core Nutation

The Interaction Between Earth's Rotation and Geophysical Processes. Nikolay S. Sidorenkov

ISBN: 978-3-527-40875-7

FTHE	Heat engine of the first type
GLONASS	Global Orbiting Navigation Satellite System, Russia
GPS	Global Positioning System
IAU	International Astronomical Union
IERS	International Earth Rotation and Reference Systems Service (formerly: International Earth Rotation Service)
IHHE	Inter-hemispheric heating engine
ILS	International Latitude Service
IPMS	International Polar Motion Service
ITCZ	Intertropical Convergence Zone
IUGG	International Union of Geodesy and Geophysics
JPL	Jet Propulsion Laboratory
LLR	Lunar Laser Ranging
LOD	Length of Day
MJD	Modified Julian Day
NCAR	U.S. National Center for Atmospheric Research
NCEP	U.S. National Centers for Environmental Prediction
NMC	National Meteorological Center (replaced by NCEP)
NSP	natural synoptic period
OAM	oceanic angular momentum
QBO	Quasi-Biennial Oscillation
SI	Système International (International System of Units)
SLR	Satellite Laser Ranging
TAI	Temps Atomique International (International Atomic Time)
TCB	Temps-coordonnée barycentrique (Barycentric Coordinate Time)
TCG	Temps-coordonnée géocentrique (Geocentric Coordinate Time)
TDB	Barycentric Dynamical Time
TDT	Terrestrial Dynamical Time
TGB	Tide Generating Potential
UKMO	U.K. Meteorological Office
UT, UT0, UT1, UT2	Universal Time
UTC	Coordinated Universal Time
VLBI	Very Long Baseline Interferometry

Appendix D

Table D.1 Series of the observed mean pole coordinates ν_1 and ν_2, the Earth's rotation velocity ν_3, and the calculated increments of specific water mass in Ocean (ζ_O), Antarctica (ζ_A), Greenland (ζ_G), and rest of land (ζ_C), g/cm^2.

Year	$\nu_1 \times 10^9$ (rad)	$\nu_2 \times 10^9$ (rad)	$\nu_3 \times 10^{10}$	ζ_O	ζ_A	ζ_G	ζ_C
1891	−102	165	73	−11	232	−49	7
1892	−78	131	33	−5	104	−15	3
1893	−87	126	−17	2	−56	31	−2
1894	−78	112	−67	10	−217	76	−7
1895	−78	82	−118	18	−381	122	−12
1896	−73	73	−168	26	−541	168	−17
1897	−73	63	−219	34	−705	215	−22
1898	−68	63	−269	42	−866	261	−28
1899	−39	78	−319	50	−1026	306	−33
1900	−19	53	−370	58	−1190	350	−38
1901	−5	34	−431	68	−1386	405	−45
1902	0	15	−471	74	−1514	441	−49
1903	5	0	−478	75	−1536	446	−49
1904	0	−48	−468	74	−1504	434	−48
1905	−39	−97	−455	71	−1462	423	−46
1906	−29	−87	−451	71	−1450	419	−46
1907	−19	−102	−452	71	−1453	418	−46
1908	−10	−145	−457	72	−1468	420	−47
1909	−29	−204	−458	72	−1471	419	−46
1910	−39	−233	−444	70	−1426	405	−45
1911	−5	−242	−416	65	−1336	375	−42
1912	−14	−291	−379	59	−1217	339	−38
1913	−24	−335	−349	54	−1120	310	−35
1914	−15	−378	−327	51	−1049	286	−32
1915	5	−378	−311	48	−997	269	−30
1916	34	−398	−288	45	−923	243	−28
1917	29	−427	−264	41	−845	220	−25
1918	82	−388	−239	37	−764	193	−23
1919	141	−301	−232	36	−742	186	−23
1920	208	−252	−231	36	−738	181	−23

(continued)

The Interaction Between Earth's Rotation and Geophysical Processes. Nikolay S. Sidorenkov

ISBN: 978-3-527-40875-7

Table D.1 (*Continued*).

Year	$\nu_1 \times 10^9$ (rad)	$\nu_2 \times 10^9$ (rad)	$\nu_3 \times 10^{10}$	ζ_O	ζ_A	ζ_G	ζ_C
1921	276	−281	−226	35	−721	168	−23
1922	310	−305	−204	32	−649	143	−20
1923	330	−301	−174	27	−553	113	−17
1924	271	−344	−143	22	−453	87	−14
1925	262	−368	−110	17	−347	56	−10
1926	208	−398	−74	11	−232	26	−6
1927	194	−378	−41	6	−126	−1	−3
1928	204	−398	−21	3	−62	−22	−1
1929	204	−499	−16	2	−45	−32	−0
1930	204	−567	−20	2	−58	−32	−0
1931	155	−635	−18	2	−51	−33	0
1932	150	−669	−9	0	−22	−43	1
1933	121	−713	3	−1	15	−54	2
1934	87	−727	9	−2	34	−57	3
1935	73	−732	7	−2	28	−54	3
1936	87	−669	−8	0	−20	−37	1
1937	145	−688	−30	4	−90	−24	−0
1938	165	−679	−60	8	−186	2	−4
1939	189	−688	−92	13	−288	29	−7
1940	199	−679	−125	19	−394	59	−10
1941	228	−654	−151	23	−478	82	−13
1942	267	−601	−164	25	−519	93	−15
1943	271	−572	−166	25	−526	96	−15
1944	305	−596	−160	24	−506	86	−15
1945	310	−616	−159	24	−503	84	−14
1946	339	−591	−161	25	−509	84	−15
1947	373	−572	−166	25	−525	86	−15
1948	412	−591	−163	25	−514	79	−15
1949	451	−621	−153	23	−482	64	−14
1950	465	−654	−138	21	−433	47	−12
1951	427	−664	−128	19	−401	41	−11
1952	359	−668	−122	18	−383	40	−10
1953	296	−693	−120	18	−377	44	−10
1954	276	−708	−115	17	−361	41	−9
1955	267	−762	−116	17	−364	40	−9
1956	281	−800	−86	12	−267	8	−6
1957	325	−834	−135	20	−424	48	−11
1958	349	−887	−165	25	−520	71	−14
1959	344	−945	−146	22	−458	50	−12
1960	305	−994	−140	21	−439	46	−11
1961	276	−984	−128	19	−401	38	−10
1962	252	−979	−148	22	−466	59	−12
1963	208	−1028	−173	26	−547	84	−14
1964	184	−1067	−223	34	−707	131	−19
1965	165	−1082	−255	39	−810	162	−22
1966	131	−1115	−281	43	−894	188	−25

Table D.1 (*Continued*).

Year	$\nu_1 \times 10^9$ (rad)	$\nu_2 \times 10^9$ (rad)	$\nu_3 \times 10^{10}$	ζ_O	ζ_A	ζ_G	ζ_C
1967	102	−1159	−273	41	−869	181	−24
1968	8	−1139	−289	44	−921	206	−25
1969	−39	−1086	−308	47	−984	231	−27
1970	−34	−1110	−315	48	−1006	236	−28
1971	−5	−1125	−335	51	−1070	251	−30
1972	39	−1139	−362	55	−1156	271	−33
1973	68	−1154	−353	54	−1126	259	−32
1974	82	−1183	−314	48	−1001	220	−28
1975	78	−1236	−311	47	−991	214	−27
1976	82	−1290	−336	51	−1071	234	−30
1977	87	−1319	−321	49	−1022	218	−28
1978	97	−1324	−334	51	−1064	229	−29
1979	112	−1357	−301	46	−957	195	−26
1980	131	−1416	−267	40	−847	158	−22
1981	150	−1440	−249	37	−789	138	−20
1982	165	−1440	−250	37	−792	138	−20
1983	179	−1450	−263	39	−834	148	−22
1984	199	−1450	−175	26	−550	64	−13
1985	218	−1454	−168	24	−528	55	−12
1986	228	−1464	−142	20	−444	29	−9
1987	213	−1503	−157	23	−492	43	−10
1988	208	−1532	−152	22	−476	37	−10
1989	223	−1551	−177	26	−556	58	−12
1990	233	−1595	−226	33	−713	100	−17
1991	228	−1619	−235	35	−742	108	−18
1992	204	−1614	−257	38	−813	131	−20
1993	179	−1668	−274	41	−868	146	−22
1994	160	−1682	−254	38	−803	129	−20
1995	170	−1677	−267	40	−845	140	−21
1996	179	−1619	−211	31	−665	90	−16
1997	160	−1605	−213	31	−672	94	−16
1998	160	−1605	−159	23	−498	44	−11
1999	160	−1605	−114	16	−353	−4	−6
2000	216	−1644	−84	11	−257	−33	−3
2001	217	−1644	−66	8	−199	−50	−1
2002	227	−1648	−55	7	−160	−63	0
2003	245	−1668	−32	3	−89	−86	3
2004	240	−1673	−37	4	−105	−81	2
2005	256	−1687	−50	6	−147	−72	1
2006	280	−1696	−95	13	−291	−32	−4
2007	280	−1696	−99	14	−304	−28	−4

Table D.2 Monthly mean of the average zonal wind velocity $\bar{u}$ in the layer of 19 to 31 km, m/s.

Year	I	II	III	IV	V	VI	VII	VIII	IX	X	XI	XII
1954	−8.5	−9.4	−10.3	−10.1	−10.1	−13.1	−12.1	−10.8	−9.6	−6.2	−3.0	−1.3
1955	0.4	2.8	5.0	7.4	9.5	10.1	10.9	9.5	8.4	6.3	4.1	1.3
1956	−0.8	−3.9	−7.3	−8.9	−10.2	−12.6	−15.5	−17.0	−17.8	−15.2	−12.7	−12.7
1957	−12.6	−8.4	−3.3	0	4.1	6.9	10.0	10.6	10.5	9.5	8.4	8.2
1958	8.1	2.3	−2.6	−4.7	−6.1	−9.1	−11.4	−12.6	−13.8	−17.5	−18.3	−17.1
1959	−15.2	−12.7	−12.0	−10.2	−6.4	−4.7	3.2	6.7	11.0	8.9	8.7	4.9
1960	2.1	−0.9	−3.0	−3.5	−3.7	−7.5	−6.8	−8.9	−10.8	−11.4	−10.7	−9.7
1961	−9.1	−5.7	−2.3	1.1	6.9	7.8	9.0	8.1	7.3	6.5	5.6	4.0
1962	2.2	−0.5	−3.2	−5.1	−6.2	−7.4	−8.0	−10.0	−11.9	−12.4	−12.7	−13.1
1963	−14.6	−15.3	−17.4	−14.5	−12.5	−10.2	−7.5	0	1.5	2.9	4.4	2.6
1964	1.2	1.5	1.6	0.9	0.6	−1.6	−3.3	−4.0	−4.3	−4.3	−4.4	−4.5
1965	−4.5	−5.0	−5.2	−6.8	−8.1	−10.9	−14.2	−16.5	−19.0	−20.4	−20.2	−18.9
1966	−17.8	−13.6	−9.5	−5.2	0	4.0	8.6	10.2	11.3	8.9	6.6	5.7
1967	4.7	4.0	3.0	1.5	0.4	−1.9	−3.5	−4.7	−5.8	−8.3	−10.4	−11.8
1968	−12.9	−14.1	−14.8	−14.2	−13.6	−11.3	−9.1	−9.6	−11.2	−10.9	−10.7	−9.4
1969	−6.4	−4.3	−1.6	1.7	5.6	11.4	14.9	14.0	10.7	8.3	5.9	3.1
1970	0.3	−2.4	−5.2	−8.1	−10.4	−13.7	−16.6	−18.4	−19.6	−17.0	−13.5	−10.0
1971	−6.9	−2.2	1.7	5.2	8.3	11.3	11.4	11.6	11.3	10.4	9.0	7.8
1972	7.1	6.6	3.5	−1.7	−6.2	−10.4	−15.1	−18.0	−20.6	−19.9	−16.3	−9.9
1973	−5.2	−0.9	2.6	5.2	8.3	7.5	5.4	0.9	0.3	0.7	4.0	1.7
1974	−0.6	−0.7	−1.4	−2.6	−5.2	−12.3	−14.7	−14.7	−13.9	−17.3	−15.9	−12.5
1975	−8.7	−11.3	−10.7	−6.7	−2.1	5.9	7.8	7.4	9.5	11.8	11.4	9.9
1976	10.6	10.1	10.6	9.4	8.0	2.6	−2.9	−7.1	−6.8	−8.5	−9.7	−10.4
1977	−10.9	−14.2	−17.1	−16.7	−12.9	−14.6	−14.4	−9.4	−7.8	−5.9	−2.3	2.1
1978	5.2	8.3	12.6	10.4	8.5	3.8	0.4	0.8	2.8	4.0	−0.1	−1.0

1979	−1.9	−1.9	−2.4	−4.3	−6.2	−10.9	−17.3	−18.2	−18.5	−18.2	−16.7	−13.3
1980	−8.0	−6.9	−5.4	−1.4	1.2	10.0	4.9	7.8	9.2	11.6	15.4	9.0
1981	6.6	9.0	6.1	3.1	2.3	−3.5	−8.5	−10.6	−12.1	−13.7	−14.9	−13.7
1982	−13.0	−16.1	−15.1	−15.2	−19.6	−17.0	−10.4	−10.6	−6.4	−1.4	3.8	7.6
1983	18.0	12.1	12.8	8.7	6.4	3.5	1.0	−4.3	−7.8	−7.8	−9.5	−11.6
1984	−12.1	−13.3	−13.9	−12.8	−15.4	−18.7	−22.5	−18.5	−17.3	−17.3	−18.0	−13.9
1985	−7.8	0	6.6	8.7	9.9	11.3	12.3	11.3	9.5	8.1	6.6	6.1
1986	6.7	7.3	4.3	5.9	0.4	−3.5	−6.9	−8.7	−9.9	−10.9	−11.4	−12.3
1987	−13.0	−15.2	−15.4	−16.6	−16.5	−15.8	−13.9	−9.5	−6.1	1.4	6.9	11.8
1988	11.6	7.6	3.5	2.1	−0.4	−2.6	−2.9	−2.8	−2.6	−1.9	−2.1	−2.3
1989	−2.6	−3.1	−9.2	−6.9	−9.5	−15.0	−16.2	−16.0	−19.3	−18.1	−15.3	−10.6
1990	−6.3	−4.7	−0.1	2.7	7.4	10.4	6.9	7.5	8.0	7.9	6.2	4.1
1991	2.8	4.8	2.7	−0.3	−5.6	−7.0	−15.5	−15.3	−12.4	−9.7	−10.7	−9.7
1992	−11.6	−10.2	−10.2	−9.9	−12.2	−12.4	−9.2	−6.5	−4.3	−1.1	2.0	7.6
1993	10.7	12.3	11.6	7.5	4.2	−0.9	−3.6	−4.5	−5.2	−7.1	−8.2	−8.6
1994	−10.7	−13.9	−12.6	−15.3	−18.2	−21.1	−20.3	−18.6	−14.7	−7.6	−1.9	5.2
1995	8.7	6.4	5.8	6.8	8.6	7.5	6.0	3.7	1.4	−1.5	−6.2	−6.3
1996	−8.7	−8.8	−10.9	−11.6	−12.4	−17.9	−19.5	−19.7	−18.7	−14.5	−10.5	−4.9
1997	−2.8	−0.9	0.1	6.8	9.2	12.1	11.9	11.2	8.7	4.9	0.4	−4.3
1998	−4.9	−5.7	−8.2	−10.4	−16.8	−18.6	−20.0	−18.4	−15.8	−13.4	−7.7	−3.6
1999	−3.2	0.6	4.2	8.2	7.8	9.0	7.0	7.5	6.3	5.5	2.4	0.9
2000	0.4	0.8	1.7	−0.8	−4.1	−7.7	−9.4	−9.7	−11.2	−12.1	−13.0	−13.9
2001	−10.6	−11.5	−11.2	−12.0	−12.8	−14.8	−16.2	−16.1	−13.7	−8.7	−3.7	0.5
2002	0.4	4.2	10.4	10.3	9.2	5.5	5.1	4.9	4.5	3.5	−1.6	−3.8
2003	−3.9	−3.6	−6.2	−10.2	−11.4	−11.3	−13.5	−14.0	−12.3	−11.8	−11.7	−6.1
2004	−4.4	−1.4	0.8	6.4	9.5	8.5	7.5	5.5	3.3	3.5	−0.2	−1.7
2005	−5.0	−2.9	−1.8	−9.4	−11.6	−16.7	−18.1	−20.4	−20.0	−19.1	−15.2	−12.4
2006	−9.3	−3.5	−0.2	−3.6	1.3	4.3	5.6	4.5	6.4	5.0	2.4	1.4
2007	−1.2	−1.6	−2.9	−6.2	−10.5	−14.8	−17.1	−18.2	−18.4	−17.6	−14.3	−11.2
2008	−9.0	−4.2	−2.1	−0.8	3.6	8.1	8.5	5.9	4.4	3.6	1.3	1.5

References

Allan, R.J., Nicholls, N., Jones, P.D. and Butterworth, I.J. (1991) A further extension of the Tahiti-Darwin SOI, Early ENSO and darwin pressure. *J. Climate.*, **4**, 743–749.

Anderson, D.L. and Hart, R.S. (1978) Q of the Earth. *J. Geophys. Res.*, **83** (B12), 5869–5882.

Astaf'eva, N.M. and Sonechkin, D.M. (1995) Multimasshtabny analis indeksa Yuzhnogo kolebaniya. [Multi-scale analysis of Index of Southern oscillation]. *Doklady Akademii Nauk*, **344** (4), 539–542 [in Russian].

Avsyuk, Yu.N. (1996) *Prilivnyye Sily i Prirodnyye Processy. [Tide-Generating Forces and Natural Processes]*, O.Yu. Schmidt Institut Fiziki Zemli Rossiyskoy Akademii, Nauk, Moscow, p. 188 [in Russian].

Barkin, Yu.V. (1996) K dinamike tverdogo yadra Zemli. [On dynamics of solid core of the Earth]. *Trudy Gosudarstvennogo Astronomicheskogo Instituta, imeni P.K. Shternberga*, **65**, 107–129 [in Russian].

Barkin, Yu.V. (2000) A mechanism of variations of the Earth rotation at different timescales, in *Polar Motion: Historical and Scientific Problems*, ASP Conference Series, **208** (eds S. Dick, D. McCarthy and B. Luzum), ASP, San Francisco, pp. 373–379.

Barnes, R.T.H., Hide, R., White, A.A. and Wilson, C.A. (1983) Atmospheric angular momentum fluctuations, length-of-day changes and polar motion. *Proc. Roy. Soc. London, Ser. A.*, **387**, 31–73.

Barsukov, O.M. (2002) Godichnye variazii seismichnosti i skorosti vrashenija Zemli [Yearly variations of seismicity and velocity of the Earth rotation]. *Fizika Zemli.*, (4), 96–98 [In Russian with English text In *Izvestiya, Physics of the Solid Earth*, 38 (4).

Batyayeva, T.F. (1960) Charts of multiyear monthly average values of air temperature at sea level for the globe. meteorologicheskiy byulleten (prilozhenie). *Tsentralny Institut Prognozov.*, 19–22 [in Russian].

Bizouard, C., Brzezinski, A. and Petrov, S. (1998) Diurnal atmospheric forcing and temporal variations of the nutation amplitudes. *J. Geodesy.*, **72**, 561–577.

Bjerknes, J. (1966) A possible response of the atmospheric Hadley circulation to equatorial anomalies of ocean temperature. *Tellus*, **18** (4), 820–829.

Bjerknes, J. (1969) Atmospheric teleconnections from the equatorial Pacific. *Mon. Wea. Rev.*, **97** (3), 163–172.

Bonchkovskaya, T.V. (1956) Nekotoryye kharakteristiki mussonnoy aktivnosti atmosfery. [Some characteristics of the monsoon activity of the atmosphere]. *Trudy Morsk. Gidrofiz. Inst.*, **4**, 102–142 [in Russian].

Braginsky, S.I. (1970) Magnitogidrodinamicheskie krutilnyye kolebaniya v zemnom yadre I variatsii dliny sutok. [Magnetohydrodynamic torsional oscillations in the Earth core and variations in the length of a day]. *Geomagnetizm i Aeronomiya.*, **10** (2), 221–232 [in Russian].

Braginsky, S.I. (1972) Proiskhozhdenie magnitnogo polya Zemli i ego vekovykh variatsy. Origin of the magnetic field of the Earth and its secular variations. *Izvestiya*

The Interaction Between Earth's Rotation and Geophysical Processes. Nikolay S. Sidorenkov

ISBN: 978-3-527-40875-7

Akademii Nauk SSSR, Fizika Zemli, (10), 3–14 [in Russian].

Bretagnon, P. and Francou, G. (1988) Planetary theories in rectangular and spherical variables. *Astron. Astrophys.*, **202**, 309–315.

Brower, D. (1952) A stydy of the changes in the rate of rotation of the Earth. *Astron. J.*, **57** (5), 125–146.

Brown, E.W. (1905) Theory of the motion of the Moon. Part 4. *Memoirs Royal Astron. Soc.*, **57**, Part 2, 51.

Brown, E.W. (1919) *Tables of the Motion of the Moon*, Yale University Press, New Haven.

Brown, E. (1926) The evidence for changes in the rate of rotation of the Earth and their geophisical consequences, with a summary and discussion of the devitations of the moon and sun from their gravitational orbits. *Trans. Astron. Obs. Yale Univ.*, **3**, Part 6, 207.

Bryazgin, N.N. (1990) Atmosfernyye osadki v Antarktide i ikh mnogoletnyaya izmenchivost [Atmospheric precipitations in Antarctic and their multi-year change], *Collection of Scientific Papers Meteorologichskie Issledovaniya v Antarktike*, vol. 1, Gidrometeoizdat, Leningrad, pp. 30–34 [in Russian].

Brzezinski, A. (1994) Polar motion excitation by variations of the effective angular momentum function, II: extended-model. *Manuscripta Geodaetica.*, **19**, 157.

Brzezinski, A., Bizouard, Ch. and Petrov, S.D. (2002) Influence of atmosphere on Earth rotation: what new can be learned from recent atmospheric angular momentum estimates? *Surv. Geophys.*, **23**, 33–69.

Bullard, E.C., Freedman, C., Gellman, H. and Nixon, J. (1950) The westward drift of the Earth's magnetic field. *Philos. Trans. Roy. Soc. London*, **A243**, 67–92.

Büllesfeld, F.J. (1985) Ein Beitrag zur harmonischen Darstellung des gezeitenerzeugenden Potentials. Deutsche Geodätische Komission Reihe C, Heft 314, München.

Butler, S.T. and Small, K.A. (1963) The excitation of atmospheric oscillations. *Proc. Roy. Astron. Soc.*, **A274**, 91–121.

Byzova, N.L. (1947) Vliyaniye sezonnogo perenosa mass vozdukha na dvizhenie zemnoy osi. [Influence of seasonal air-mass transport on the Earth's axis motion]. *Doklady Akademii Nauk SSSR*, **58** (3), 393–396 [in Russian].

Cain, J.C., Schmitz, D.R. and Kluth, C. (1985) Eccentric geomagnetic dipole drift. *Phys. Earth Planet. In.*, **39** (4), 237–242.

Cartwright, D.E. and Edden, A.C. (1973) Corrected tables of tidal harmonics. *Geophys. J. R. Astron. Soc.*, **33**, 253–264.

Cartwright, D.E. and Tayler, R.I. (1971) New computations of the tide-generating potential. *Geophys. J. Roy. Astron. Soc.*, **23** (1), 45–74.

Catalogue (1964) *Catalogue of the Macrosynoptic Processes on G.J. Vangengejma's Classification. 1891–1962*, Arctic and Antarctic Research Institute, GUGMSSM SSSR, Leningrad, p. 158 [in Russian].

Chandler, S. (1891) On the variation in latitude. *Astron. J.*, **11**, 83.

Chapman, S. and Lindzen, R.S. (1970) *Atmospheric Tides*, D. Reidel, Norwell, Mass.

Chapront-Touzé, M. and Chapront, J. (1988) ELP2000-85: a semi-analytical lunar ephemeris adequate for historical times. *Astron. Astroph.*, **190**, 342–352.

Chuchkalov, B.S. (1972) Osobennosti razvitiya kvazidvukhletnego tsikla v svyazi s perenosom massy vozdukha v ekvatorialnoy stratosfere. [Features of development of a quasi-two-year cycle in relation to air mass transport in the equatorial stratosphere]. *Trudy Gidromettsentra SSSR*, **107**(107), 3–17 [in Russian].

Chuchkalov, B.S. (1989) *Operativnyy Kontrol Obshchey Tsirkulyatsii Atmosfery. [Monitoring of General Circulation of Atmosphere]. Collection of Scientific Papers: Shestdesyat let Tsentru Gidrometeorologicheskikh Prognozov*, Gidrometeoizdat, Leningrad [in Russian].

Chuikova, N.A. and Maksimova, T.G. (2005) Izostaticheskoe ravnovesie kory i verchnei mantii Zemli [Isostatic equilibrium of the crust and upper mantle of the Earth]. Vestnik of the Moscow State University.

Ser. 3. Physics and Astronomy, 4, 64–72 [in Russian].

Climate Diagnostics Bulletin. publ. monthly by NOAA. Climate Prediction Center. Washington. D.C. 20223. – 1990–2008.

Coleman, P. (1971) Solar wind torque on the geomagnetic cavity. *J. Geophys. Res.*, **76** (16), 3800–3805.

De Sitter, W. (1927) On the secular accelerations and the fluctuations of the longitudes of the moon, the sun, Mercury and Venus. *Bull. Astr. Inst. Netherlandes*, **4** (124), 21–38.

Deacon, E.L. and Webb, E.K. (1962) Microscale interaction, in *The Sea. Ideas and Observations on Progress in the Study of the Seas*, vol. 1 (ed. M.N. Hill), Interscience Publishers, John Wiley & Sons, New York–London, pp. 5–57.

Dikiy, L.A. (1969) *Teoriya Kolebaniy Zemnoy Atmosfery. [Theory of Oscillations of the Earth Atmosphere]*, Gidrometeoizdat, Leningrad, p. 196 [in Russian].

Doodson, A.T. (1922) The harmonic development of tide-generating potential. *Proc. Royal Soc. London*, **A100**, 305–329.

Dymnikov, V.P., Petrov, V.L. and Lykosov, V.N. (1979) Gidrodinamicheskaya zonalnaya model obshchey tsirculatsii atmosfery. [Hydrodynamical zonal model of general circulation of atmosphere]. *Izv. AN SSSR. Fizika atmosfery i okeana.*, **15** (5), 484–497 [in Russian].

Eckart, C. (1960) *Hydrodynamics of Oceans and Atmospheres*, Pergamon Press, New York, p. 290.

Efimov, V.V., Kulikov, E.A., Rabinovich, A.B. and Fayn, I.V. (1985) *Volny v Pogranichnykh Oblastyakh Okeana. [Waves in the Boundary Areas of Ocean]*, Gidrometeoizdat, Leningrad, p. 280 [in Russian].

Eubanks, T.M. (1993) Variations in the orientation of the Earth, in *Contributions of Space Geodesy to Geodynamics: Earth Dynamics. Geodynamics 24* (ed. D.E. Smith), American Geophysical Union, Washington, DC, pp. 1–54.

Fedorov, E.P., Korsun, A.A., Mayor, S.P., Panchenko, N.I., Taradiy, V.K. and Yatskiv, Ya. (1972) *Dvizheniye Polyusa Zemli s 1890. 0 po 1969. 0. [Motion of the Earth's Pole from 1890.0 to 1969.0]*, Naukova dumka, Kiev, p. 264 [in Russian].

Gandin, L.S., Laykhtman, D.L., Matveev, L.T. and Yudin, M.I. (1955) *Osnovy Dinamicheskoy Meteorologii. [Principles of Dynamic Meteorology]*, Gidrometeoizdat, Leningrad, p. 648 [in Russian].

Gaposchkin, E.M. (1972) Analysis of pole positions from 1846 to 1970, *Rotation of the Earth*, (eds P. Melchior and S. Yumi), D. Reidel Publishing Co., Dordrecht, Holland, pp. 19–32.

Garratt, J. (1977) Review of drag coefficients over oceans and continents. *Mon. Weather Rev.*, **105** (7), 915–929.

Gill, A.E. (1982) *Atmosphere – Ocean Dynamics*, Academic Press, New York.

Girs, A.A. (1971) *Mnogoletnie Kolebaniya Atmosphernoy Tsirkulatsii i Dolgosrochnye Gidrometeorologicheskie Prognozy. [Long-Term Fluctuations of Atmospheric Circulation and Long-Term Hydrometeorologic Forecasting]*, Gidrometeoizdat, Leningrad, p. 280 [in Russian].

Gorkaviy, N.N., Trapeznikov, Yu.A. and Fridman, A.M. (1994b) O globalnoy sostavlyayushchey seysmicheskogo protsessa i ee svyazi s nablyudaemymi osobennostyami vrashcheniya Zemli. [On the global component of seismic process and its connection with the observable features of the Earth's rotation]. *Doklady Rossiyskoy Akademii Nauk.*, **338** (4), 525–527 [in Russian].

Gorkaviy, N.N., Levitzky, L.S., Taidakova, T.A., Trapeznikov, Ju.A. and Fridman, A.M. (1994a) O vyjavlenii trech component seismicheskoi aktivnosti Zemli [About detection of three components of a seismic activity of the Earth]. *Fizika Zemli*, (10), 23–32 [In Russian with English text In *Izvestiya, Physics of the Solid Earth*, 30 (10)].

Gross, R.S. (1993) The effect of ocean tides on the Earth's rotation as predicted by the results of an ocean tide model. *Geophys. Res. Lett.*, **20**, 293–296.

Gu, D. and Philander, S.G.H. (1995) Secular changes of annual and interannual variability in the tropics during the past century. *J. Climate.*, **8**, 864–876.

Guterman, I.G. (ed.) (1975) *Aeroklimaticheskiy Spravochnik Kharakteristik Tsirkulatsii Atmosfery v Uzlakh Koordinatnoy Setki Severnogo Polushariya, [Book of Reference on the Aeroclimatic Characteristics of Circulation of Atmosphere in the Nodes of a Grid in the Northern Hemisphere]*, Gidrometeoizdat, Moscow, p. 61 [in Russian].

Guterman, I.G. (1976) Srednyaya tsirculatsiya atmosfery i gorizontalny mezhshirotny obmen. [Mean circulation of atmosphere and horizontal inter-latitude exchange]. *Trudy VNIIGMI MCD*, **29**, 3–37 [in Russian].

Guterman, I.G. (ed.) (1978) *Aeroklimaticheskiy Atlas-Spravochnik Kharakteristik Tsirkulatsii Atmosfery nad Yuzhnym Polushariem, [Atlas-Directory of Aeroclimatic Characteristics of Atmosphere Circulation in the Southern Hemisphere]*, vol. 1, Part 2, Gidrometeoizdat, Moscow, p. 56 [in Russian].

Guterman, I.G. and Khanevskaya, I.V. (1972) Osnovnye cherty raspredeleniya temperatury i vozdushnykh techeniy v svobodnoy atmosphere. [Main features of distribution of temperatures and air currents in open atmosphere of the Earth], *Trudy V Vsesoyuznogo Meteorologicheskogo Kongressa*, vol. 3, Gidrometeoizdat, Leningrad, pp. 132–151 [in Russian].

Haltiner, G.J. and Martin, F.L. (1957) *Dynamical and Physical Meteorology*, McGraw-Hill, New York-Toronto-London.

Hamilton, K. and Garsia, R.R. (1986) Theory and observations of the short-period normal mode oscillations of the atmosphere. *J. Geophys. Res.*, **91** (D11), 11867–11875.

Hartmann, T. and Wenzel, H.-G. (1994) The harmonic development of the Earth tide generating potential due to the direct effect of the planets. *Geophys. Res. Lett.*, **21**, 1991–1993.

Hartmann, T. and Wenzel, H.-G. (1995) The HW95 tidal potential catalogue. *Geophys. Res. Lett.*, **22**, 3553–3556.

Hide, R. (1989) Fluctuations in the earth's rotation and the topography of the core-mantle interface. *Philos. Trans. Roy. Soc.*, **A328**, 351–363.

Hide, R., Dickey, J.O., Marcus, S.L., Rosen, R.D. and Salstein, D.A. (1997) Atmospheric angular momentum fluctuations in global circulation models during the period 1979–1988. *J. Geophys. Res.*, **102**, 16423–16438.

Hirshberg, J. (1972) Upper limit of the torque of the solar wind on the Earth. *J. Geophys. Res.*, **77** (25), 4855–4857.

Holton, J.R. (1975) Dynamical meteorology of the stratosphere and mesosphere. *Meteor. Monogr.*, **15** (37), American Meteor. Soc., Boston.

Holton, J.R. and Lindzen, R.S. (1972) An updated theory for the quasi-biennial cycle of the tropical stratosphere. *J. Atmos. Sci.*, **29**, 1076–1080.

Hough, S.S. (1897) On the application of garmonic analysis to the dynamical theory of tides. Part I. On Laplace's "Oscillations of the first species", and on the dynamics of ocean currents. *Philos. Trans. Roy. Soc. London*, **A189**, 210–257.

Hough, S.S. (1898) On the application of garmonic analysis to the dynamical theory of tides. Part II. On the general integration of the Laplace's dynamical equations. *Philos. Trans. Roy. Soc. London*, **A191**, 139–185.

Hsu, H.H. and Hoskins, B.J. (1989) Tidal fluctuations as seen in ECMWF data. *Quart. J. R. Meteorol. Soc.*, **115**, 247–264.

IERS Conventions (2003), IERS Technical Note No. 32. Verlag des Bundesamts für Kartographie und Geodäsic, Frankfurt a. Main, 2004, p. 127.

IERS Standards (1989) IERS Technical Note 3. Observatoire de Paris, 76.

IERS (2000) Annual Report, International Earth Rotation Service, Observatoire de Paris, 144.

Izvekov, B.I. and Kochin, N.E. (eds) (1937) *Dynamicheskaya Meteorologiya*, Chast 2, [Dynamic Meteorology]. Part 2, Gidrometeoizdat, Moscow – Leningrad, pp. 280 [in Russian].

Jacobs, G.A. and Mitchell, J.L. (1996) Ocean circumpolar variations associated with the

Antarctic Circumpolar Wave. *Geophys. Res. Lett.*, **23** (21), 2947–2950.

Jean-Pierre, R. and Jolanta, N. (1992) Sur les irregularites de rotation de la Terre couplees a sa temperature globale. *C. R. Acad. Sci. – Ser. 2*, **315** (6), 667–672.

Jeffreys, H. (1916) Causes contributory to the annual variation of latitude. *Mon. Not. R. Astron. Soc.*, **76** (6), 499–525.

Jeffreys, H. (1926) On the dynamics of geostrophic winds. *Quart. J. Roy. Meteorol. Soc.*, **52**, 85–104.

Jeffreys, H. (1928) Possible tidal effects on accurate time-keeping. *Mon. Not. Roy. Astron. Soc., Geophys.*, **2** (Suppl.), (1), 56–58.

Jeffreys, H. (1959) *The Earth*, Cambridge University Press, Cambridge, p. 323.

Jeffreys, H. (1963) On the hydrostatic theory of the figure of the Earth. *Geophys. J. Roy. Astron. Soc.*, **8** (2), 196–202.

Jeffreys, H. (1968) The variations of latitude. *Mon. Not. R. Astron. Soc.*, **141** (2), 255–268.

Jeffreys, H. and Swirles, B. (1966) *Methods of Mathematical Physics*, Cambridge University Press, Cambridge, p. 323.

Jin, R.S. and Thomas, D.M. (1977) Spectral line similarity in the geomagnetic dipole field variations and length of day fluctuations. *J. Geophys. Res.*, **82**, 828–834.

Jones, P.D. and Mann, M.E. (2004) Climate over past millennia. *Rev. Geophys.*, **42** (2), RG2002, DOI: 10.1029/2003RG000143.

Kalnay, E., Kanamitsu, M., Kistler, R., Collins, W., Deaven, D., Gandin, L., Iredell, M., Saha, S., White, G., Woollen, J., Zhu, Y., Chelliah, M., Ebisuzaki, W., Higgins, W., Janowiak, J., Mo, K.C., Ropelewski, C., Wang, J., Leetmaa, A., Reynolds, R., Jenne, R. and Joseph, D. (1996) The NMC/NCAR 40-year reanalysis project. *Bull. Am. Meteor. Soc.*, **77**, 437–471.

Kidson, J.W., Vincent, D.G. and Newell, R.E. (1969) Observational studies of the general circulation of the Tropics: long term mean values. *Q. J. Roy. Meteor. Soc.*, **95** (404), 258–287.

Kikuchi, N. (1971) Annual and early variations of the atmospheric pressure and the Earth' rotation. *Proc. Int. Latitude Observatory of Mizusawa*, (11), 11–28.

Klige, R.K. (1985) *Izmeneniya Globalnogo Vodoobmena. [Variations in the Global Water Exchange]*, Moscow, Nauka, p. 248 [in Russian].

Kligue, R.K. (1980) *Uroven Okeana v Geologicheskom Proshlom. [Sea Level in the Geological Past]*, Nauka, Moscow, [in Russian].

Klyashtorin, L.B. and Sidorenkov, N.S. (1996) Dolgoperiodnye klimaticheskiye izmeneniya i flyuktuatsii chislennosti pelagicheskikh ryb Patsifiki. [Long-term climatic changes and fluctuations of population of the Pacific pelagic fish]. *Izvestiya Tikhookeanskogo Nauchno-Issledovatelskogo, Instituta Rybnogo Khozyaystva I Okeanografii. Rybokhozyaystvennogo Tsentra*, **119**, 33–54 [in Russian].

Korsun, A.A., Mayor, S.P. and Yatskiv, Ya.S. (1974) *O godovom Dvizhenii Polyusov Vrashcheniya i Inertsii Zemli. [On Annual Motion of the Poles of the Earth's Rotation and Inertia. Collection of Scientific Papers: "Astrometriya i Astrofizika"*, Naukova dumka, Kiev, No. 24, pp. 26–45 [in Russian].

Kotlyakov, V.M. (1968) *Snezhnyy Pokrov Zemli i Ledniki. [Snow Cover and Glaciers of the Earth]*, Gidrometizdat, Leningrad, p. 479 [in Russian].

Kudryavtsev, S.M. (1999) Accurate and quick account of the tidal effects by the new analytical method. *J. Braz. Soc. Mech. Sci.*, **XX**, 552–557.

Kudryavtsev, S.M. (2002) Precision analytical calculation of geodynamical effects on satellite motion. *Celest. Mech. Dyn. Astron.*, **82** (4), 301–316.

Kudryavtsev, S.M. (2004) Improved harmonic development of the Earth tide generating potential. *J. Geodesy*, **77** (12), 829–838.

Kulikov, K.A. and Sidorenkov, N.S. (1977) *Planeta Zemlya. [Planet Earth]*, Nauka, Moscow, p. 192 [in Russian].

Lamb, H. (1932) *Hydrodynamics*, Cambibge University Press, Cambridge, p. 728.

Lambeck, K. (1980) *The Earth's Variable Rotation: Geophysical Causes and Consequences*, Cambridge University Press, Cambridge, p. 450.

Lambeck, K. (1988) *Geophysical Geodesy*, Clarendon Press, Oxford.

Landau, L.D. and Lifshitz, E.M. (1959) *Fluid Mechanics*, Pergamon Press.

Landau, L.D. and Lifshitz, E.M. (1965) *Mechanics*, Nauka, Moscow.

Landscheidt, T. (2003) New little ice age instead of global warming. *Energy and Environment*, **14**, 4.

Lappo, S.S., Gulev, S.K. and Rozhdestvenskiy, A.E. (1990) *Krupnomasshtabnoe Teplovoe Vzaimodeystviye v Sisteme Okean-Atmosfera i Energoaktivnye Oblasti Mirovogo Okeana. [Large-Scale Thermal Interaction in the Ocean-Atmosphere System and Energetically Active Zones of Ocean*, Gidrometeoizdat, Leningrad, p. 336 [in Russian].

Latif, M. and Villwock, A. (1990) Interannual variability as simulated in coupled ocean-atmosphere models. *J. Mar. Systems*, **1**, 51–60.

Lindzen, R.S. and Holton, J.R. (1968) A theory of the quasi-biennial oscillation. *J. Atmos. Sci.*, **25**, 1095–1107.

Livshits, M.A., Sidorenkov, N.S. and Starkova, L.I. (1979) On the possible relation between the irregularity of the earth rotation and the high-latitude magnetic field of the Sun. *Sov. Astron.-AJ*, **23** (3), 584–589.

Lliboutry, L. (1965) *Traite de Glaciologie*, vol. 11, Masson, Paris, p. 610.

Longuet-Higgins, M.S. (1968) The eigenfunctions of Laplace's tidal equation over a sphere. *Philos. Trans. Roy. Soc. London*, **A262**, 511–607.

Lorenz, E.N. (1967) *The Nature and Theory of the General Circulation of the Atmosphere*, World Meteorological Organization, Geneva.

Love, A. (1927) *A treatise on the Mathematical Theory of Elasticity*, Dover Publications, New York.

Mac Donald, G.J.F. (1963) The deep structure of oceans and continents. *Rev. Geophys.*, **1**, 587–665.

Magnitskiy, V.A. (1965) *Vnutrennee Stroeniye i Fizika Zemli. [The Internal Structure and Physics of the Earth]*, Nedra, Moscow, [in Russian].

Malakhov, S.G. (1971) O perenose radioaktivnykh aerozoliy v atmosfere iz polushariya v polusharie. [On atmospheric transfer of radioactive aerosols from one hemisphere to the other]. *Meteorologiya i Gidrologiya*, **1971** (9), 40–48 [in Russian].

Massey, B.S. (1983) *Mechanics of Fluids*, 5th edn, Van Nostrand Reinhold, UK, 0-442-30552-4.

Matveev, L.T. (1965) *Osnovy Obshchey Meteorologii. Fizika Atmosfery. [Principles of General Meteorology. Physics of Atmosphere]*, Gidrometeoizdat, Leningrad, p. 876 [in Russian].

McCarthy, D.D. and Babcock, A.K. (1986) The length of day scince 1656. *Phys. Earth Plan. Int.*, **44**, 281–292.

McKenzie, D.P. (1966) The viscosity of the Lower Mantle. *J. Geophys. Res.*, **71** (16), 3995–4010.

Mechoso, C.R., Ma, C.C., Farrara, J.D., Spahr, J. and Moore, R.W. (1993) Parallelization and distribution of a coupled atmosphere-ocean general circulation model. *Mon. Wea. Rev.*, **121**, 2062–2076.

Melchior, P. (1971) *Physique et Dynamique Planetaires*, Vander–editeur, Bruxelles.

Melchior, P. (1983) *The Tides of the Planet Earth*, 2nd edn, Pergamon Press, Oxford, p. 641.

Monin, A.S. and Yaglom, A.M. (1965) *Statisticheskaya Gidromekhanika. Chast I. [Statistical Hydromechanics. Part I]*, Nauka, Moscow, p. 710 [in Russian].

Moritz, H. and Mueller, I. (1987) *Earth Rotation. Theory and Observation*, The Unger Publishing Company, New York, p. 512.

Morrison, L.V. (1973) Rotation of the Earth from A.D. 1663–1972 and the constancy of G. *Nature*, **241** (5391), 519–520.

Mul'tanovskii, B.P. (1933) *Basic Positions of a Synoptic Method of Long-Term Weather Forecasts*, Publishing house TsUEGMS, Moscow, p. 139 [in Russian].

Munk, W. and Hassan, E.S.H. (1961) Atmospheric excitation of the Earth's wobble. *Geophisical Journal of the R.A.S.*, **4**, 339–358.

Munk, W.H. and Macdonald, G.J.F. (1960) *The Rotation of the Earth, a Geophysical Discussion*, Cambridge University Press, Cambridge, p. 323.

Nastula, J., Ponte, R.M. and Salstein, D.A. (2000) Regional signals in atmospheric and oceanic excitation of polar motion, in *Polar Motion: Historical and Scientific Problems*, ASP Conference Series, vol. 208 (eds S. Dick, D. McCarthy and B. Luzum), ASP, San Francisco, pp. 463–472.

Neelin, J.D. (1990) A hybrid coupled general circulation model for studies. *J. Atmos. Sci.*, **47**, 674–693.

Newcomb, S. (1891) On the periodic variations of latitude and the observations with the Washington Prime-Vertical Transit. *The Astron. J.*, **11**, 81–82.

Newton, R.R. (1970) *Ancient Astronomical Observations and Acceleration of the Earth and Moon*, Johns Hopkins Press, London, p. 309.

Newton, R.R. (1972) The historical acceleration of the Earth, *Rotation of the Earth*, (eds P. Melchior and S. Yumi), D. Reidel Publishing Co., Dordrecht, Holland, pp. 160–161.

Nikolaevskiy Victor, N. (2003) *Angular Momentum in Geophysical Turbulence: Continuum Spatial Averiging Method*, Kluwer Academic Publishers, p. 256.

Pagava, S.T., Aristov, N.A., Bljumina, L.I. and Turketti, Z.L. (1966) *Bases of a Synoptic Method of the Seasonal Weather Forecasts*, Gidrometeoizdat, Leningrad, p. 362 [in Russian].

Panchev, S. (1967) *Sluchaynye Funktsii I Turbulentnost. [Stochastic Functions and Turbulence]*, Gidrometeoizdat, Leningrad, p. 448 [in Russian].

Pariiski, N.N. (1954) *Neravnomernost Vrashcheniya Zemli. [Irregularity in the Velocity of the Earth's Rotation]*, Izd. AN SSSR, Moscow, p. 48.

Pariiski, N.N. and Berlyand, O.S. (1953) Vliyaniye sezonnykh izmeneniy atmosfernoy tsirkulyatsii na skorost vrashcheniya Zemli. [Influence of the seasonal changes of the atmospheric circulation on the velocity of the Earth's rotation]. *Trudy Geofizicheskogo Instituta AN SSSR*, **19** (19), 103–122 [in Russian].

Parker, D.E., Folland, C.K. and Jackson, M. (1995) Marine surface temperature: observed variations and data requirements. *Climatic Change*, **31**, 559–600.

Petersen, R.G. and White, W.B. (1998) Slow oceanic teleconnections linking the Antarctic Circumpolar Wave with the tropical El Nino–Southern Oscillation. *J. Geophys. Res.*, **103** (C11), 24573–24583.

Petrov, V.N. (1975) *Atmosfernoe Pitaniye Lednikovogo Pokrova Antarktidy. [Atmospheric Nourishment of Antarctic Ice Sheet]*, Gidrometeoizdat, Leningrad, [in Russian].

Philander, S.G.H. (1990) *El Nino, La Nina, and the Southern Oscillation*, Academic Press, San Diego, CA, p. 293.

Philander, S.G.H., Pacanowski, R.C., Lau, N.C. and Nath, M.J. (1992) Simulation of ENSO with a global atmospheric GCM coupled to a highresolution, tropical Pacific ocean GCM. *J. Climate*, **5**, 308–329.

Pushkov, A.N. and Chernova, T.A. (1972) *Osobennosti Prostranstvenno Vremennoy Struktury Vekovoy Variatsii Geomagnitnogo Polya [Features of the Space-Time Structure of the Secular Variation of the Geomagnetic Field]*, Preprint No. 18., IZMIRAN, Moscow, [in Russian].

Rao, V.P. (1964) Interhemispheric circulation. *Q. J. Roy. Meteor. Soc.*, **90** (384), 190–194.

Rasmusson, E.M. (1991) Observational aspects of ENSO cycle teleconnections, in *Teleconnectios Linking Worldwide Climate Anomalies* (eds M.H. Glantz, R.W. Katz and N. Nicholls) Cambridge University Press, New York, pp. 309–343.

Rasmusson, B.M. and Carpenter, T.H. (1982) Variation in tropical sea surface temperature and surface wind fields associated with the Southern oscillation/El-Nina. *Mon. Wea. Rev.*, **110**, 354–384.

Reed, R.J. (1964) A tentative model of the 26-month oscillation in tropical latitudes. *Quart. J. Roy. Meteor. Soc.*, **90**, 441–466.

Ricard, Ya., Dogliony, C. and Sabadini, R. (1991) Differential rotation between lithosphere and mantle: a consequence of lateral mantle viscosity variations. *J. Geophys. Res.*, **96** (B5), 8407–8415.

Roosbeek, F. (1996) RATGP95: a harmonic development of the tide-generating potential using an analytical method. *Geophys. J. Int.*, **126**, 197–204.

Ropelewski, C.F. and Jones, P.D. (1987) An Extension of the Tahiti-Darwin southern oscillation index. *Mon. Wea. Rev.*, **115**, 2161–2165.

Rosenhead, L. (1929) The annual variation of latitude. *Mon. Not. Roy. Astron. Soc.*, Geophys. Suppl., **2** (3), 140–170.

Rykhlova, L.V. (1971) Godovaya sostavlyayushchaya dvizheniya polyusa za 119 let [Annual component of the pole motion for 119 years]. *Soobshcheniya Gos. Astronomicheskogo instituta im. P.K., Shternberga*, **169** (169), 22–26 [in Russian].

Salstein, D.A. (2000) Atmospheric excitation of polar motion, in *Polar Motion: Historical and Scientific Problems*, ASP Conference Series, vol. 208 (eds S. Dick, D. McCarthy and B. Luzum), ASP, San Francisco, pp. 437–446.

Salstein, D.A. and Rosen, R.D. (1997) Global momentum and energy signals from reanalysis systems. Preprints, 7th Conference on Global Climate Variations, American Meteorological Society, Boston, MA, pp. 344–348.

Salstein, D.A., Kann, D.M., Miller, A.J. and Rosen, R.D. (1993) The sub-bureau for atmospheric angular momentum of the International Earth Rotation Service. A Meteorological data center with geodetic applications. *Bull. Am. Meteor. Soc.*, **74**, 67–80.

Schneider, U. and Schonwiese, C.D. (1989) Some statistical characteristics of El Nino/southern oscillation and North Atlantic oscillation indices. *Atmosfera.*, **2** (3), 167–180.

Schopf, P.S. and Suarez, M.J. (1988) Vacillations in a coupled ocean-atmosphere model. *J. Atmos. Sci.*, **45**, 549–566.

Shaw, N. (1936) *Manual of Meteorology*, 2nd edn, vol. II, Cambridge University Press, London.

Shuleykin, V.V. (1965) Analiz slozhnykh teplovykh usloviy v oblasti zamknutykh tsiklicheskikh techeniy Atlanticheskogo okeana. [Analysis of complex thermal conditions in the area of circulating currents of the Atlantic ocean]. *Izv. AN SSSR. Fizika atmosferyi okeana.*, **1** (4), 413–425 [in Russian].

Shuleykin, V.V. (1968) *Fizika Morya. [Physics of the Sea]*, Nauka, Moscow, p. 1083 [in Russian].

Sidorenkov, N.S. (1961) O vliyanii neravnomernosti vrashcheniya Zemli na protsessy v ee atmosfere i gidrosfere. [On influence of irregularity in the velocity of the Earth's rotation on the Earth's atmosphere and hydrosphere]. *Problemy Arktiki i Antarktiki.*, **9** (9), 45–49 [in Russian].

Sidorenkov, N.S. (1963) K voprosu o prirode neravnomernosti vrashcheniya Zemli. [On the issue of the nature of irregularity in the velocity of the Earth's rotation]. *Izvestiya AN SSSR, Seriya geofizicheskaya*, (5), 730–739. [in Russian].

Sidorenkov, N.S. (1968) Methods for estimating the influence of atmospheric circulation on the Earth's rotational velocity. *Sov. Astron.-AJ*, **12** (2), 303–308.

Sidorenkov, N.S. (1969) The influence of atmospheric circulations on the Earth's rotational velocity, 1956.8-1964.8. *Sov. Astron.-AJ*, **12** (4), 706–714.

Sidorenkov, N.S. (1973) The inertia tensor of the atmosphere, the annual variations of its components, and the variations of the earth's rotation. *Bull. (Izv.), Acad. Sci. USSR, Atmospheric and Oceanic, Physics*, **9** (4), 185–192.

Sidorenkov, N.S. (1975) O mezhdupolusharnoy teplovoy mashine v atmosfere Zemli. [On the inter-hemispheric heat engine in the atmosphere of the Earth]. *Doklady AN SSSR.*, **221** (4), 835–838 [in Russian].

Sidorenkov, N.S. (1976a) Nonuniformities in the earth's rotation from astronomical observations during 1971.0-1974.5. *Sov. Astron.-AJ*, **19** (5), 665–667.

Sidorenkov, N.S. (1976b) A study of angular momentum of the atmosphere. *Bull. Izv. Acad. Sci. USSR, Atmospheric and Oceanic, Physics*, **12** (6), 351–356.

Sidorenkov, N.S. (1978) Neravnomernost vrashcheniya Zemli i protsessy v atmosfere. [The Earth's rotation irregularity and the atmospheric processes]. *Trudy Gidromettsentra SSSR*, **205** (205), 48–66 [in Russian].

Sidorenkov, N.S. (1979) An investigation of the role of the atmosphere in the excitation of multiyear variations in the earth's rotation rate. *Sov. Astron.-AJ*, **23** (1), 102–109.

Sidorenkov, N.S. (1980a) Irregularities of the Earth's rotation as possible indices of global water exchange. *Sov. Meteorol. Hydrol.*, (1), 39–45.

Sidorenkov, N.S. (1980b) Mean annual zonal circulation of the atmosphere. *Sov. Meteorol. Hydrol.*, (8), 22–26.

Sidorenkov, N.S. (1982a) K voprosu o prirode zonalnoy tsirkulyatsii atmosfery. [On the issue of the nature of the zonal circulation of the atmosphere]. *Trudy Gidromettsentra SSSR*, (248), 66–75 [in Russian].

Sidorenkov, N.S. (1982) Nekotorye parametry globalnogo vodoobmena po dannym o vekovom dvijenii polusa i neravnomernosti vrasheniya Zemli. [Estimation of some parameters of the global water exchange based on data on the secular polar motion and nonuniformity of the Earth's rotation]. *Vodnye Resursy*, (3), 39–46 [in Russian].

Sidorenkov, N.S. (1983) *Vekovoe Dvizheniye Polyusa kak Geofizicheskaya Problema. [Secular Motion of the Poles as a Geophysical Problem]. Collection of Scientific Papers: Vrashcheniye Zemli i Geodinamika*, FAN, Tashkent, pp. 25–29 [in Russian].

Sidorenkov, N.S. (1987) The nature of seasonal and interannual variations of the Earth's rotation. Proc. Int. Symp. "Figure and Dynamics of the Earth, Moon, and Planets." Prague, pp. 947–960.

Sidorenkov, N.S. (1988) Tsirkulyatsiya vozdukha mezhdu severnym i yuzhnym polushariyami. [Air circulation between the northern and southern hemispheres]. *Trudy Gidromettsentra SSSR*, (297), 3–17 [in Russian].

Sidorenkov, N.S. (1991a) *Atmosfera. [The Atmosphere.] Collection of Scientific Papers: Aktualnye Problemy Geodinamiki*, Nauka, Moscow, pp. 105–129 [in Russian].

Sidorenkov, N.S. (1991b) Fizika zonalnoy tsirkulyatsii atmosfery. [Physics of zonal circulation of the atmosphere]. *Trudy Gidromettsentra SSSR*, (316), 3–18 [in Russian].

Sidorenkov, N.S. (1991c) Kharakteristiki yavleniya El-Nino – Yuzhnoe kolebaniye. [Characteristics of the El Nino phenomenon – Southern oscillation]. *Trudy Gidromettsentra SSSR*, (316), 31–44 [in Russian].

Sidorenkov, N.S. (1997) The effect of the El Nino Southern oscillation on the excitatation of the chandler motion of the Earth's pole. *Astron. Rep.*, **41** (5), 705–708.

Sidorenkov, N.S. (1998) Planetarnye atmosfernye protsessy. [Planetary atmospheric processes], *Atlas Vremennykh Variatsiy Prirodnykh, Antropogennykh i Sotsialnykh Protsessov. Tsiklicheskaya Dinamika v Prirode i Obshchestve*, vol. 2, Nauchnyy mir, Moscow, pp. 274–281 [in Russian].

Sidorenkov, N.S. (1999) Chandlerovskoe dvizheniye polyusov kak chast nutatsii sistemy atmosfera-okean-Zemlya. [The Chandler polar motion as a component of nutation of the atmosphere–ocean–earth system]. *Kinematics and Physics of Celestial Bodies*, Appendix No. 1 (Nauchno-tekhnicheskiy zhurnal NAN Ukrainy), 55–61 [in Russian].

Sidorenkov, N.S. (2000a) Chandler wobble of the poles as part of the nutation of the atmosphere – ocean – earth system. *Astron. Rep.*, **44** (6), 414–419.

Sidorenkov, N.S. (2000b) Mekhanizmy mezhgodovoy izmenchivosti atmosfery i okeana. [Mechanisms of interannual variability of atmosphere and ocean]. *Trudy Gidromettsentra Rossii*, (335), 26–41 [in Russian].

Sidorenkov, N.S. (2000c) Monitoring obshchey tsirkulyatsii atmosfery. [Monitoring of general circulation of the atmosphere. *Trudy Gidromettsentra Rossii*, (331), 12–41 [in Russian].

Sidorenkov, N.S. (2000d) Prilivnye kolebaniya atmosfernoy tsirkulyatsii. [Tidal oscillations of atmospheric circulation]. *Trudy Gidromettsentra Rossii*, (331), 49–63 [in Russian].

Sidorenkov, N.S. (2000) Excitation of the Chandler wobble, in *Polar Motion: Historical and Scientific Problems*, ASP Conference Series, vol. 208 (eds S. Dick, D. McCarthy and B. Luzum), ASP, San Francisco, pp. 455–462.

Sidorenkov, N.S. (2001) Chandler wobble as part of the inter-annual oscillation of the atmosphere-ocean-Earth system, *Journees 2001. Systemes de Reference Spatio-Temporels*, Observatoire royal de Belgique, Brussels, 2003, pp. 74–80.

Sidorenkov, N.S. (2002a) *Physics of the Earth's Rotation Instabilities*, Nauka, Fizmatlit, Moscow, p. 384 [in Russian].

Sidorenkov, N.S. (2002b) *Atmospheric Processes and the Earth Rotation*, Gidrometeoizdat, St. Peterburg, p. 366 [in Russian].

Sidorenkov, N.S. (2003a) Changes in the Antarctic ice sheet mass and the instability of the Earth's rotation over the last 110 years, in *International Association of Geodesy Symposia*, vol. 127, (ed. F. Sanso), V Hotine-Marussi symposium on Mathematical geodesy, Matera, Italy, June 17–21, Springer-Verlag, Berlin, Heidelberg, Germany, pp. 339–346.

Sidorenkov, N.S. (2003b) Influence of the atmospheric tides on the Earth rotation. *Celest. Mech. Dyn. Astr.*, **87** (1–2), 27–38.

Sidorenkov, N.S. (2004) Influence of the atmospheric and oceanic circulation on the plate tectonics. Journees 2003. Astrometry, Geodynamics and Solar System Dynamics: from milliarcseconds to microarcseconds, St. Petersburg, September 22–25 (eds A. Finkelstein and N. Capitaine), pp. 225–230.

Sidorenkov, N.S. (2005) Physics of the Earth's rotation instabilities. *Astron. Astrophys. Trans.*, **24** (5), 425–439.

Sidorenkov, N.S. (2006) Long term changes of the variance of the Earth orientation parameters and of the excitation functions. Proceedings of the "Journees 2005 Systemes de Reference Spatio-Temporels" Space Research Centre PAS, Warsaw, Poland, 2006 (eds A. Brzezinski, N. Capitaine and B. Kolaczek), pp. 225–228.

Sidorenkov, N.S. (2008a) On peculiarity of movement of geophysical and astrophysical continuums. *Radio Physics and Radio Astronomy*, **13** (3), 124–129.

Sidorenkov, N.S. (2008b) Lunar-solar tides and atmospheric processes. *Nature (Russian)*, (2), 23–31 [in Russian].

Sidorenkov, N.S. and Orlov, I.A. (2008) Atmosphere circulation epochs and changing of climate. *Russ. Meteorol. Hydrol.*, (9), 22–29.

Sidorenkov, N.S. and Shveikina, V.I. (1994) Changes of the climatic regime over the Volga Basin and the Caspian Sea during last century, in *Contemp. Climatol.* (eds R. Brazdil and M. Kolar), Brno, pp. 525–530.

Sidorenkov, N.S. and Shveikina, V.I. (1996) Izmeneniye klimaticheskogo rezhima basseyna Volgi i Kaspiyskogo morya za poslednee stoletie. [Climatic regime shift in the catchment areas of the River Volga and the Caspian sea for the last century]. *Vodnye Resursy*, **23** (4), 401–406 [in Russian].

Sidorenkov, N.S. and Sidorenkov, P.N. (2002) *Russ. Patent, Russian State Invention Register* (10 May 2002), No. 2182344.

Sidorenkov, N.S. and Stekhnovsky, D.I. (1971) The Total Mass of the Atmosphere and Its Seasonal Redistribution. *Bull. (Izv.), Acad. Sci. USSR, Atmospheric and Oceanic, Physics*, **7** (9), 646–648 [in Russian].

Sidorenkov, N.S. and Stekhnovsky, D.I. (1972) Seasonal redistribution of air masses between the continents and oceans of the northern and southern hemispheres. *Trudy Gidromettsentra SSSR*, (107), pp. 80–84. [in Russian].

Sidorenkov, N.S. and Svirenko, P.I. (1983) Long-term fluctuations in atmospheric circulation. *Sov. Meteorol. Hydrol.*, (11), 12–15.

Sidorenkov, N.S. and Svirenko, P.I. (1988) Diagnoz nekotorykh parametrov globalnogo vodoobmena po dannym o nepravilnostyakh vrashcheniya Zemli. [Diagnosis of some parameters of global water exchange based on the data on irregularities of the Earth's rotation]. *Izv. AN SSSR, Seriya Geograficheskaya*, **5**, 16–23 [in Russian].

Sidorenkov, N.S. and Svirenko, P.I. (1991) Monitoring momenta impulsa zonalnykh vetrov atmosfery. [Monitoring of the angular momentum of zonal winds of the atmosphere]. *Trudy Gidromettsentra SSSR*, (316), 19–25 [in Russian].

Sidorenkov, N.S. and Wilson, I.R. (2009) The decadal fluctuations in the Earth's rotation and in the climate characteristics, *In Journees 2008. Systemes de Reference Spatio-Temporels*, Technische Universitet. Lohran Observatorium, Dresden, (in press).

Sidorenkov, N.S., Lutsenko, O.V., Bryazgin, N.N. *et al.* (2005) Variation of the mass of the sheet of Antarctica and instability of the Earth's rotation. *Russ. Meteorol. Hydrol.*, (8), 1–8.

Siebert, M. (1961) Atmospheric tides, *Advances in Geophysics*, vol. 7, Academic Press, New York, pp. 105–187.

Simon, J.L., Bretagnon, P., Chapront, J., Chapront-Touzé, M., Francou, G. and Laskar, J. (1994) Numerical expressions for precession formulae and mean elements for the Moon and planets. *Astron. Astrophys.*, **282**, 663–683.

Sonechkin, D.M. and Burlutskiy, R.F. (2005) *Khaos i Poryadok v Dinamike Pogody i Klimata. Vodnyy i Teplovoy Balansy Troposfery. [Chaos and Order in the Weather and Climate Dynamics. Water and Thermal Balances of Troposphere]*, Gidrometeoizdat, St. Peterburg, p. 256 [in Russian].

Spitaler, R. (1901) Die periodischen Luftmassen verschiebungen und ihr Einfluss auf die Langenanderungen der Erdachse (Breitenschwankungen). *Petermanns Mitt., Erganzungsband.*, **29**, 137.

Standish, E.M. (1998) JPL planetary and lunar ephemerides DE405/LE405. JPL IOM 312.F-98-048, Pasadena.

Standish, E.M. and Williams, J.G. (1981) Planetary and lunar ephemerides DE200-LE200. URL: ftp:.

Starr, V.P. (1968) *Physics of Negative Viscosity Phenomena*, McGraw-Hill, New York.

Stekhnovsky, D.I. (1960) Maps of Longterm Mean Monthly Values of the Atmospheric Pressure at Sea Level and Some Characteristics of the Pressure Field of the Earth. *Meteorologicheskiy Byulleten, Prilozheniye*, Tsentr. Inst. Prognozov. 3–18 [in Russian].

Stekhnovsky, D.I. (1962) *Baricheskoye Pole Zemnogo Shara. [The Pressure Field of the Earth]*, Gidrometeoizdat, Moscow, p. 147 [in Russian].

Stekhnovsky, D.I. and Sidorenkov, N.S. (1977) Sredniy mnogoletniy rezhim pereraspredeleniya mass vozdukha na zemnom share ot iyulya k yanvaryu. [Normal annual regime of redistribution of air mass on the globe from July to January]. *Trudy Gidromettsentra SSSR*, (171), 38–40 [in Russian].

Stephenson, F.R. (1997) *Historical Eclipses and Earth's Rotation*, Cambridge University Press, Cambridge, p. 557.

Stephenson, F.R. and Morrison, L.V. (1984) Long term changes in the rotation of the Earth: 700 B.C. to A.D. 1980. *Philos. Trans. Roy. Soc. London*, **A313**, 47.

Sutton, O.G. (1953) *Micrometeorology*, McGraw-Hill, New York-Toronto-London.

Tamura, Y. (1987) A harmonic development of the tide-generating potential. *Bull. Info. Marées Terrestres*, **99**, 6813–6855.

Tamura, Y. (1995) Additonal terms to the tidal harmonic tables, in *Proc. of the 12th Int. Symp. on Earth Tides* (ed. H.T. Hsu), Science Press, Beijing/NewYork, pp. 345–350.

Thomson Sir, W. (1882) *Collected Works*, vol. III, Cambridge University Press, p. 337.

Tom, I.I. (1953) *Morskoy Atlas*, [Sea atlas. Vol. II.], Izd. Glavnogo Shtaba VMS, Moscow, [in Russian].

Trenberth, K.E. (1976) Spatial and temporal variations of the Southern oscillations. *Quart. J. Roy. Met. Soc.*, **102** (433), 639–653.

Trenberths, K.E. (1991) General characteristics of El Nino-Southern Oscillation, in *Teleconnectios Linking Worldwide Climate Anomalies* (eds M.H. Glantz, R.W. Katz and N. Nicholls), Cambridge University Press, New York, pp. 13–42.

Troup, A.J. (1965) The southern oscillation. *Quart. J. Roy. Met.Soc.*, **91**, 490–506.

Trubitsyn, V.P. (2000) Principles of the tectonics of floating continents. *Izvestiya Phys. Solid Earth*, **36** (9), 708–741.

Trubitsyn, V.P. and Rykov, V.V. (1998) Global tectonics of floating continents and oceanic lithospheric plates. *Doklady. Ross. Akad. Nauk*, **359** (1), 109–111 [In Russian.].

Vangengeim, G.Ya. (1935) *Opyt Primeneniya Sinopticheskikh Metodov k Izucheniyu i Kharakteristike Klimata. [Data on Synoptic Methods Used to Study and Describe Climate]*, TsUGMS, Moscow, p. 112 [in Russian].

Vestine, E.H. and Kahle, A.B. (1968) The westward drift and geomagnetic secular change. *Geophys. J. R. Astron. Soc.*, **15**, 29–37.

Vlasova, I.L., Sonechkin, D.M. and Chuchkalov, B.S. (1987) Kvazidvukhletnie kolebaniya nizhnestratosfernykh ekvatorialnykh vetrov za 30 let. [Quasi-two-year oscillations of lower stratospheric equatorial winds during 30 years]. *Meteorologiya i Gidrologiya*, (5), 47–55 [in Russian].

Volland, H. (1988) *Atmospheric Tidal and Planetary Waves*, Kluwer Academic Publishers, Dordrecht, The Netherlands, p. 348.

Volterra, V. (1895) Sulla teoria dei movimenti del polo-terrestre. *Astronomische Nachrichten*, **138**, 33.

Vondrak, J. (1999) Earth rotation parameters 1899.7–1992.0 after reanalysis within the Hipparcos frame. *Surv. Geophys.*, **20**, 169–195.

Vondrak, J. and Pejovic, N. (1988) Atmospheric excitation of polar motion: Comparison of polar motion spectrum with spectra of atmospheric effctive angular momentum functions. *Bull. Astron. Inst. Czechosl.*, **39** (3), 172–185.

Wahr, J.M., Sasao, T. and Smith, M.L. (1981) Effect on the fluid core on changes in the length of day due to long period tides. *Geophys. J. R. Astron. Soc.*, **64**, 635–650.

Walker, G.T. (1924) Correlation in seasonal variations of weather, IX: A further study of world weather (World Weather II). *Memoirs India Meteorol. Dept.*, **24**, 275–332.

Walker, A. and Young, A. (1957) Further results on the analysis of the variation of latitude. *Mon. Not. R. Astron. Soc.*, **117** (2), 119–141.

Wang, B. and Fang, Z. (1996) Chaotic oscillations of tropical climate: a dynamic system theory for ENSO. *J. Atmos. Sci.*, **53** (19), 549–566.

Wang, B. and Wang, Y. (1996) Temporal structure of the Southern Oscillation as revealed by waveform and wavelet analysis. *J. Climate.*, **9**, 1586–1598.

White, W.B. and Peterson, R.G. (1996) An Antarctic circumpolar wave in surface pressure, wind, temperature and sea-ice extent. *Nature*, **380**, 699–702.

Wilkinson, J.H. and Reinsch, C. (1976) *Handbook for Automatic Computation*. Vol. II: *Linear algebra*, Springer-Verlag, New York.

Wilson, C.R. (2000) Excitation of polar motion, in *Polar Motion: Historical and Scientific Problems*, ASP Conference Series, vol. 208 (eds S. Dick, D. McCarthy and B. Luzum), ASP, San Francisco, pp. 411–419.

Wilson, C.R. and Haubrich, R.A. (1976) Meteorological excitation of the Earth's wobble. *Geophysical Journal of the Roy. Astron. Soc.*, **46**, 707–743.

Wilson, C.R. and Haubrich, R.A. (1976) Atmospheric contributions to the excitation of the Earth's wobble. 1901–1970. *Geophysical Journal of the Roy. Astron. Soc.*, **46**, 745–760.

Wilson, I.R.G., Carter, B.D. and Waite, I.A. (2008) Does a spin–orbit coupling between the sun and the jovian planets govern the solar cycle? *Publ. Astron. Soc. Aust.*, **25**, 85–93.

Wolff, E.W. and Peel, D.A. (1985) The record of global pollution in polar snow and ice. *Nature*, **313** (6003), 535–540.

Woolard, E. (1959) Inequalities in mean solar time from tidal variations in the rotation of the Earth. *Astron. J.*, **64** (1269), 140–142.

Wright, P.B. (1989) Homogenised long-period southern oscillation indices. *Int. J. Climatol.*, **9**, 33–54.

Xi, Q.W. (1987) A new complete development of the tide-generating potential for the epoch J2000.0. *Bull. Info. Marées Terrestres*, **99**, 6766–6812.

Xi, Q.W. (1989) The precision of the development of the tidal generating potential and some explanatory notes. *Bull. Info. Marées Terrestres*, **105**, 7396–7404.

Yatskiv, Ya.S., Mironov, N.T., Korsun, A.A. and Taradiy, V.K. (1976) Dvizheniye polyusov i neravnomernost vrashcheniya Zemli. [Motion of the poles and Irregularity in the velocity of the Earth's rotation], in *"Astronomiya" ("Itogi nauki i tekhniki")*, vol. 12, (Parts 1 and 2), VINITI, Moscow, [in Russian].

Yoder, C.F., Williams, J.G. and Parke, M.E. (1981) Tidal variations of Earth rotation. *J. Geophys. Res.*, **86** (B2), 881–891.

Young, A. (1953) The effect of the movement of surface masses on the rotation of the earth. *Mon. Not. R. Astron. Soc.*, Geophysical, Supplement., **6**, 482.

Zastavenko, L.G. (1972) Baricheskoe pole troposfery Severnogo polushariya. [Pressure-field of troposphere of the Northern hemisphere]. *Trudy NIIAK*, **86** (237), pp. 1–237 [in Russian].

Zastavenko, L.G. (1975) *Sredniy Absolyutnyy Geopotentsial Izobaricheskikh Poverkhnostey 100, 50, 30 mb v Uzlakh Kartograficheskoy Setki Yuzhnogo Polushariya. [Mean Absolute Geopotential of Isobaric Surfaces of 100, 50 and 30 Millibars in the Units of a Map Grid of the Southern Hemisphere]*, Gidrometeoizdat, Moscow, p. 43 [in Russian].

Zebiak, S.E. and Cane, M.A. (1987) A model El Nino-Southern Oscillation. *Mon. Wea. Rev.*, **115**, 2262–2278.

Zernov, N.V. and Karpov, V.G. (1972) *Theory of Radio Circuit*, 2 edn, Energiya leningrad., p. 816.

Zharkov, V.N. (1983) *Vnutrenneye Stroeniye Zemli i Planet. [The Internal Structure of the Earth and Planets]*, Nauka, Moscow, p. 416 [in Russian].

Zharkov, V.N. and Trubitsyn, V.P. (1978) *Fizika Zemli i Planet*, Nauka, Moscow, p. 384 [in Russian].

Zharkov, V.N. and Trubitsyn, V.P. (1980) *Fizika Planetnykh Nedr. [Physics of Planetary Interiors]*, Nauka, Moscow, p. 448 [in Russian].

Zharkov, V.N., Molodensky, S.M., Brzezinski, A., Groten, E. and Varga, P. (1996) *The Earth and its Rotation: Low Frequency Geodynamics*, Wichmann, Heidelberg, p. 501.

Zharov, V.E. (1996a) Vliyaniye atmosfernykh prilivov na vrashcheniye Zemli, Venery i Marsa. [Influence of atmospheric tides on the rotation of the Earth, Venus and Mars]. *Astronomicheskiy Vestnik*, **30** (4), 321–330. [in Russian].

Zharov, V.E. (1996b) Sutochnye atmosfernye prilivy i ikh vliyaniye na vrashcheniye Zemli. [Diurnal atmospheric tides and their influence on the rotation of the Earth]. *Vestnik Mosk. Universiteta. Ser.3, Fizika. Astronomiya.*, **37** (1), 75–82 [in Russian].

Zharov, V.E. (1997a) Vliyaniye vrashcheniya atmosfery na nutatsiyu Zemli. [Influence of rotation of the atmosphere on the Earth's nutation]. *Vestnik Mosk. Universiteta. Ser. 3, Fizika. Astronomiya*, (6), 65–67 [in Russian].

Zharov, V.E. (1997b) Vrashcheniye Zemli i atmosfernye prilivy. [Rotation of the Earth and atmospheric tides]. *Astron. Vestnik.*, **31** (6), 558–565 [in Russian.].

Zharov, V.E. and Gambis, D. (1996) Atmospheric tides and rotation of the Earth. *J. of Geodesy*, **70**, 321–326.

Zharov, V.E., Konov, A.S. and Smirnov, V.B. (1991) Variatsii parametrov vrashcheniya Zemli. [Variations of parameters of the rotation of the Earth and their connection with the strongest earthquakes]. *Astronomicheskiy Zhurnal*, **68** (1), 187–196 [in Russian].

Zharov, V.E., Konov, A.S. and Smirnov, V.B. (1991) Variazii parametrov vrashenija Zemli i ich svjaz s silneishimi zemletrjasenijami mira [Variations of parameters of the Earth rotation and their connection with the strongest earthquakes of the world]. *Astronomicheskii Zhurnal*, **68** (1), 187–196 [In Russian with English text In *Astronomy Reports*, 35(1).

Zhou, Y.H., Salstein, D.A. and Chen, J.L. (2006) Revised atmospheric excitation function series related to Earth variable rotation under consideration of surface topography. *J. Geophys. Res.*, **111**, D12108, DOI: 10.1029/2005JD006608.

Zilitinkevich, S.S. (1970) *Dinamika Pogranichnogo Sloya Atmosfery. [Dynamics of the Boundary Layer of the Atmosphere]*, Gidrometeoizdat, Leningrad, p. 292 [in Russian].

Index

The Interaction Between Earth's Rotation and Geophysical Processes. Nikolay S. Sidorenkov

ISBN: 978-3-527-40875-7